GENETIC ENGINEERING APPLICATIONS FOR INDUSTRY

GENETIC ENGINEERING APPLICATIONS FOR INDUSTRY

Edited by J.K. Paul

Based on research prepared at:

Genex Corporation
Massachusetts Institute of Technology
Poly-Planning Services

NOYES DATA CORPORATION
Park Ridge, New Jersey, U.S.A.
1981

Library of Congress Catalog Card Number: 81-14028
ISBN: 0-8155-0869-7
ISSN: 0198-6880
Printed in the United States

Published in the United States of America by
Noyes Data Corporation
Noyes Building, Park Ridge, New Jersey 07656

Library of Congress Cataloging in Publication Data
Main entry under title:

Genetic engineering applications for industry.

(Chemical technology review ; 197)
"Four working papers which were used in the preparation of Impacts of applied genetics--micro-organisms, plants, and animals, by the Office of Technology Assessment (OTA) of the Congress of the United States."
Bibliography: p.
Includes index.
1. Genetic engineering--Industrial applications.
I. Paul, J. K.. II. Series.
TP248.G46 575.1 81-14028
ISBN 0-8155-0869-7 AACR2

Foreword

The documents that make up this book are four working papers which were used in the preparation of *Impacts of Applied Genetics—Micro-Organisms, Plants, and Animals* by the Office of Technology Assessment (OTA) of the Congress of the United States. However, **the documents presented in the book contain considerable additional technical information that is not found in the OTA report.** This additional material is of great value and, therefore, Noyes is publishing the papers in their entirety.

Genetic engineering, with all of its implications and ramifications, is the most exciting area of scientific research in the world today. Biotechnological advances could bring substantial changes, equal to or greater than those brought by the introduction of the computer.

Although research has been in progress for many years, the issuance of a patent ("Process for Producing Biologically Functional Molecular Chimeras," assigned to the Board of Trustees of Stanford University) in December, 1980 signaled the official opening of the industrial genetic age. The debate as to where basic genetic research should be done—in the academic or the industrial domain—may continue for some time. Meanwhile, the possibilities for applications of genetic engineering are extremely interesting and offer almost unlimited challenges to anyone involved in future planning in such industries as chemicals, foods, pharmaceuticals (human and veterinary), agricultural products, and energy, and, of course, in medical research.

We are aware that there is an occasional omission of references, a figure, or a table and that the quality of some figures and tables is not perfect; but the papers were available in this form only and it was thought that the occasional omission would not detract from the overall value of the documents. It should once again be emphasized that significantly more technical data is contained in these working papers than in the corresponding sections of the OTA report.

Following are the titles of the working papers used in the book:

> *Impact of Genetic Engineering Technology on Industry*, prepared by Genex Corporation for the Office of Technology Assessment of the Congress of the United States.
>
> *Chemical Industry* (second draft), prepared for the Office of Technology Assessment of the Congress of the United States at the Massachusetts Institute of Technology.
>
> *Implications of Molecular Genetics for Medicine* (revised version) prepared by W.P. O'Neill, Ph.D., M.B.A., of Poly-Planning Services for the Office of Technology Assessment of the Congress of the United States.
>
> *The Use of Applied Genetics in the Development of Pharmaceuticals and Medical Research Technology*, prepared for the Office of Technology Assessment of the Congress of the United States at the Massachusetts Institute of Technology.

The table of contents is organized in such a way as to serve as a subject index and provides easy access to the information contained in the book.

Glossary

The Glossary is from *Impacts of Applied Genetics—Micro-Organisms, Plants, and Animals* (OTA-HR-132), prepared by the Office of Technology Assessment of the Congress of the United States, April 1981.

Aerobic—Growing only in the presence of oxygen.

Alkaloids—A group of nitrogen-containing organic substances found in plants; many are pharmacologically active, e.g., nicotine, caffeine, and cocaine.

Allele—Alternate forms of the same gene. For example, the genes responsible for eye color (blue, brown, green, etc.) are alleles.

Amino acids—The building blocks of proteins. There are 20 common amino acids; they are joined together in a strictly ordered "string" which determines the character of each protein.

Anaerobic—Growing only in the absence of oxygen.

Antibody—A protein component of the immune system in mammals found in the blood.

Antigen—A large molecule, usually a protein or carbohydrate, which when introduced in the body stimulates the production of an antibody that will react specifically with the antigen.

Aromatic chemical—An organic compound containing one or more six-membered rings.

Aromatic polymer—Large molecules consisting of repeated structural units of aromatic chemicals.

Artificial insemination—The manual placement of sperm into the uterus or oviduct.

Bacteriophage (or phage)—A virus that multiplies in bacteria. Bacteriophage lambda is commonly used as a vector in recombinant DNA experiments.

Bioassay—Determination of the relative strength of a substance (such as a drug) by comparing its effect on a test organism with that of a standard preparation.

Biomass—Plant and animal material.

Biome—A community of living organisms in a major ecological region.

Biosynthesis—The production of a chemical compound by a living organism.

Biotechnology—The collection of industrial processes that involve the use of biological systems. For some of these industries, these processes involve the use of genetically engineered microorganisms.

Blastocyst—An early developmental stage of the embryo; the fertilized egg undergoes several cell divisions and forms a hollow ball of cells called the blastocyst.

Callus—The cluster of plant cells that results from tissue culturing a single plant cell.

Carbohydrates—The family of organic molecules consisting of simple sugars such as glucose and sucrose, and sugar chains (polysaccharides) such as starch and cellulose.

Catalyst—A substance that enables a chemical reaction to take place under milder than normal conditions (e.g., lower temperatures). Biological catalysts are enzymes; nonbiological catalysts include metallic complexes.

Cell fusion—The fusing together of two or more cells to become a single cell.

Cell lysis—Disruption of the cell membrane allowing the breakdown of the cell and exposure of its contents to the environment.

Cellulase—An enzyme that degrades cellulose to glucose.

Cellulose—A polysaccharide composed entirely of several glucose units linked end to end; it constitutes the major part of cell walls in plants.

Chimera—An individual composed of a mixture of genetically different cells.

Chloroplast—The structure in plant cells where photosynthesis occurs.

Chromosomes—The thread-like components of a cell that are composed of DNA and protein. They contain most of the cell's DNA.

Clone—A group of genetically identical cells or organisms asexually descended from a common ancestor. All cells in the clone have the same genetic material and are exact copies of the original.

Conjugation—The one-way transfer of DNA between bacteria in cellular contact.

Crossing-over—A genetic event that can occur during cellular replication, which involves the breakage and reunion of DNA molecules.

Cultivar—An organism developed and persistent under cultivation.

Cytogenetics—A branch of biology that deals with the study of heredity and variation by the methods of both cytology (the study of cells) and genetics.

Cytoplasm—The protoplasm of a cell, external to the cell's nuclear membrane.

Diploid—A cell with double the basic chromosome number.

DNA (deoxyribonucleic acid)—The genetic material found in all living organisms. Every inherited characteristic has its origin somewhere in the code of each individual's complement of DNA.

DNA vector—A vehicle for transferring DNA from one cell to another.

Dominant gene—A characteristic whose expression prevails over alternative characteristics for a given trait.

Embryo transfer—Implantation of an embryo into the oviduct or uterus.

Endotoxins—Complex molecules (lipopolysaccharides) that compose an integral part of the cell wall, and are released only when the integrity of the cell is disturbed.

Enzyme—A functional protein that catalyzes a chemical reaction. Enzymes control the rate of metabolic processes in an organism; they are the active agents in the fermentation process.

Escherichia coli—A bacterium that commonly inhabits the human intestine. It is a favorite organism for many microbiological experiments.

Estrogens—Female sex hormones.

Estrus ("heat")—The period in which the female will allow the male to mate her.

Eukaryote—A higher, compartmentalized cell characterized by its extensive internal structure and the presence of a nucleus containing the DNA. All multicellular organisms are eukaryotic. The simpler cells, the prokaryotes, have much less compartmentalization and internal structure; bacteria are prokaryotes.

Exotoxins—Proteins produced by bacteria that are able to diffuse out of the cells; generally more potent and specific in their action than endotoxins.

Fermentation—The biochemical process of converting a raw material such as glucose into a product such as ethanol.

Fibroblast—A cell that gives rise to connective tissues.

Gamete—A mature reproductive cell.

Gene—The hereditary unit; a segment of DNA coding for a specific protein.

Gene expression—The manifestation of the genetic material of an organism as specific traits.

Gene mapping—Determining the relative locations of different genes on a given chromosome.

Genetic code—The biochemical basis of heredity consisting of codons (base triplets along the DNA sequence) that determine the specific amino acid sequence in proteins and that are the same for all forms of life studied so far.

Genetic drift—Changes of gene frequency in small population due to chance preservation or extinction of particular genes.

Genetic engineering—A technology used at the laboratory level to alter the hereditary apparatus of a living cell so that the cell can produce more or different chemicals, or perform completely new functions. These altered cells are then used in industrial production.

Genome—The basic chromosome set of an organism—the sum total of its genes.

Genotype—The genetic constitution of an individual or group.

Germ cell—The sex cell of an organism (sperm or egg, pollen or ovum). It differs from other cells in that it contains only half the usual number of chromosomes. Germ cells fuse during fertilization.

Germplasm—The total genetic variability available to an organism, represented by the pool of germ cells or seed.

Glycopeptides—Chains of amino acids with attached carbohydrates.

Glycoprotein—A conjugated protein in which the nonprotein group is a carbohydrate.

Haploid—A cell with only one set (half of the usual number) of chromosomes.

Heterozygous—When the two genes controlling a particular trait are different, the organism is heterozygous for that trait.

Homozygous—When the two genes controlling a particular trait are identical for a pair of chromosomes, the organism is said to be homozygous for that trait.

Hormones—The "messenger" molecules of the body that help coordinate the actions of various tissues; they produce a specific effect on the activity of cells remote from their point of origin.

Hybrid—A new variety of plant or animal that results from cross-breeding two different existing varieties.

Hydrocarbon—All organic compounds that are composed only of carbon and hydrogen.

Immunoproteins—All the proteins that are part of the immune system (including antibodies, interferon, and cytokines).

In vitro—Outside the living organism and in an artificial environment.

In vivo—Within the living organism.

Leukocytes—The white cells of blood.

Lipids—Water insoluble biomolecules, such as cellular fats and oils.

Lipopolysaccharides—Complex substances composed of lipids and polysaccharides.

Lymphoblastoid—Referring to malignant white blood cells.

Lymphokines—The biologically active soluble factor produced by white blood cells.

Maleic anhydride—An important organic chemical used in the manufacture of synthetic resins, in fungicides, in the dyeing of cotton textiles, and to prevent the oxidation of fats and oils during storage and rancidity.

Messenger RNA—Ribonucleic acid molecules that serve as a guide for protein synthesis.

Metabolism—The sum of the physical and chemical processes involved in the maintenance of life and by which energy is made available.

Mitochondria—Structures in higher cells that serve as the "powerhouse" for the cell, producing chemical energy.

Monoclonal antibodies—Antibodies derived from a single source or clone of cells which recognize only one kind of antigen.

Mutants—Organisms whose visible properties with respect to some trait differ from the norm of the population due to mutations in its DNA.

Mutation—Any change that alters the sequence of bases along the DNA, changing the genetic material.

Myeloma—A malignant disease in which tumor cells of the antibody producing system synthesize excessive amounts of specific proteins.

n-alkanes—Straight chain hydrocarbons—the main constituents of petroleum.

Nif genes—The genes for nitrogen fixation present in certain bacteria.

Nucleic acid—A polymer composed of DNA or RNA subunits.

Nucleotides—The fundamental units of nucleic acids. They consist of one of the four bases—adenine, guanine, cytosine, and thymine (uracil in the case of RNA)—and its attached sugar-phosphate group.

Organic compounds—Chemical compounds based on carbon chains or rings, which contain hydrogen, and also may contain oxygen, nitrogen, and various other elements.

Parthenogenesis—Reproduction in animals without male fertilization of the egg.

Pathogen—A specific causative agent of disease.

Peptide—Short chain of amino acids.

pH—A measure of the acidity or basicity of a solution; on a scale of 0 (acidic) to 14 (basic); for example, lemon juice has a pH of 2.2 (acidic), water has a pH of 7.0 (neutral), and a solution of baking soda has a pH of 8.5 (basic).

Phage—(See *bacteriophage.*)

Phenotype—The visible properties of an organism that are produced by the interaction of the genotype and the environment.

Plasmid—Hereditary material that is not part of a chromosome. Plasmids are circular and self-replicating. Because they are generally small and relatively simple, they are used in recombinant DNA experiments as acceptors of foreign DNA.

Plastid—Any specialized organ of the plant cell other than the nucleus, such as the chloroplast.

Ploidy—Describes the number of sets of chromosomes present in the organism. For example, humans are diploid, having two homologous sets of 23 chromosomes (one set from each parent) for a total of 46 chromosomes; many plants are haploid, having only one copy of each chromosome.

Polymer—A long-chain molecule formed from smaller repeating structural units.

Polysaccharide—A long-chain carbohydrate containing at least three molecules of simple sugars linked together; examples would include cellulose and starch.

Progestogens—Hormones involved with ovulation.

Prostaglandin—Refers to a group of naturally occurring, chemically related long-chain fatty acids that have certain physiological effects (stimulate contraction of uterine and other smooth muscles, lower blood pressure, affect action of certain hormones).

Protein—A linear polymer of amino acids; proteins are the products of gene expression and are the functional and structural components of cells.

Protoplast—A cell without a wall.

Protoplast fusion—A means of achieving genetic transformation by joining two protoplasts or joining a protoplast with any of the components of another cell.

Recessive gene—Any gene whose expression is dependent on the absence of a dominant gene.

Recombinant DNA—The hybrid DNA produced by joining pieces of DNA from different sources.

Restriction enzyme—An enzyme within a bacterium that recognizes and degrades DNA from foreign organisms, thereby preserving the genetic integrity of the bacterium. In recombinant DNA experiments, restriction enzymes are used as tiny biological scissors to cut up foreign DNA before it is recombined with a vector.

Reverse transcriptase—An enzyme that can synthesize a single strand of DNA from a messenger RNA, the reverse of the normal direction of processing genetic information within the cell.

RNA (ribonucleic acid)—In its three forms—messenger RNA, transfer RNA, and ribosomal RNA-it assists in translating the genetic message of DNA into the finished protein.

Somatic cell—One of the cells composing parts of the body (e.g., tissues, organs) other than a germ cell.

Tissue culture—An in vitro method of propagating healthy cells from tissues, such as fibroblasts from skin.

Transduction—The process by which foreign DNA becomes incorporated into the genetic complement of the host cell.

Transformation—The transfer of genetic information by DNA separated from the cell.

Vector—A transmission agent; a DNA vector is a self-replicating DNA molecule that transfers a piece of DNA from one host to another.

Virus—An infectious agent that requires a host cell in order for it to replicate. It is composed of either RNA or DNA wrapped in a protein coat.

Zygote—A cell formed by the union of two mature reproductive cells.

Contents and Subject Index

PART I: INDUSTRIAL IMPACTS

PART II: CHEMICAL INDUSTRY

Part I

Industrial Impacts

The information in Part I is from *Impact of Genetic Engineering Technology on Industry,* prepared by Genex Corporation for the Office of Technology Assessment of the Congress of the United States, April 1980.

1.0 Introduction

1.1 Objective and Scope

The objective of this study was to provide the Office of Technology Assessment (OTA) with data on the current and potential uses of genetic engineering in the pharmaceutical, food, agricultural, chemical, and energy industries. Data include market and technical information pertaining to individual compounds whose production is most likely to benefit from the application of genetic engineering. Also considered is the impact of genetic engineering on each of the aforementioned industries as a whole as well as on supporting industries.

This report utilizes both text (Sections 2.0-4.0)and charts (Section 5.0) to assess each industry. To aid in the selection process of compounds, each industry was divided into the product categories listed in the Genex report to OTA, dated December 31, 1979.

The assessment of the various product categories was based on the parameters described in the "Statement of Work" (for period ending April 15, 1980) given to Genex by OTA. These parameters limit the data compiled to compounds that have the largest market value in particular product groupings or that play a key role in the economy. Also, through genetic engineering's inherent relationship to biology, many biological compounds (mostly pharmaceuticals) have been evaluated in this study, that do not have a market as yet but could be very important compounds both in economic and social terms.

The Genex report of December 31, 1979 concluded that recombinant DNA technology might have a significant impact on all the aforementioned industries within a twenty year period. Included in the current report is a 20 year projection of market values, in 1980 dollars, of various product categories where the application of genetic engineering may have its greatest impact (in terms of expanding or initiating a market for the appropriate products).

The current report examines market and technical information on each compound, grouped according to product category. The report was prepared this way in order to present a reasonably complete picture of each product category of interest. Since many product categories are spread throughout more than one industry, if the data had been reviewed solely by industry, the product categories would have been scattered and the information would have been presented less coherently.

1.2 Explanation of the Charts

The charts present the compounds from two points of view. Charts 1-7 group the compounds by industry, subgrouped by product category. Charts 8-31 group the compounds by product category irrespective of industry.

The charts based on industry present end use data for each compound; e.g., in the pharmaceutical industry aspirin is listed as an aromatic used as an analgesic, whereas in the chemical industry aniline is listed as an aromatic used as a cyclic intermediate. Thus, the similarities and differences between compounds of similar origin, i.e., product category, are revealed.

The charts based solely on product category are divided into two types; one type pertaining to market data (Charts 8, 9 and the subsequent odd numbered charts through Chart 31), and the other pertaining to technical data (the even numbered Charts from Chart 10 through Chart 30).

The market data were obtained both from published sources and from prior proprietary studies performed by Genex. In the latter case, data are used that are not proprietary although the sources must remain confidential. Market values were estimated by multiplying the market volume (total amount of product sold in 1978) by the bulk cost (unit bulk selling price in 1980). Except for aromatics and aliphatics, all market data represent world-wide estimates. Market data for aromatics and aliphatics are restricted to the U.S. Data that could not be found were marked not available (N/A). Compounds with a high market value were identified, and those compounds that could be produced biologically were selected for this report. High market values were relative to the industry and end use listing of each compound. For example, with respect to chemicals, normally only cyclic intermediates with production volumes (which differ from market volumes) exceeding 50 million pounds were selected, but in the case of flavor and perfume materials, compunds with production values generally exceeding 1 million pounds were selected. In the case of many pharmaceuticals, clinical importance was weighed heavily in their selection process.

The technical data were also obtained both from published and proprietary sources. In regard to the timetable for commercial production, the stated length of time is the time required to develop existing technology (including both genetic and biochemical engineering) to the point where it may be applied to appropriate manufacturing facilities for large scale production of the desired compounds. These time intervals should be sufficient for undertaking all the research and development, starting from our current knowledge base, in order to demonstrate that the desired compounds can be biologically produced first in the laboratory and then in the pilot plant at commercially desirable yields and reaction efficiencies. The timetable does not consider delays caused by construction of new facilities, nor does it consider delays required for government regulatory approval of new products.

It should be noted that in the technical data charts, when glucose is listed as an alternate precursor by fermentation, other carbohydrates, e.g., cellulose and corn starch, could be used. Moreover, if glucose were the precursor of choice, the actual feedstock would probably be a commodity like molasses as opposed to pure glucose.

1.3 Summary

Chart 8 lists over 100 compounds representing 17 different product categories that span the five industries under evaluation. The current market value of all these products exceeds $27 billion. One compound, however, methane, accounts for over $12 billion. The even numbered charts from Chart 10 to Chart 30 project that within 20 years all these products could be manufactured in genetically engineered microbial strains on a more economical basis than by means of today's conventional technologies. In many cases the time required to apply genetically engineered strains in commercial fermentations could be reduced to as little as 5 years.

Chart 31 illustrates the impact of genetic engineering on selected markets. Only five product categories are considered here, and in one of these categories, amino acids, only a few of the compounds comprising it are evaluated. The products represented in the five categories currently have a total market value exceeding $800 million. However, within 20 years this market value could rise to over $5 billion (in 1980 dollars) due largely to the application of genetic engineering. In a number of cases the desired products would very likely not be available in significant quantities if it were not for the application of genetic engineering technology.

2.0 Product Assessment

2.1 Amino Acids

Amino acids are the constituents of protein and play many key roles in all organisms. In the past two decades there have been great market demands for amino acids, and it appears that the amazing growth of these products will continue. The three industries that have strong demands for amino acids are the pharmaceuticals, foods, and agricultural products.

In the pharmaceutical industry, essential amino acids such as phenylalanine and tryptophan (which the body cannot synthesize at all) are used as nutrients for infusions (Chart 1). When research disclosed that non-essential amino acids (which the body can synthesize) also have nutritional significance, the demand for these increased dramatically.[25] Some amino acids are also utilized as therapeutic agents. Arginine is effective against hyperammonemia and liver disease by increasing arginase activity in the liver. Cysteine is used for treatment of bronchitis and nasal catarrh.[25]

In the foods industry, amino acids are used as flavoring agents and nutritional additives (Chart 3). Monosodium glutamate (MSG), the amino acid with the largest market, is utilized almost universally as a flavor enhancer.[26] Aspartic acid is also a flavor enhancer but is not employed as widely as MSG. The amino acid cysteine, with its reducing power, is used in the baking of breads to improve their quality and in fruit juices as an anti-oxidant.[25]

In the agricultural industry, amino acids are employed as feed additives to improve the efficiency of protein utilization (Chart 4). Lysine and methionine are the amino acids primarily used as feed additives.[26] Other more expensive amino acids, such as tryptophan and threonine, could be marketed extensively as feed additives if their costs of manufacturing were reduced.[17]

Current prices of amino acids range from $1.00 to over $50.00 per pound, depending on the type of amino acid, its purity, and the quantity required (Chart 9). Total world demand for all of the amino acids is climbing towards a billion pounds per year. This demand has kept abreast of cost-reducing technological advances.

For example, in 1954 world consumption of methionine was only a few million pounds at a cost of almost $3.00 per pound.[17] By 1960, as technological advances permitted, the cost fell to $1.35 per pound, and consumption was substantially increased. By 1965, the cost was again reduced by half, and world consumption reached 30 million pounds. In 1974, methionine consumption exceeded 70 million pounds, yet the price of $0.70 per pound remained the same as in 1965, regardless of continual inflation and devaluation of the dollar. In recent years methionine production has soared over 200 million pounds per year. The same is true for glutamic acid, the most widely produced amino acid. As far back as 1965, annual output of glutamate exceeded 250 million pounds worldwide (output is 600 million pounds today). Such a large demand was satisfied even though the technology for economic production had only been developed a few years earlier.

Conventional technology for producing glutamic acid economically is based on pioneering work that was subsequently applied to other amino acids.[(25)] The production methods employed microbial strains to overproduce amino acids beyond their normal biosynthetic capabilities. This was accomplished by using two distinct methods: (a) induction of wild type strains, and (b) the isolation of naturally occuring mutants. In the first method, the addition of inducers, e.g., penicillin or polyoxyethylene-monopalmitate, creates an environment in the culture medium whereby the permeability of the cell wall is increased, thereby resulting in enhanced excretion of the amino acid by the microorganism. In the second method, a microorganism is isolated that overproduces an amino acid through an auxotrophic mutation, e.g., oleic acid mutations that result in weak cell walls and increased permeability with respect to glutamate. Also utilized are regulatory mutants that possess feedback-insensitive key enzymes that cause the microorganism to accumulate amino acids in large amounts.

Through recombinant DNA technology, microorganisms can readily be constructed with specific mutations at regulatory points along a desired biosynthetic pathway. Such genetic engineering substitutes for an otherwise statistically improbable phenomenon in wild type organisms (random mutations occur at a rate of 10^{-7}). Furthermore, the gene dosage of the genes coding for the enzymes of the desired biosynthetic pathway can be increased by placing the pertinent genes on a plasmid which replicates in multiple copies in the host organism. Enzyme biosynthesis increases relative to gene dose.[12a,24a] Thus the amount of regulation on the synthesis and excretion of an amino acid may be minimized, and the yields increased dramatically. Also, through genetic engineering, alternate, less expensive precursors of the end product may now be employed.

2.2 Vitamins

Vitamins are organic substances that function in organisms in trace amounts as cofactors for enzymatic reactions. Although these substances are vital for many forms of life, their biological importance was first recognized because some organisms cannot synthesize them and must acquire them from the environment.

The vitamins have been shown to have properties other than their nutritional significance.[32] In the pharmaceutical industry, vitamin E has been shown to function as an infertility drug[10] and vitamin B_{12} is used as a hematopoeitic factor.[21]

In the food industry, vitamins C and D are used as food enrichment agents. In the agricultural industry, vitamins E, C, B_2, and nicotinic acid are used as feed additives. Of these, vitamins C and E are the most important with about 98% of the market.

Vitamins are produced in various ways (Chart 12). Most are produced by microbial fermentation except nicotinic acid, which is chemically synthesized, and vitamin E, which is extracted from wheat germ oil.[52] Vitamin C is produced by a semi-synthetic process from glucose (involving both fermentation and chemical synthesis), while vitamin D is produced by the fermentation of various yeasts.[52] Riboflavin fermentation processes have become very effective in producing relatively large yields of the vitamin.[38] In 1947, Meade developed a process with Clostridium acetobutylicum that produced 97 milligrams per liter. In 1973, the processes had improved, through the selection of mutants, to a point where Szczeniak applied a procedure which yielded 6.42 grams per liter. Vitamin B_{12}, cyanocobolamin, is a vitamin where fermentation has drastically changed the market.[13] Between 1951 and 1963, the time where research into the fermentation brought about new processes, the bulk selling price of cyanocobolamin dropped from $500 to $15 per gram (the current price is $15.40 per gram) (Chart 11).

Genetic engineering, through increasing the gene dosage of these minor metabolic routes, could produce strains of microorganisms that should still further increase the yields of the desired vitamin in the fermentation process.

2.3 Antibiotics

Antibiotics are chemical substances produced by various species of microorganisms, that suppress the growth of other microorganisms and may eventually kill them. Over the 35 years since antibiotics were first applied therapeutically, roughly 4000 antibiotics have been discovered, and the industry producing them has become highly competitive. Of these 4000, approximately 100 are prepared on a commercial scale, most of which are separated into three groups: penicillins, tetracyclines, and cephalosporins.[37]

In addition to their obvious utilization as antimicrobial agents in the pharmaceutical industry, antibiotics have other applications in the agricultural and food industries.[37] Erythromycin is used as a growth promoter in poultry and animals, and nisin, an erythromycin polypeptide from Streptococcus cremoris, is used as a food preservative.

All the antibiotics that are produced commercially are made by aerobic fermentation.[20,43] In these industrial processes, the products may account for up to 30% by weight of the nutrient supplied. These high yields come from the selection of mutant strains of microorganisms that normally produce the antibiotic in very small quantities.

For example, the yield of benzylpenicillin (penicillin G) was improved to well over 20 mg/ml by the use of strains selected from Penicillium chrysogenum NRRC 1951, a culture producing only 120 μg/ml. A problem that arises with mutant penicillin strains is instability.[37] The greater the productivity of the mutant strain, the less stable it is.

With the applications of recombinant DNA technology, strains that have larger increases in productivity of the antibiotic may be developed.[3a] These strains will be more stable because the mechanism through which the productivity is increased is different than that of the naturally occuring mutant strain.

2.4 Steroid hormones

A steroid hormone is a polycyclic substance secreted by a specific tissue and transported to a cellular receptor site on another tissue where it exerts its effects. The steroids include the hormones of the adrenal cortex (cortisone, prednisone, prednisolone, and aldosterone), the androgens (testosterone), and the estrogens (estradiol). They are all derived from the tetracyclic hydrocarbon perhydrocyclopentanophenanthrene and differ primarily in the number, type, and location of the substituent functional groups and in the number and position of the double bonds.

All the steroid hormones are manufactured by the pharmaceutical industry as therapeutic agents and prophylactics. The adrenal cortex hormones serve as anti-inflammatory agents except aldosterone, which is used to control electrolyte imbalance.[23] The androgens serve as infertility drugs and are administered in the treatment of hypogonadism and hypopituitarism.[18] The estrogens are used as a contraceptives and in the treatment of vaginitis.[19]

In 1949, when cortisone was found to be a capable agent in the treatment of arthritis, the demand for the drug could not be met since no practical method for large scale production existed.[42] The chemical synthesis was complicated and very expensive. In the early and mid-1950s, many investigators reported the microbial transformation, in high yields, of several intermediates to compounds that corresponded to the chemical synthetic scheme. Thus, by saving many chemical steps and achieving higher yields, manufacturers were able to reduce the price of steroids to a level to where they were a marketable commodity.

For example, the 11 α-hydroxylation of progesterone brought the 1949 price of cortisone in a very short time from $200 to $6 per gram.(17) Through further improvements, the price dropped to less than $1.00 per gram (1980 price is $0.46 per gram).

Still, the production of steroids remains a demanding, time-consuming, relatively expensive process (Chart 15). Many different microorganisms are needed to carry out the elaborate transformations. Recombinant DNA technology could enable the construction of a microorganism in which many steps could be carried out by that single strain. Such genetic engineering would further decrease the price of the particular steroid of interest.

2.5 Enzymes

Enzymes are proteins that serve as catalysts of biochemical reactions. The enzyme industry currently produces 30 million pounds of 50 different enzymes per year, excluding therapeutic enzymes, which amount to approximately $200 million in sales.(17) Charts 13 and 14 list only a few of these enzymes, which while representing a small percentage of the total market volume, comprise the bulk of the sales. Thus, malt amylase, a plant enzyme which is produced at a level of 20 million pounds per year, is conspicuously absent from these charts.

It should be noted that a number of manufacturing processes in major industries are absolutely dependent on the action of certain enzymes, e.g., rennin in cheese manufacturing. The availability of these enzymes in turn control several billion dollars worth of business. There is such increasing interest in developing enzyme products that there are patents for over 1000 different enzymes.[2] Microbial, animal, and plant enzymes are all utilized commercially with the microbial enzymes being the most important. Charts 2-4,6, and 7 cite some of the diverse industrial applications.

With respect to proteases, *Bacillus* protease production alone amounts to approximately 1 million pounds annually. The proteases are mainly used in detergents, but they have other applications such as the removal of gelatin from photographic film to facilitate silver recovery and the processing of casein and soybean protein. Proteases are also used to solubilize fish and tenderize meat.

α-Amylase and amyloglucosidase together are produced at a level of 1.2 million pounds annually.[2] These enzymes are utilized commercially for alcohol fermentation, glucose production, and starch processing and removal.

Other important enzymes include glucose isomerase, rennin, glucose oxidase, papain, and pepsin. Glucose isomerase converts glucose to a mixture of fructose and glucose, this mixture being sweeter than sucrose (table sugar) or glucose alone. Microbial rennin, used in cheese production, has to a large extent replaced the expensive calf rennin.[2] Glucose oxidase converts glucose to gluconic acid, an industrial chemical. It is also used in approximately one million diagnostic tests per day worldwide for measurement of blood sugar. Papain, a plant-derived enzyme, is added to animal feeds as a meat tenderizer. Pepsin, an animal protease, is used in manufacturing as the microbial proteases are.

Enzymes that are therapeutics are relatively new products and include urokinase, asparaginase and tyrosine hydroxylase. Urokinase, a very expensive pharmaceutical (several thousand dollars per dose), is used as an antithrombotic agent in the treatment of acute pulmonary embolisms.[33] Asparaginase is used in cancer chemotherapy.[5b] Tyrosine hydroxylase converts tyrosine to DOPA, which is therapeutic against Parkinson's disease.

Strain development in the enzyme industry is of extreme importance.[2] An optimal strain is one that grows on an inexpensive medium and gives a constant, high yield of enzyme in a short time. Secondary enzyme activities and the content of metabolites in the fermented broth should be minimal. Optimization of strains by mutations has been the method used till now because wild type strains often have undesirable side activities which need to be removed in the recovery processes. An example of this is the enzyme transglucosidase in glucoamylase preparations.

Through recombinant DNA technology, the construction of stable strains that contain all the properties listed above can be achieved. Regulation of the gene that codes for the desired enzyme can be altered by introducing a stronger promoter and a mutation which causes the inactivation of any repressor molecules. These changes in the host organism, together with an increase in gene dose for the desired enzyme, will enable it to produce high concentrations of the enzyme with no detrimental by-products.

2.6 Short Peptides

Short peptides, oligomers with less than 10 amino acid residues, have various uses and although the market for these is small, new compounds show great promise. Today most peptides are used in research as reference materials in protein structure or enzyme specificity studies. Other uses are chemotactic agents for immunological studies and as additives in tissue culture.

For example, in 1973 Thaler reported the use of the tripeptide Gly-His-Lys as an agent in tissue culture. It was found that this tripeptide, added in nanogram quantities, increased the survival time, cell growth, and both DNA and RNA synthesis of either normal or transformed rat liver cell cultures.[(49)]

Clearly, any short peptide could be made by fermentation by means of genetically engineered organisms. By polymerizing the peptide's DNA sequence and placing this on a plasmid, this gene (in high dosage) would cause the organism to produce a polypeptide in very large quantities.

One very short peptide that could be extremely important in the food industry is aspartame. This C-terminal methyl ester of the dipeptide aspartyl-phenylalanine, has been found to be attractive as an artificial sweetener. While more powerful synthetic sweeteners have been prepared (aspartame is 100 - 200 times as sweet as sucrose), aspartame has the advantage of having been well studied and has exhibited very low toxicity. It contains hardly any calories and causes virtually no aftertaste. Aspartame is currently produced chemically at $110 per pound, this high cost due largely to the cost of phenylalanine (current price is 38.18 per pound).

Fermentation of a genetically engineered microorganism to make large quantities of phenylalanine (wild type microorganisms have strong regulatory pathways for phenylalanine) could help the economics of chemical synthesis of aspartame by decreasing the price of phenylalanine. Another approach would be to construct a microorganism containing a poly (aspartate-phenylalanine) gene that would enable the organism to produce large quantities of the repeating dipeptide. The latter could be enzymatically cleaved and methylated to aspartame by chemical or enzymatic means. This second approach might be more economical since the microorganism could condense the amino acids in quantitative yields and in the right stereospecific configuration, a problem that arises in chemical synthesis.

2.7 Peptide Hormones

Peptide hormones are a diverse class of hormones that are synthesized and excreted in various sites within the body. The peptides listed in Charts 17 and 18 are hormones of the pancreas (insulin and glucagon), the anterior pituitary (human, bovine, porcine and ovine growth hormones), the posterior pituitary (vasopressin), the adrenal cortex (adrenocorticotropic hormone:ACTH), and the brain (endorphins and enkephalins).

All of the peptide hormones are current or potential products in the pharmaceutical industry except the bovine, porcine and ovine growth hormones (Charts 2 and 4). These are not currently marketed but have potential as growth promoters in the agricultural industry. The pancreatic hormones are therapeutic agents with respect to glucose regulation. Insulin is used to control diabetes, and glucagon is used in the treatment of insulin-induced hypoglycemia.[24] The brain hormones, the endorphins and enkephalins, are natural opiate receptor antagonists and may eventually be used as narcotics and prophylactics. In regard to the pituitary hormones, human growth hormone is currently in short supply and therefore very expensive to use against dwarfism. Vasopressin is an antidiuretic agent.[4] ACTH is a diagnostic agent in adrenal instability.

Peptides hormones are currently produced by extraction from the organs in which they originate. This process is quite expensive, and in the case of extraction from human tissue, availability of the organs is the limiting factor. When non-human hormones are used as pharmaceuticals, the possible risk of allergic reaction is greatly enhanced. Persons who need daily injections of the hormone, e.g. insulin, run the risk of serum sickness that brings constant malaise and ultimately could lead to anaphylaxis if the allergic reaction is strong enough.

Like the short peptides, the peptide hormones could be produced in large quantities in genetically engineered microorganisms. This could be done in two ways. First, a chemically synthesized gene coding for the desired peptide could be placed on a multicopy plasmid under the influence of a strong promoter. The resulting recombinant DNA molecule would be inserted into a microorganism that could be readily grown in large scale fermentation. Second, a naturally occurring gene could be utilized by isolating its messenger RNA. The RNA would serve as a template for synthesis of its gene. By means of the viral enzyme reverse transcriptase and then DNA polymerase, the DNA itself could be obtained in its normal double stranded configuration. This gene could then be placed on a plasmid with a strong promoter, much the same way as the synthetic gene. An example of the former application is the chemical synthesis of the human insulin gene. Insulin, the peptide hormone with the largest market share ($183 million), has already been produced in by a microorganism in low yields.[19a,49a] Improvement of the strain should increase productivity to a point where commercial fermentation would be economically feasible. The gene for human growth hormone, a $75 million pharmaceutical (in 1978), has been cloned into an microorganism using the latter technique.[19b,34a,42a]

2.8 Nucleotides

Nucleotides are the monomeric units of polymerized nucleic acids that are present in all living organisms. The most important of these compounds commercially are 5'-IMP and 5'-GMP.[36] These, as well as other minor commercial forms of nucleotides, e.g., inosine, are produced by four different processes: enzymatic hydrolysis of yeast RNA, chemical hydrolysis of yeast RNA to nucleosides and their subsequent phosphorylation, fermentative production of 5'-nucleotides, and fermentative production of nucleosides and their subsequent phosphorylation.

Of these, the most widely used process is the enzymatic hydrolysis method.[36] Since 1961, much emphasis has been placed on isolating and identifying microbial enzymes that hydrolyze RNA in many different ways, e.g., exonucleases and endonucleases. For a specific nucleotide preparation many microbes are needed, thereby creating a multistep process. Genetic engineering would enable the use of a microorganism that contained many enzymes needed for a particular preparation through recombinants of several microbes.

In the fermentative processes, naturally occuring mutants have been isolated that are genetically blocked in the biosynthetic pathways to the nucleic acids. Thus, the nucleotide pathways are released from negative feedback regulation by the end products of the polymerization reactions. By means of genetic engineering the dosage of genes coding for the enzymes of the nucleotide pathways could be increased by placing the pertinent genes on a multicopy plasmid. This would increase the yields of nucleotides resulting from fermentation. By using a microbial strain with appropriate regulatory mutations as the host organism of choice, the increase in yields of nucleotides could be substantial.

2.9 Pesticides

Pesticides include fungicides, herbicides, insecticides, rodenticides, and related products such as plant growth regulators, seed disinfectants, soil conditioners, and soil fumigants. Control of insects has the largest market (roughly $500 million annually) and is broken into two categories, chemical and microbial. Although microbial insecticides have been known for years, only 5% of the market value is spent for this type.(27) Recent successes in developing entomopathogenic viruses and bacteria, and the negative publicity brought on the chemical insecticide industry has encouraged a revival in microbial insecticides.

Of the 15,000 known species of insects, only 300 are sufficiently harmful to warrant control and/or destruction.

Fortunately, most of these 300 species have pathogenic microorganisms associated with them. In regard to these pathogens, they are required to possess lttle or no toxicity to man, non-target animals, and plants to be used as commercial insecticides.

There are approximately 100 known species of bacteria that are pathogenic to insects, but only three have been developed into commercial insecticides: Bacillus popilliae, B.thuringiensis, and B.moritai.(27) B.popilliae is produced only in the larvae of Japanese beetles. The other Bacillus species are produced by conventional fermentation techniques. These latter species have been most useful because they form spores that can be easily mass-produced and are stable enough to be handled commercially. The actual substances that cause toxicity to the insect are toxins synthesized by the microbes.

Although there are only two viral insecticides produced commercially, most experts believe that viruses have the greatest potential for development as insecticides. Entomopathogenic viruses are specific and are active against many economically important insects. Because of their specificity and their property as obligate parasites and consequent production only in their hosts, insect viruses are more difficult to produce on an industrial scale than bacteria and fungi. Just as the commercially acceptable bacteria are protected by spores, the most useful viral insecticides are protected by proteinaceous inclusion bodies.

The two viral insecticides that have been commercially produced are the Heliothis NPV (a nuclear polyhedrosis virus) produced in the U.S., and Dendrolimus CPV (a cytoplasmic polyhedrosis virus) produced in Japan.[27] Other viral types under experimental investigation include NPV of Lymantria, Orgyia, and Prodenia.

Genetic engineering should enable the construction of bacterial strains that are more potent insecticides, through increasing the dosage of the genes that code for the synthesis of the toxins involved. Also, synergists could be produced due to the mixture of genes capable of directing synthesis of various toxins acting on a broad spectrum of insects.

It should also be noted that bioconversions resulting in chemical insecticides may be achieved by means of microbial enzymes. Genetic engineering could be aplied here similarly to the way described in Section 2.5.

2.10 Viral Antigens

Viral antigens are specific proteins on the capsid of a virus that can act as inducers of antibody formation. The manufacturing of viral antigens for vaccines involves the production of the desired virus by infection of a host cell line (usually embryonic tissue culture) and subsequent quantitative inactivation. The inactivated virus is then purified and administered parenterally, creating antibodies without infection.

There are, however, certain disadvantages to the use of inactivated viruses as vaccines.[28] Extreme care is required in the inactivation step and the handling of the tissue culture to make certain that no residual, live, virulent virus is present in the vaccine. The immunity acquired is often brief and must be boostered. This not only involves the logistic problem of repeatedly reaching the individuals in need of the vaccine, but also raises the probability of hypersensitivity reactions by the individuals through repeated administration of foreign proteins. In addition, parenteral administration of the vaccine gives varying results upon infection by a wild type virulent strain. Local resistance is not induced in significant amounts at the natural portal of infection or the primary site of multiplication of the wild type virus.

Live attenuated vaccines have been developed that give better results.(28) This type of vaccine, because it has the advantage of acting like the natural infection, tends to stimulate longer lasting antibody and also induces resistance at the portal of infection. These viruses, although they grow and replicate in the host, have mutations that cause them to lose their toxicity to the host. This is obviously desirable. A major problem with this type of vaccine, however, is the fact that the mutant viruses have a tendency of reverting back to the wild type virulent strain in the host.

Human viruses that have vaccines associated with them are rubella, poliomyelitis, varicella, smallpox, and influenza. Some of these vaccines have proven very effective, while some have had only mild success. An example of a vaccine that has not proven effective is the influenza virus. The problem is that the virus itself undergoes many mutations which alter the toxic viral proteins. Antibodies that are induced by the vaccine cannot bind to these proteins.

Vaccines associated with animal viruses could also

claim a large market, especially in the poultry industry where large populations of chickens are susceptible to viral infections, e.g., avian leukemia virus.

Through genetic engineering, a very pure antigen of a virus could be produced (Charts 2 and 4). A major protein in the coat of the virus could be identified, and the gene that codes for the protein isolated. The gene could be placed on a plasmid in a bacterium, thereby enabling the bacterium to produce the viral protein in large quantities through conventional fermentation. A vaccine could be made from this protein that would react with the immune system of the host to create very specific antibody against that virus. Viral antagens of hepatitis B virus has already been cloned into bacteria [5a], but commercial production has not yet been achieved.

2.11 Miscellaneous Proteins

Proteins other than enzymes are also of commercial interest. Such proteins include human serum albumin, interferon, and monoclonal antibodies.

Human serum albumin, the smallest but most abundant of the plasma proteins, is used therapeutically in the treatment of various clinical conditions including shock, hypoproteinemia, and erythroblastosis fetalis.[26a] It is currently prepared from human plasma, making its supply both limited and expensive.

Interferon, the protein that has attracted so much attention lately, is found in miniscule amounts in cells infected by almost any animal virus. It has been used therapeutically, in experimental settings, as an antiviral drug and in the treatment of several types of cancer.[45a]

Interferon is currently extracted from peripheral blood leukocytes or produced in cultures of either lymphoid cells or fibroblasts. Since the protein is produced in such small quantities in the cell, the production is very laborious and thus very expensive. The demand for interferon is also driving up the cost that currently runs as high as $10 billion per pound.

Monoclonal antibodies are products of a novel cell hybridization technique which permits the production of these antibodies that are directed at a single determinant of an antigen. The hybridized cells, called hybridomas, are the hybrid of two cells: a single antibody-producing cell that may be obtained from primary immunologic tissue, and a cell which can be grown indefinitely in culture, e.g., an oncogenic cell line that has lost its ability to produce its own antibody. Monoclonal antibodies substitute for conventional techniques for raising antibody levels.

For example, immunization with an antigen derived from microorganisms or cell components results in the production of antibodies directed against many different determinants of the antigen. Separation of the antibodies with respect to specificity for the different determinants from the serum of the immunized individual presents a major task. Also, impurities in the sera are difficult to separate from the antibodies, and these contaminants limit the usefulness of the antibodies.

Monoclonal antibiodies have near perfect specificity and homogeniety and can be consistently produced in large quantities at reasonable costs. These hybridomas are used mainly in research, but applications in diagnostics and therapeutics are definite possibilities, especially since these cells can be also used in radioimmunoassays. A hepatitis monoclonal antibody has already been produced and is just being marketed as a diagnostic.

Competition in the monoclonal antibody market could be very tight. The U.S. radioimmunologic market, already approaching $200 million annuallly, could expand very rapidly. Cancer diagnostics, a definite application of monoclonal antibodies, will be the main reason for this growth.

By the use of recombinant DNA technology, it should eventually be possible to construct bacterial strains that synthesize monoclonal antibodies in quantities large enough to make production by fermentation feasible. The strategy involved in construction of such strains would be much like that of the peptide hormones: isolation of the messenger RNA and preparation of the DNA from the messenger by reverse transcriptase and DNA polymerase, followed by placing this gene on a multicopy plasmid for insertion into the host bacterium. Costs of production would be dramatically reduced, and quantity control problems in handling hybridomas would be eliminated.

Human serum albumin is very compatible to this type of technique, since it is one of the major proteins produced in the liver. Thus, the messenger should be readily isolated. In fact, ovalbumin, a related ovian protein, has already been expressed in genetically engineered microorganisms. With respect to interferon, its messenger is relatively hard to obtain. Nevertheless, its messenger has recently been isolated and utilized for preparation of the pertinent recombinant DNA. Interferon has now been expressed in bacteria although in relatively small amounts. (35a)

2.12 Gene Preparations

A wide variety of hereditary diseases plague mankind. The application of gene therapy is hardly suitable for all such diseases but may be applied to some. In order for a hereditary illness to be responsive to gene therapy the following conditions should be met. First, all clinical symptoms should result from expression of a single pair of recessive genes, e.g., sickle cell anemia. Thus, in each target cell only one of the alleles need to be repaired or replaced by a normal nucleotide sequence. Second, the molecular difference between the normal and abnormal alleles should be as small as possible in order to optimize the chances for insertion of the normal nucleotide sequence and replacement of the abnormal base sequence. Third, the allele responsible for the genetic disorder should code for a gene product that functions at a terminal stage of differentiation. Fourth, the primary etiology should be confined to a single tissue or a limited number of tissues in which the abnormal allele is expressed. Finally, the genetic disease should be one in which as few target cells as possible need to be altered by the donor DNA in order to yield therapeutically promising results. That is, only some of the cells of the population that synthesize the defective gene product should have to be transformed by the donor DNA.

Certain diseases of the blood may eventually be treatable by gene therapy. Those disorders that meet all the criteria discussed above include sickle cell anemia, the thallasemias, and perhaps the hemophilias. The first two disorders involve alterations in the hemoglobin protein. The hemophilias involve deficiencies of factors involved in blood coagulation.

Various laboratories have already established the concept of gene therapy for hereditary diseases such as sickle cell anemia by demonstrating the expression of exogenous mammalian hemoglobin genes in recipient cells.[26b,34aa,34c,53] The demonstration of gene transfer in live animals by transformation of their bone marrow cells by fragments of DNA has also been accomplished.[9]

2.13 Chemicals

2.13.1 Introduction

The chemical industry is one of the largest and most important in the world today. The U.S. market just for synthetic organic chemicals, excluding primary products from petroleum, natural gas, and coal tar, exceeded $35 billion in 1978.[47] (This value includes the bulk market for medicinal chemicals and pesticides). As Chart 8 shows, genetic engineering technology could have a profound impact on the chemical industry. For aromatics and aliphatics (excluding methane) alone, commercial fermentation of these products by means of genetically engineered strains may result in close to $4 billion of sales (in 1980 dollars) within 20 years (Charts 8,25 and 27). As explained in Section 1.2, this value represents U.S. sales (not world-wide). Sales of fermentation-derived methane and inorganics could also be quite substantial by the year 2,000 (Charts 8,29 and 30).

The advantages of using commercial fermentation over chemical reactions to synthesize these compounds are many. The major advantages include the use of renewable resources, the application of enzymes as catalysts, one-step production processes, and decreased pollution. These four advantages will be discussed in the following sections.

The application of genetic engineering technology to the chemical indsutry may be successful in both producing the compounds at lower costs and employing production processes that are safer to the environment. This technology will enable the utilization of renewable resources by microbial strains that have their normal content of enzymes altered. The concentrations of the enzymes involved in a biosynthetic pathway for the production of a particular compound can be increased by placing the genes that code for these enzymes on a multicopy plasmid. Also, this microbial process of transforming cheap, renewable precursors to product occurs in one step, thus requiring only one purification procedure to obtain the final product.

The enzymatic processes that carry out the chemical transformations involved in chemical synthesis produce no side-products (through an enzyme's very high specificity to its substrate) or by-products (through an enzyme's strong catalytic power). Since these side-products and by-products are the major causes of pollution, the potential for these hazards is removed.

2.13.2 Resources

The chemical industry has been based on the coal tar and petroleum industries for its raw materials since its inception nearly 100 years ago. Great strides in chemical technology plus the availability and very low cost of these materials caused tremendous expansion of the chemical industry in the middle of this century. This growth pattern seemed endless with many technological advances occuring each year.

Serious problems arose at the start of the 1970s when skyrocketing prices and severe cutbacks in availability of petroleum-based products sharply curtailed the industry's expansion. Further escalation of the prices of these raw materials in the future could make matters even worse.

Genetic engineering technology has the potential for releasing the chemical industry from dependence upon petroleum based products through the use of renewable resources in commercial production. All microorganisms are able to metabolize carbohydrates to various end products. These carbohydrates are found throughout the plant world and constitute what is otherwise known as biomass. There has been increasing research and development on the conversion of cellulose, the major carbohydrate in plants, to products of commercial interest by genetically engineered strains. The monomer of cellulose, glucose, can, of course, be readily used as a feedstock for fermentation. Other sources of carbohydrates include corn starch, molasses, and lignin. Lignin is an aromatic polymer found in wood, whose monomeric units resemble coniferyl alcohol. This monomer could be used as a precursor for aromatic substances, ultimately making biosynthetic processes to aromatic chemicals much easier and economical.

2.13.3 Catalysts

An important task in chemical production is to find

optimum velocity and efficiency in converting a precursor to product, i.e., chemical transformation. Substances that affect the rate of a reaction without being altered themselves are called catalysts.

Most chemical catalysts are metals or metal complexes. The effectiveness of a chemical catalyst appears to depend upon the adsorption of one or more of the reactants on the surface of the metal. Energy released as a result of the bonding of reactant molecules to the catalyst surface may serve to weaken a bond which is to be broken, therefore lowering the energy of activation. The catalyst surface may also act as a matrix to hold the reactant molecule in an orientation particularly favorable to reaction with another molecule, thus increasing the probability that a reaction will take place.[35] A problem that arises is that although there is some specificity of the catalyst to the substrate, it is not always enough to create reaction conditions that progress quantitatively to a single product.

Biological catalysts, called enzymes, have this specificity. Only certain substrates are acted upon and only a single-type reaction takes place, without side reactions or by-products, in nearly quantitative yields. Enzymes are proteins, polymers of amino acids with a complex three dimensional structure. This structure is formed when the hydrophilic groups of the amino acids conform to the outside of the protein and the hydrophobic groups conform to the inside of the molecule. This conformation creates indentations or pockets in the protein structure, that serve as "active sites" which bind specific substrates by covalent bonding (Figure 1). An active site orients the substrate so that only a certain reaction takes place. Proteins can have more than one active site for binding more than one substrate. They also have the property whereby the binding of a substrate to one active site on a protein will increase the binding efficiency of the other active sites. This type of enzyme, called an allosteric enzyme, is exemplified by hemoglobin molecules that have increased efficiency in binding oxygen once an oxygen atom is bound to one active site.

FIGURE 1

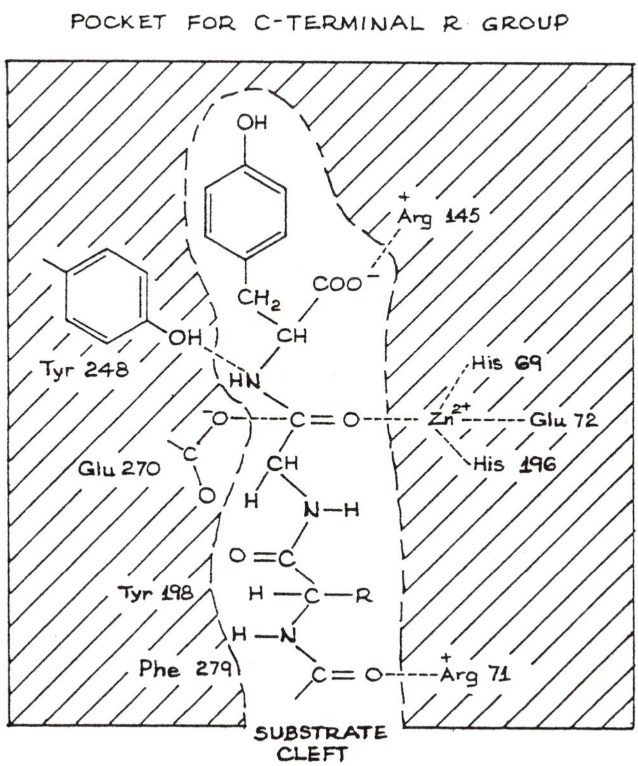

SCHEMATIC REPRESENTATION OF THE ACTIVE SITE OF THE ENZYME CARBOXYPEPTIDASE AND ITS SUBSTRATE

Reprinted from Leninger A.L. (1975) *Biochemistry, 2nd ed.*, New York: Worth Publications, p. 233.

Although enzymes are proteins and thus are relatively fragile molecules, they exert their extraordinary catalytic activities in dilute aqueous solutions at biological pH and moderate temperature, in sharp contrast to the rather extreme conditions often required to accelerate industrial organic reactions.

2.13.4 Processes

In the chemical synthesis of compounds, each reaction that takes place in a synthetic scheme is usually a separate entity. Because most chemical reactions are not quantitative, the product of each individual reaction will probably need to be purified before it can be utilized in the next step in the reaction scheme. This type of approach is very time consuming and expensive. For example, a synthetic scheme starts with ethylene, a petroleum-based product. If the scheme requires ten steps with each step in 90% yields (very optimistic yields in chemical syntheses), only about one-third of the ethylene would be recovered as the final end product. Purification may be costly, and often the chemicals involved , e.g., organic solvents for extractions, as well as side products of the reaction are toxic and require special disposal.

In biological systems, microorganisms are able to complete entire synthetic schemes at one time. Their enzymes transform the precursor through the intermediates to the desired end product in quantitative yields; therefore no purification steps are needed until the end product has been made.

2.13.5 Pollution

Typically, conventional chemical processes have particular features which relate to potential pollution hazards. These are:

1. side-product formation
2. by-product formation
3. incomplete conversion
4. mechanical and accidental loss.

It is assumed that the last characteristic is relatively insensitive to the type of product synthesis route.
The first three characteristics all represent inefficiencies in utilization of precursor. As such, they each contribute to the complexity of plant process design by necessitating recycle and separation. Separation can be particularly complicated by the fact that side products are often similar in their chemical and physical properties. Most separation techniques are based upon differences in physical properties (e.g., density, volatility, size). The more similar these properties, the more energy required to achieve a separation.

To the extent that by- and side-products have no value or unconverted precursor cannot economically be recycled, the problems of waste disposal and pollution will occur. Solution of these problems in the current chemical industry requires ingenuity, vigilance, energy, and dollars. A genetically engineered organism is designed to be precursor and product specific, with each enzyme having essentially 100 per cent conversion efficiency. As a result, biological processes can be expected to eliminate many conventional process waste and disposal problems at the front end of the system--in the reactor (fermentor). The enzymatic specificity of biologically performed reactions will dramatically reduce the numbers of undesired compounds which must later be separated from product and disposed of. Their high conversion efficiency will reduce precursor costs due to incomplete conversion losses and production costs by elimination of recycle. In addition, specificity and efficiency of the biological conversion process will generally result in simplification of product recovery with reduced capital and operating costs.

Specific comparisons of the environmental hazards produced by a conventional system as compared to a biological one are difficult. There is an abundance of data detailing the pollution parameters for various current processes but much less information regarding fermentation processes. The problem is compounded by the fact that there are few compounds currently produced by both methods. However, in most beverage distilling operations pollution has been essentially reduced to zero with complete recovery of still slops to provide animal feeds of high nutritional value. Such control procedures are generally applicable to fermentation (or bioconversion) reorders. Table 1 gives an example of the range of values of conventional pollution parameters produced by current processes and those expected from biologically based processes. The values reported for biological processes are based on typical data for alcohol and antibiotic production.

Significant reductions in a number of polluting variables can be seen. These reductions are primarily the result of the expected precursor and product specificity of the engineered organism. One parameter that is of special note is five day biochemical oxygen demand (BOD_5). This refers to the oxygen required over a five day period by organisms which will consume degradable organics in the waste stream. If the oxygen demand is too high, discharge of such a stream into a body of water will deplete the dissolved oxygen to the extent that it threatens aquatic life. An important variable that must be considered along with BOD is COD (chemical oxygen demand). Large differences between the COD and BOD of a waste stream can indicate the presence of non-biodegradables. Although the conventional process stream in Table 1 has less BOD_5 than the biological process stream, its COD content probably means that non biodegradable substances are present and specialized waste treatment is necessary.

TABLE 1

WASTE STREAM POLLUTION PARAMETERS:
CURRENT PROCESSES VS. BIOLOGICAL PROCESSES

COMPOUNDS: Mixed chemicals, including ethylene oxide, propylene oxide, glycols, amines and ethers

POLLUTION PARAMETERS	CURRENT PROCESSES	BIOLOGICAL PROCESSES
Alkalinity (mg/l)	4,060	~ 0
BOD_5 (mg/l)	1,950	4,000- 12,000
Chlorides (mg/l)	430- 800	~ 0
COD (mg/l)	7,970- 8540	5,000- 13,000
Oils (mg/l)	547	~ 0
pH	9.4- 9.8	4- 7
Sulfates (mg/l)	655	~ 0
Total Nitrogen (mg/l)	1,160- 1,253	50- 200
Phosphates (mg/l)	0	50- 200

NOTE: See text for explanation of BOD_5 and COD.

BOD is one area where traditional fermentation based processes have posed pollution problems. Batch fermentation processes typically generate large quantities of dead cells and residual nutrients which result in very large BOD if dumped directly into a dynamic aquatic environment. This difficulty can be circumvented by the use of spent cell material as an animal feed supplement or as fertilizer. These applications have been intensively investigated and have met with success. Also, with the renewed interest developed by the potentials of genetic engineering, some traditional fermentation systems are being revamped. Immobilized cell techniques are beginning to appear in commercial operations. Immobilization allows the reuse of cells that would otherwise have to be discarded. These systems can perform under continuous use for many months as compared to the usual fermentation time in a batch process of about one week or less. Immobilized operations create waste cells much less often than batch systems and therefore generate less BOD.

The contributions of genetic engineering to pollution control are not limited to direct displacement of current chemical synthetic routes. In many sectors implementation of genetic engineering technology will not take place for decades. In others, biological techniques may never be able to compete economically; however, genetic engineering may still find application in the field of waste disposal. Biological treatment of waste material is already a burgeoning science. Organisms engineered with more varied appetites and digestion efficiency will expand the range of application and use.

The implementation of genetic engineering technology for large-scale chemical synthesis can be expected to displace a number of currently non-biological processes. Concommitant to a switchover in production technology is a change in the characteristics of those waste materials generated during synthesis, separation, and purification that pose environmental pollution problems. Among the products that have been identified in this report, current processes for aliphatics and aromatics pose the major large volume pollution problems. Alleviation of these problems by biological processes should proceed along the same timetable set out for the development of genetically engineered strains to synthesize the compounds.

By their nature, biologically based chemical processes tend to create waste products that are biodegradable and valuable as a nutrient source. On the other hand many current chemical processes create wastes of no utility, which require elaborate degradation procedures to make them environmentally acceptable. In 1980 the chemical industry is expected to spend $883 million on capital outlays for pollution control, and well over $200 million on research and development of new control techniques and replacement products.[30] These figures do not include the millions of dollars that have been spent in recent years to clean up toxic chemical dumps and for compensation to persons harmed by these poorly disposed wastes. Also they do not include the cost of energy and labor required to operate pollution control systems.

The Environmental Protection Agency has estimated that the U.S. government and industry together will spend over $360 billion to control air and water pollution from 1977 through 1986. The share to be applied to the chemical and allied industries is about $26 billion.(8) Other than their positive effects upon our environment these expenditures have no return and represent a tremendous loss to the nation's economy. Genetic engineering technology may help alleviate this burden by developing cleaner synthetic routes to compounds and better biological waste treatment systems. The monetary impact could be tremendous. If just 10% of the current chemical industry were affected, capital, and research and development spending on pollution could be reduced by upwards of $100 million per year.

3.0 Case Studies

3.1 Introduction

This section presents the interdependence of both genetic and biochemical engineering in the application of commercial fermentation to the production of two compounds, ethanol and pullulan.

In the case of ethanol production, fermentation of corn syrup to ethanol by the conventional method costs $1.08 per gallon. By the use of genetically engineered thermophilic bacteria (microorganisms capable of growth at very high temperatures) that are fixed onto a solid support, i.e., immobilized, ethanol production can be reduced to $0.96 per gallon. The genetic engineering enables the organism to produce ethanol at a higher efficiency and at a temperature where ethanol vaporizes, thus creating a reactor where both production and purification are accomplished. The biochemical engineering allows for the continuous flow of precursor to the bacteria. Such a process decreases the operating cost by reducing down-time of the fermentor, which normally occurs in batch fermentations. Also, the stability of the microorganisms are increased by virtue of their being immobilized.

The interdependence between genetic and biochemical

engineering is again seen in the case of pullulan, a polysaccharide that could be utilized in the plastics industry. Recombinant DNA technology would enable the construction of an organism capable of producing high levels of pullulan, thereby increasing precursor transformation efficiency. The biochemical engineering, i.e., cell immobilization, creates a reduction in the operating cost. This increased transformation efficiency plus the decreased operating cost could lower the total production cost to a point where pullulan would be competitive with other plastics on the market.

3.2 Ethanol production

Extreme Thermophilic Ethanol Fermentation

Thermophilic ethanol producers have already been described in the genus Clostridium (i.e., *C. thermocellum*).[11] In addition, genetically engineered organisms, described as "a cross between yeasts and thermophilic bacteria" are able to ferment at 70°C.[14]

Advantages of such thermophilic fermentations are significant:

- Fermentation time is significantly reduced.
- The risk of contamination is nearly eliminated.
- Cooling requirements are lower due to the higher temperature of the fermenting broth.

It should be possible to engineer organisms able to ferment at even higher temperatures (82°C to 85°C) in such a way that fermentation and part of the distillation would be combined in the same reactor.

The carbon dioxide evolved at the same time as ethanol would have a stripping effect and would cause vaporization of ethanol at a lower temperature than in pure water-ethanol systems. The heat of vaporization of ethanol would be partly provided by the heat of fermentation.

For optimum results, the cells should be immobilized in a fixed film on column packing as described recently by Sitton et.al..[44] Such a configuration would permit the fermentation of highly concentrated sugar solutions, since the ethanol is removed as quickly as it is formed and does not accumulate in the beer to inhibit fermentation.

At the same time, the top part of the column could be used to enrich the ethanol concentration of the gas output and to provide an internal recycle of water. This distillation is driven by two heat sources. The first is the exothermic heat of reaction from fermentation of glucose to ethanol. The second comes from a reboiler to recover ethanol from a liquid bleed that prevents accumulation of toxic metabolites not volatile at the operating conditions of the system (such as non-volatile organic acids). A condenser drives the external recycle and separates CO_2 from ethanol and water. The CO_2 can be recovered for sale to beverage industries, or for partial recycling to increase the stripping rate in a column. A concentration of 90% ethanol would be achieved, and thus one distillation column would be eliminated.

Removal of ethanol in the gas phase has already been described by Cysewaki and Wilke for mesophilic fermentation by the use of vacuum.[12] Rates of fermentation of 82g/liter-hr were achieved with a feed solution containing 33% glucose, in a stirred reactor with cell recycle. A vacuum fermenation process has also been described by Ramalingham and Finn in which 50% sugar solutions were fermented.[39]

The use of extreme thermophilic organisms will make these processes feasible at atmospheric pressure. Coupled with the immobilized cell configuration and the high metabolic rate associated with thermophilic fermentations, one can predict that such a system should be able to produce up to 400g ethanol/liter-hr of fermenting broth. A combination fermentor-stripping column is sketched in Figure 2 and a comparative evaluation of the system is indicated in Table 2.

A total saving of 12 cents per gallon of ethanol is the result of a smaller capital investment and lower operating costs.

FIGURE 2. ETHANOL FERMENTATION SYSTEM

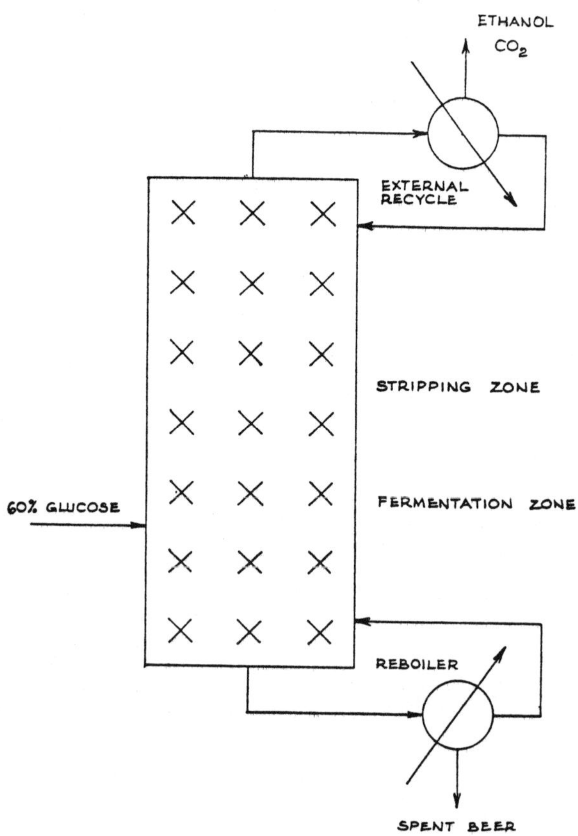

TABLE 2. ETHANOL FERMENTATION:

COMPARATIVE ECONOMIC EVALUATION OF

EXTREME THERMOPHILIC ETHANOL FERMENTATION (ETEF)

Basis: Fermentation of Corn Syrup

State of the Art: ADM process (15)

	ADM	ETEF
Raw Material Cost per Gallon Ethanol	\$0.91	\$0.91
By Product Credit	\$0.31	\$0.31
Net Raw Material Cost	\$0.60	\$0.60
Operating Cost	\$0.33	\$0.26 (17)
Cost of Capital	\$0.15	\$0.10 (17)
TOTAL	\$1.08	\$0.96

3.3 Pullulans

Most plastic materials, such as polyethylene, nylon, etc., are based on petrochemical feedstocks. As these non-renewable resources get more valuable with time, one investigation of interest would be the biological production of feedstocks, such as ethylene, which would permit the utilization of an existing and vast infrastructure. Another approach, however, could become more economical in the long run; that is, the direct synthesis of polymers by biological routes.

Such a material might be pullulan[22], a product of starch conversion by the yeast <u>Pullularia pullulans</u>. Actually, not only starch, but other carbohydrates can be used as a substrate. The product is a polymer of maltotriose, sub-units linked by α-1-4 bonds.[6]

(maltotriose)

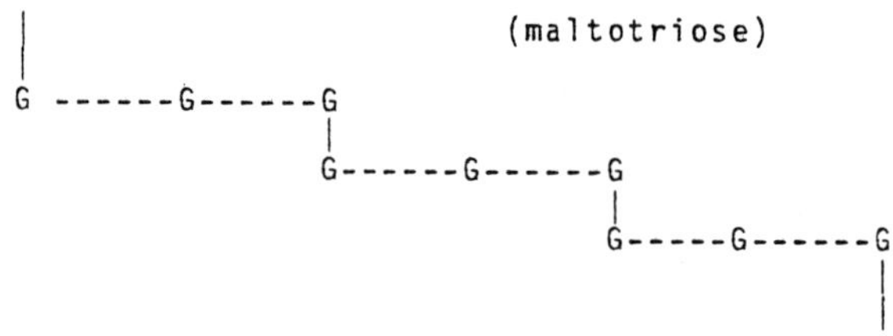

The mode of biosynthesis seems to be by trans-glycosylation of UDPG. ATP is required although its role is not clear.[48]

The material, a white powder when precipitated from the fermentation broth by aqueous ethanol, has an average molecular weight of 50,000 and can be thermally processed without the need of a plasticizer. After processing, it becomes insoluble and has a transparency and consistency similar to commercial food wraps, nylon, etc. The gas permeability of the plastic is low. One of the major advantages of the polymer is its biodegradability, and, as a corollary, it is not a polluting material. It is also totally nontoxic and can be incinerated without yielding noxious gases.

Many applications have been patented: Flocculating agents, textile yarns, low calorie foods, gauze, adhesives, molded products, packaging film, pharmaceutical and cosmetic applications.

Recombinant DNA technology would be advantageous in increasing the gene dosage and, consequently, the enzyme concentration to increase the rate of transglycosylation. If starch is used as raw material, the amylase gene should also be amplified to speed up glucose generation.

Growth of the genetically engineered yeast on a column packing would permit the use of a semicontinuous system to increase throughput and payload.

A possible process is shown in Figure 3.

FIGURE 3. PULLULAN FERMENTATION SYSTEM

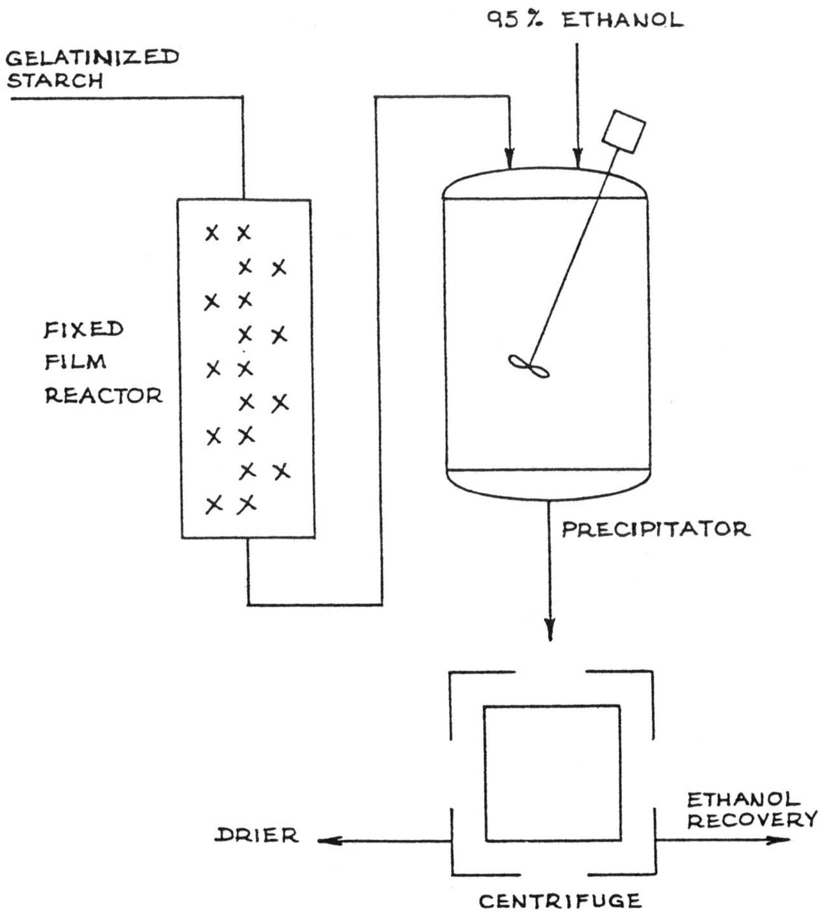

4.0 References

1. Aunstrup K., Industrial Approach to Enzyme Production (1977), in Z. Bohak and N. Sharon, eds., *Biotechnological Applications of Proteins and Enzymes*, New York: Academic Press, pp. 39-50.

2. Aunstrup K., O. Andresen, E.A. Falch, and T.K. Nielsen, Production of Microbial Enzymes (1979), in H.J. Peppler and D. Perlman, eds., *Microbial Technology: Microbial Processes*, *Vol. 1*, New York: Academic Press, pp.282-311.

3. Bernard Wolnak and Associates, in *Food Prod.*, July, 1978, p.41.

3a. Bibb M., J.L. Schottel, and S.N. Cohen (1980) A DNA Cloning System for Interspecies Gene Transfer in Antibiotic Producing *Streptomyces*, *Nature* *284*: 526-531.

4. Brazeau P., Agents Affecting the Renal Conservation of Water, (1975) in L. S. Goodman and A. Gilman, eds., *The Pharmacological Basis of Therapeutics*, 5th Ed., New York: MacMillan Publishing Co. Inc., pp.898-859.

4a. Bunn H.F., B.C. Forget, and H.M. Ranney (1977) *Human Hemoglobins*, Philadelphia: Sanders.

5. Burns J. J., Water Soluble Vitamins; Ascorbic Acid, (1975) in L. S. Goodman and A. Gilman, eds., *The Pharmacological Basis of Therapeutics*, 5th Ed., New York:MacMillan Publishing Co. Inc., pp. 1564-1590.

5a. Burrell, et.al. (1979) Expression in *Escherichia coli* of Hepatitis B Virus DNA Sequences *Cloned in pBR322*, *Nature* *279*:43-47.

5b. Calabresi P. and R.E. Parks, Chemotherapy of Neoplastic Diseases (1975) in L. S. Goodman and A. Gilman, eds., *The Pharmacological*

Basis of Therapeutics, 5th Ed., New York: MacMillan Publishing Co. Inc., pp. 1248-1308.

6. Catley B. J. and W. J. Whelan (1971) Archives of Biochemistry and Biophysics 141: 138.

7. Chemical and Engineering News, April 14, 1980, p.6.

8. Chemical Engineering, October 22, 1979.

9. Chemical Marketing Reporter, March and April, 1980.

9a. Cline, et.al. (1980) Gene Transfer in Intact Animals, Nature 284:422-425.

10. Cohn V. H., Fat-Soluble Vitamins; Vitamin K and Vitamin E (1975) in L.S. Goodman and A. Gilman, eds., The pharmacological Basis of Therapeutics, 5th ed., New York: MacMillan Publishing Co., Inc. pp.1591-1600.

11. Cooney, C., Massachusetts Institute of Technology, personal communication.

12. Cysewaki and Wilke (1977), Biotechnology and Bioengineering 19:1125.

12a. Davis M.G. and J.M. Calvo (1977) Relationship Between Messenger Ribonucleic Acid and Enzyme Levels Specified by the Leucine Operon of Escherichia coli K-12, J. Bacteriol. 131: 997-1007.

13. Florent J. and L. Nitiet, Vitamin B_{12} (1979), in H.J. Peppler and D. Perlman, eds., Microbial Technology: Microbial Processes, Vol. 1, New York: Academic Press, pp.497-520.

14. Food Engineering, February, 1980, p.68.

15. F. Eberstadt and Company, Inc., October 18, 1979.

16. Genex Corporation, Impact of Recombinant DNA Technology on Various Industries: Final Report to OTA, December 31, 1979.

17. Genex Corporation, proprietary information.

18. Gilman A.G. and F. Muriad, Androgens and Anabolic Steroids (1975) in L. S. Goodman and A. Gilman, eds., The Pharmacological Basis of Therapeutics, 5th Ed., New York: MacMillan Publishing Co. Inc., pp. 1451-1471.

19. Gilman A.G. and F. Muriad, Estrogens and Progestins (1975) in L. S. Goodman and A. Gilman, eds., The Pharmacological Basis of Therapeutics, 5th Ed., New York: MacMillan Publishing Co. Inc., pp. 1423-1450.

19a. Goeddel D.V., et.al. (1979) Expression in Escherichia coli of Chemically Synthesized Genes for Human Insulin, Proc. Natl. Acad. Sci. 76:106-110.

19b. Goeddel D.V., et.al. (1979) Direct Expression in Escherichia coli of a DNA Sequence Coding for Human Growth Hormone, Nature 281:544-548.

20. Gorman M. and F.M. Huber, β-Lactam Antibiotics (1978) in G.T. Tsao and D. Perlman, eds., Annual Reports on Fermentation Processes, Vol. 2, New York:Academic Press, pp. 203-222.

21. Greengard P., Water Soluble Vitamins, The Vitamin B Complex (1975) in L. S. Goodman and A. Gilman, eds., The Pharmacological Basis of Therapeutics, 5th Ed., New York: MacMillan Publishing Co. Inc., pp. 1549-1563.

21a. Guyton A.C. (1966) Textbook of Medical Physiology, Philadelphia: Sanders.

22. Hayashubara Biochemical Research Institute, Okayama City, Japan.

23. Haynes R.C. and J. Larner, Adrenocorticotropic Hormone: Adrenocorticol Steroids and Their Synthetic Analogs: Inhibitors of Adrenocorticol Steroid Biosynthesis (1975) in L. S. Goodman and A. Gilman, eds., The Pharmacological Basis of Therapeutics, 5th Ed., New York: MacMillan Publishing Co. Inc. pp. 1472-1506.

24. Haynes R.C. and J. Larner, Insulin and Oral Hypoglycemic Drugs: Glucagon (1975) in L. S. Goodman and A. Gilman, eds., The Pharmacological Basis of Therapeutics, 5th Ed., New York: MacMillan Publishing Co. Inc. pp. 1507-1533.

24a. Hershfield V., H.W. Boyer, C. Yanofsky, M.A Lovett, and D.R. Helinski (1974) Plasmid ColE1 as a Molecular Vehicle for Cloning and Amplification of DNA, Proc. Natl. Acad. Sci. 71: 3455-3459.

25. Hirose Y. and H. Okada, Microbial Production of Amino Acids (1979) in H.J. Peppler and D.

Perlman, eds., Microbial Technology: Microbial Processes, Vol. 1, New York: Academic Press, pp.211-240.

26. Hirose Y., K. Sano and H. Shibai, Amino Acids (1978) in G.T. Tsao and D. Perlman, eds., Annual Reports on Fermentation Processes, Vol. 2, New York: Academic Press, pp. 155-190.

26a. Huff B.,ed. (1968) Physicians Desk Reference, 22nd, Oradell, New Jersey: Chapman-Reinhold Inc..

26b. Huttner, K. M., G. A. Scangos, and F. H. Ruddle (1979) DNA-Mediated Gene Transfer of a Circular Plasmid into Murine Cells, Proc. Natl. Acad. Sci. 76: 5820-5824.

27. Ignoffo C.M. and R.F. Anderson, Bioinsecticides (1979) in H.J. Peppler and D. Perlman, eds., Microbial Technology: Microbial Processes, Vol. 1, New York: Academic Press, pp.211-240.

28. Jawetz E., J.L. Melnick and E.A. Adelberg (1974) Reviews of Medical Microbiology, 11th ed., Los Altos: Lange Medical Publications.

29. Jones H.R. (1971) Environmental Control in the Organic and Petrochemical Industries, Pollution Control Review Number 3, Noyes Data Corporation, Park Ridge, New Jersey.

30. Journal of Water Pollution Control (1979) 51:1419.

31. Koziorowski B. and J. Kucharski (1972) Industrial Waste Disposal, Pergammon Press, Wydawnictwa Naukowo-Techniczne, Warsaw, Poland.

32. Leninger A.L. (1975) Biochemistry, 2nd ed., New York: Worth Publications.

33. Levine W.G., Anticoagulants, Antithrombotic and Thrombolytic Drugs (1975) in L. S. Goodman and A. Gilman, eds., The Pharmacological Basis of Therapeutics, 5th Ed., New York: MacMillan Publishing Co. Inc. p.1365.

34. Lockwood L.B., Production of Organic Acids by Fermentation (1979) in H.J. Peppler and D. Perlman, eds., Microbial Technology: Microbial Processes, Vol. 1, New York: Academic Press, pp.367-372.

34a. Martial J.A., et.al. (1979) Complementary DNA Cloning and Expression in Bacteria, Science 205: 602-607.

34aa. Mantei, N., W. Boll, and C. Weissmann (1979) Rabbit β-Globin mRNA Production in Mouse L Cells Transformed with Cloned Rabbit β-Globin Chromosomal DNA, Nature 281: 40-46.

34b. McKusick (1978) Mendelian Inheritances in Man: Autosomal Dominant, Autosomal Recesive, and X-Linked Phenotypes, 6th ed., Baltimore: Johns Hopkins Press, pp. 762-764.

34c. Mulligan, R. C., B. H. Howard, and P. Berg (1979) Synthesis of Rabbit β-Globin in Cultured Monkey Kidney Cells Following Infection with a SV40 β-Globin Recombinant Genome, Nature 277: 108-114.

35. Murphy D.B. and V. Rousseau (1969) Foundations of College Chemistry, New York:Ronald Press Company, p.311.

35a. Nagata, et.al. (1980) Synthesis in Escherichia coli of a Polypeptide with Human Leukocyte Inteferon Activity, Nature 284: 316-319.

36. Nakao Y., Microbial Production of Nucleosides and Nucleotides (1979) in H.J. Peppler and D. Perlman, eds., Microbial Technology: Microbial Processes, Vol. 1, New York: Academic Press, pp.312-355.

37. Perlman D., Microbial Production of Antibiotics (1979) in H.J. Peppler and D. Perlman, eds., Microbial Technology: Microbial Processes, Vol. 1, New York: Academic Press, pp.241-281.

38. Perlman D., Microbial Process for Riboflavin Production (1979) in H.J. Peppler and D. Perlman, eds., Microbial Technology: Microbial Processes, Vol. 1, New York: Academic Press, pp.521-528.

39. Ramalingham and Finn (1977) Biotechnology and Bioengineering, 20:583.

40. Roberts J.D. and M.C. Caserio (1964) Basic Principles of Organic Chemistry, New York: Benjamin Inc., p.1096.

41. Sax I.N.,ed., (1974) Industrial Pollution, New York: Van Nostrand Reinhold Company.

42. Sebek O.K. and D. Perlman, Microbial Transformation of Steroids and Sterols, (1979) in H.J. Peppler and D. Perlman, eds., Microbial Technology: Microbial Processes, Vol. 1, New York: Academic Press, pp.483-496.

42a. Seeberg, et.al. (1978) Synthesis of Growth Hormone by Bacteria, Nature 276: 795-798.

43. Shibata M. and M. Uyeda, Microbial Transformation of Antibiotics (1978) in G.T. Tsao and D. Perlman, eds., Annual Reports on Fermentation Processes, Vol. 2, New York: Academic Press, pp. 267-304.

44. Sitton, et al. (1979) Chemical Engineering Progress, December, p.52.

45. Solomons G.L., The Microbial Production of Enzymes (1977) in Z. Bohak and N. Sharon, eds., Biotechnological Applications of Proteins and Enzymes, New York: Academic Press, pp.51-61.

45a. Stewart W.E. (1979) The Interferon System, Berlin: Springer Publications.

46. Straw J.A., Fat-Soluble Vitamins; Vitamin D (1975) in L. S. Goodman and A. Gilman, eds., The Pharmacological Basis of Therapeutics, 5th Ed., New York: MacMillan Publishing Co. Inc. pp. 1579-1590.

47. Synthetic Organic Chemicals, March and April 1980.

48. Taguchi R. et al. (1978) Agricultural Biological Chemistry 37:1635.

49. Thaler (1973) Biochem. Biophys. Res. Comm. 243: 85.

49a. Villa-Komaroff A., A. Efstratiadis, S. Broome, P. Lomedico, R. Tizard, S.P. Nader, W.L. Chick, and W. Gilbert (1978) A Bacterial Clone Synthesizing Proinsulin, Proc. Natl. Acad. Sci. 75: 3727-3731.

50. Weatherall D.J. and J.B. Clegg (1979) Recent Developments in the Molecular Genetics of Human Hemoglobin. Cell 16:467-480.

51. Weinstein L., Chemotherapy of Microbial Diseases (1975) in L. S. Goodman and A. Gilman, eds.,

The Pharmacological Basis of Therapeutics, 5th Ed., New York: MacMillan Publishing Co. Inc. pp. 1090-1247.

52. Windholz M.,ed. (1976) The Merck Index, 9th ed., Rahway, New Jersey: Merck and Co..

53. Wold, B., M. Wigler, E. Lacy, T. Maniatis, S. Silverstein, and R. Axel (1979) Introduction and Expression of a Rabbit β-Globin Gene in Mouse Fibroblasts, Proc. Natl. Acad. Sci. 76: 5684-5688.

5.0 Charts

Chart 1: Pharmaceuticals—Small Molecules

Product category	End use
Amino acids	
Phenylalanine	Intravenous solutions
Tryptophan	Intravenous solutions
Arginine	Therapeutic: liver disease and hyperammonemia
Cysteine	Therapeutic: bronchitis and nasal catarrh
Vitamins	
Vitamin E	Intravenous solutions, prophylactic
Vitamin B_{12}	Intravenous solutions
Aromatics	
Aspirin	Analgesic
p-acetaminophenol	Analgesic
Steroid hormones	
Corticoids	
Cortisone	Therapeutic: anti-inflammatory agent
Prednisone	Therapeutic: anti-inflammatory agent
Prednisolone	Therapeutic: anti-inflammatory agent
Aldosterone	Therapeutic: control of electrolyte imbalance
Androgens	
Testosterone	Therapeutic: infertility, hypogonadism, and hypopituitarism
Estrogens	
Estradiol	Prophylactic, therapeutic: vaginitis
Antibiotics	
Penicillins	Control of infectious diseases
Tetracyclines	Control of infectious diseases
Cephalosporins	Control of infectious diseases
Short peptides	
Glycine-Histidine-Lysine	Manufacturing processes: tissue culture

Chart 2: Pharmaceuticals—Large Molecules

Product category	End use
Peptide hormones	
Insulin	Control of diabetes
Endorphins	Analgesics, narcotics, prophylactics
Enkephalins	Analgesics, narcotics, prophylactics
ACTH[a]	Diagnostic: adrenal instability
Glucagon	Therapeutic: diabetes-induced hypoglycemia
Vasopressin	Therapeutic: antidiuretic
Human growth hormone	Therapeutic: dwarfism
Enzymes	
Glucose oxidase	Diagnostic: measurement of blood sugar
Urokinase	Therapeutic: antithrombotic
Asparaginase	Therapeutic: antineoplastic
Tyrosine hydroxylase	Therapeutic: Parkinson's disease
Viral antigens	
Hepatitis viruses	Vaccine
Influenza viruses	Vaccine
Herpes viruses	Vaccine
Varicella virus	Vaccine
Rubella virus	Vaccine
Reoviruses	Vaccine: common cold
Epstein-Barr virus	Vaccine: infectious mononucleosis, nasopharyngeal carcinoma, Burkitt lymphoma
Miscellaneous proteins	
Interferon	Control of infectious diseases
Human serum albumin	Therapeutic: shock and burns
Monoclonal antibodies	Diagnostics: hepatitis, cancer, etc; therapeutics
Gene preparations	
Sickle-cell anemia	Control of hereditary disorder
Hemophilias	Control of hereditary disorder
Thallasemias	Control of hereditary disorder

[a]Adrenocorticotropic hormone.

Chart 3: Food Products

Product category	End use
Amino acids	
Glutamate	Food enrichment agent, flavoring agent
Cysteine	Food enrichment agent, manufacturing processes
Aspartate	Flavoring agent
Vitamins	
Vitamin C	Food additive, food enrichment agent
Vitamin D	Food enrichment agent
Aromatics	
Benzoic acid	Food preservative
Aliphatics	
Propionic acid	Food preservative
Short peptides	
Aspartame	Artificial sweetener
Enzymes	
Rennin	Manufacturing processes
Amyloglucosidase	Manufacturing processes
α-amylase	Food enrichment agent, manufacturing processes
Glucose isomerase	Manufacturing processes: sweetener
Nucleotides	
5'-IMP[a]	Flavoring agent
5'-GMP[b]	Flavoring agent

[a]5'-inosinic acid.
[b]5'-guanylic acid.

Chart 4: Agricultural Products

Product category	End use
Amino acids	
Lysine	Feed additive
Methionine	Feed additive
Threonine	Feed additive
Tryptophan	Feed additive
Vitamins	
Nicotinic acid	Feed additive
Riboflavin (B_2)	Feed additive
Vitamin C	Feed additive
Aliphatics	
Sorbic acid	Feed preservative
Antibiotics	
Penicillins	Feed additive, prophylactic
Erythromycins	Feed additive, prophylactic
Peptide hormones	
Bovine growth hormone	Growth promoter
Porcine growth hormone	Growth promoter
Ovine growth hormone	Growth promoter
Viral antigens	
Foot-and-mouth disease virus	Vaccine
Rous sarcoma virus	Vaccine
Avian leukemia virus	Vaccine
Avian myeloblastosis virus	Vaccine
Enzymes	
Papain	Feed additive
Glucose oxidase	Feed preservative
Pesticides	
Microbial	Insecticide
Aromatic	Insecticide
Inorganics	
Ammonia	Fertilizer

Chart 5: Chemicals—Aliphatics

Compound	End use
Acetic acid	Miscellaneous acyclic
Acrylic acid	Miscellaneous acyclic
Adipic acid	Miscellaneous acyclic
Bis (2-ethylhexyl) adipate	Plasticizer
Citronellal	Flavor/perfume material
Citronellol	Flavor/perfume material
Ethanol[a]	Miscellaneous acylic
Ethanolamine	Miscellaneous acyclic
Ethylene glycol	Miscellaneous acyclic
Ethylene oxide	Miscellaneous acyclic
Geraniol	Flavor/perfume material
Glycerol	Miscellaneous acyclic
Isobutylene	Miscellaneous acyclic, flavor/perfume material
Itaconic acid	Plastics/resin
Linalool	Flavor/perfume material
Linalyl acetate	Flavor/perfume material
Methane	Primary petroleum product
Nerol	Flavor/perfume material
Pentaerythritol	Miscellaneous acyclic
Propylene glycol	Miscellaneous acylic
Sorbitol	Miscellaneous acyclic
α-terpineol	Flavor/perfume material
α-terpinyl acetate	Flavor/perfume material

Chart 7: Energy Products

Product category	End use
Enzymes	
Ethanol dehydrogenase	Manufacturing processes
Hydrogenase	Manufacturing processes
Biodegradation	Petroleum byproducts removal
Aliphatics	
Methane	Fuel
Ethanol	Fuel
Inorganics	
Hydrogen	Fuel
Mineral leaching	
Uranium	Fuel

Chart 8: Total Market Values for the Various Product Categories

Product category	Number of compounds	Current value ($ millions)
Amino acids	9	$ 1,703.0
Vitamins	6	667.7
Enzymes	11	217.7
Steroid hormones	6	376.8
Peptide hormones	9	263.7
Viral antigens	9	N/A
Short peptides	2	4.4
Nucleotides	2	72.0
Miscellaneous proteins	2[a]	300.0
Antibiotics	4[b]	4,240.0
Gene preparations	3	N/A
Pesticides	2[b]	100.0
Aliphatics:		
Methane	1	12,572.0
Other	24[c]	2,737.5
Aromatics	10[c]	1,250.9
Inorganics	2	2,681.0
Mineral leaching	5	N/A
Biodegradation	N/A	N/A
Totals	107	$27,186.7[d]

[a]Only two of a number of compounds are considered here.
[b]These numbers refer to major classes of compounds; not actual numbers of compounds.
[c]These numbers refer only to those compounds representing the largest market volume in classes specified in the text.
[d]Current value excluding methane = $14,614,700,000.

Chart 9: Amino Acids—Market Information (9, 25, 26)

	Current market data		
Compound	Market volume 1,000 lb	Bulk cost $/lb	Market value ($ millions)
Arginine	900	12.73	11.46
Aspartate	3,000	2.86	8.6
Cysteine	600	22.75	13.6
Glutamate	600,000	1.80	1,080.0
Lysine	129,000	2.10	258.0
Methionine	210,000	1.40	294.0
Phenylalanine	300	38.18	11.46
Threonine	300	58.18	16.2
Tryptophan	225	43.18	9.71

Chart 6: Chemicals—Aromatics and Miscellaneous

Product category	End use
Aromatics	
Aniline	Cyclic intermediate
Benzoic acid	Cyclic intermediate
Cresols	Cyclic intermediate
Phenol	Cyclic intermediate
Phthalic anhydride	Cyclic intermediate
Cinnamaldehyde	Flavor/perfume material
Diisodecyl phthalate	Plasticizer
Dioctyle phthalate	Plasticizer
Inorganics	
Ammonia	Manufacturing processes
Hydrogen	Manufacturing processes

Product category	End use
Enzymes	
Pepsin	Manufacturing processes
Bacillus protease	Manufacturing processes
Mineral leaching	
Transition metals (cobalt, nickel, manganese, iron)	Inorganic intermediates; catalysts
Biodegradation	Removal of organic phosphates, aryl sulfonates, and haloaromatics

Chart 10: Amino Acids—Technical Information (17, 25, 26, 51)

Compound	Typical synthetic process	Typical precursor	Is precursor renewable/non-renewable limited	Alternate precursor by fermentation	Time to implement commercial fermentation by genetically engineered strain
Arginine	fermentation	glucose and NH_4^+[a]	renewable	—	5 yrs.
Aspartate	fermentation	fumaric acid and ammonia	limited	—	5 yrs.
Cysteine	extraction	protein hydrolysis	renewable	—	5 yrs.
Glutamate.......	fermentation	glucose and NH_4^+	renewable	—	5 yrs.
Lysine	fermentation	glucose and NH_4^+	renewable	—	5 yrs.
Methionine	chemical	β-methylmercapto propionaldehyde	nonrenewable	glucose and NH_4^+	10 yrs.
Phenylalanine....	chemical	α-acetamino-cinnamic acid	limited	glucose and NH_4^+	5 yrs. 5 yrs.
Threonine	fermentation	glucose and NH_4^+	renewable	—	5 yrs.
Tryptophan	chemical	α-ketoglutaric phenylhydrazone	nonrenewable	glucose and NH_4^+	5 yrs.

[a]Ammonium ion.

Chart 11: Vitamins—Market Information (9, 10, 13, 38, 47)

	Current market data		
Compound	Market volume 1,000 lb	Bulk cost $/lb	Market value ($ millions)
Nicotinic acid....	1,400	1.82	2.5
Riboflavin (B_2)....	22	15.40	0.34
Vitamin B_{12}	22	6,991.60	153.8
Vitamin C	90,000	4.50	405.0
Vitamin D	12	42.50	0.51
Vitamin E	3,641	29.00	105.6

Chart 12: Vitamins—Technical Information (5, 13, 17, 21, 38, 46)

Compound	Typical synthetic process	Typical precursor	Is precursor renewable/non-renewable limited	Alternate precursor by fermentation	Time to implement commercial fermentation by genetically engineered strain
Nicotinic Acid ...	chemical	alkyl α-subst. pyridines	nonrenewable	glucose and NH_4^+[a]	10 yrs.
Riboflavin (B_2)....	fermentation	glucose	renewable	—	10 yrs.
Vitamin B_{12}	fermentation	carbohydrates	renewable	—	10 yrs.
Vitamin C	semisynthetic	glucose or sorbitol	renewable	—	10 yrs.
Vitamin D	fermentation	glucose	renewable	glucose	10 yrs.
Vitamin E	extraction	wheat germ oil	limited	glucose	15 yrs.

[a]Ammonium ion.

Chart 13: Enzymes—Market Information (1, 2, 3, 45, 47)

	Current market data		
Compound	Market volume 1,000 lb	Bulk cost $/lb	Market value ($ millions)
α-amylase	600	19.33	11.6
Amyloglucosidase	600	.00	12.0
Asparaginase	(Information not available)		
Bacillus protease.	1,000	8.28	8.2
Ethanol dehydrogenase .	(Information not available)		
Glucose isomerase	100	400 00	40 0
Glucose oxidase .	5	160.00	0.80
Hydrogenase	(Information not available)		
Papain..........	200	59.00	11.8
Pepsin..........	10	380.00	3.8
Rennin..........	24	696.00	40.0
Tyrosine hydroxylase	(Information not available)		
Urokinase		60,000 IU[a]	89.5

[a]IU = international units.

Chart 14: Enzymes—Technical Information (1, 2, 7, 17, 33, 45, 51)

Compound	Typical synthetic process	Typical precursor	Is precursor renewable/non-renewable limited	Alternate precursor by fermentation	Time to implement commercial fermentation by genetically engineered strain
α-amylase	fermentation	molasses	renewable	—	5 yrs.
Amyloglucosidase	fermentation	molasses	renewable	—	5 yrs.
Asparaginase	extraction	tissue culture	renewable	glucose and NH_4^+	5 yrs.
Bacillus protease	fermentation	molasses	renewable	—	5 yrs.
Ethanol dehydrogenase		(Information not available)		glucose and NH_4^+	10 yrs.
Glucose isomerase	fermentation	glucose and NH_4^+ [a]	renewable	—	5 yrs.
Glucose oxidase	fermentation	molasses	renewable	—	5 yrs.
Hydrogenase		(Information not available)		glucose and NH_4^+	10 yrs.
Papain	extraction	papaya	renewable	glucose and NH_4^+	5 yrs.
Pepsin	fermentation	molasses	renewable	—	5 yrs.
Rennin	fermentation	molasses	renewable	—	5 yrs.
Tyrosine	extraction	tissue culture	renewable	glucose and NH_4^+	5 yrs.
Urokinase	extraction	tissue culture	renewable	glucose	5 yrs.

[a]Ammonium ion.

Chart 15: Steroid Hormones—Market Information (9, 17)

Compound	Current market data: Market volume 1,000 lb	Bulk cost $/lb	Market value ($ millions)
Corticoids			305.8
Cortisone	N/A	208.84	N/A
Prednisone	N/A	467.62	N/A
Prenisolone	N/A	463.08	N/A
Aldosterone	N/A	N/A	N/A
Androgens			10.8
Testosterone	(Information not available)		
Estrogens			60.2
Estradiol	(Information not available)		

Chart 16: Steroid Hormones—Technical Information (17, 18, 19, 23, 42)

Compound	Typical synthetic process	Typical precursor	Is precursor renewable/non-renewable limited	Alternate precursor by fermentation	Time to implement commercial fermentation by genetically engineered strain
Corticoids					
Cortisone					
Prednisone	semisynthetic	diosgenin or stigmasterol	renewable	glucose	10 yrs.
Predisolone					
Aldosterone					
Androgens					
Testosterone	semisynthetic	chemical modification of cholesterol	renewable	glucose	10 yrs.
Estrogens					
Estradiol	semisynthetic	chemical modification of cholesterol	renewable	glucose	10 yrs.

Chart 17: Peptide Hormones—Market Information (17)

Compound	Market volume 1,000 lb	Bulk cost $/lb	Market value ($ millions)
	Current market data		
ACTH[a]	N/A	N/A	5.6
Bovine growth hormone	0.0	0.0	0.0
Endorphins	(Information not available)		
Enkephalins	(Information not available)		
Glucagon	(Information not available)		
Human growth hormone	N/A	N/A	75.0
Insulin	N/A	N/A	183.1
Ovine growth hormone	0.0	0.0	0.0
Porcine growth hormone	0.0	0.0	0.0
Vasopressin	(Information not available)		

[a]Adrenocorticotropic hormone.

Chart 18: Peptide Hormones—Technical Information (4, 17, 24)

Compound	Typical synthetic process	Typical precursor	Is precursor renewable/non-renewable limited	Alternate precursor by fermentation	Time to implement commercial fermentation by genetically engineered strain
ACTH[a]	extraction	adrenal cortex	limited	glucose and NH_4^+ [b]	5 yrs.
Bovine growth hormone	extraction	anterior pituitary	limited	glucose and NH_4^+	5 yrs.
Endorphins	extraction	brain	limited	glucose and NH_4^+	5 yrs.
Enkephalins	extraction	brain	limited	glucose and NH_4^+	5 yrs.
Glucagon	extraction	pancreas	limited	glucose and NH_4^+	5 yrs.
Human growth hormone	extraction	anterior pituitary	limited	glucose and NH_4^+	5 yrs.
Insulin	extraction	pancreas	limited	glucose and NH_4^+	5 yrs.
Ovine growth hormone	extraction	anterior pituitary	limited	glucose and NH_4^+	10 yrs.
Porcine growth hormone	extraction	anterior pituitary	limited	glucose and NH_4^+	10 yrs.
Vasopressin	extraction	posterior pituitary	limited	glucose and NH_4^+	5 yrs.

[a]Adrenocorticotropic hormone.
[b]Ammonium ion.

Chart 19: Viral Antigens—Market Information (17)

Compound	Market volume 1,000 lb	Bulk cost $/lb	Market value ($ millions)
	Current market data		
Avian leukemia virus	(Information not available)		
Avian myeloblastosis virus	(Information not available)		
Epstein-Barr virus	0.0	0.0	0.0
Hepatitis virus	0.0	0.0	0.0
Herpes virus	0.0	0.0	0.0
Hoof and mouth disease virus	0.0	0.0	0.0
Influenza virus	(Information not available)		
Reoviruses	0.0	0.0	0.0
Rous sarcoma virus	(Information not available)		
Rubella virus	(Information not available)		
Varicella virus	(Information not available)		

Chart 20: Viral Antigens—Technical Information (17, 28)

Compound	Typical synthetic process	Typical precursor	Is precursor renewable/non-renewable limited	Alternate precursor by fermentation	Time to implement commercial fermentation by genetically engineered strain
Avian leukemia... virus		(Information not available)		glucose and NH_4^+ [a]	5 yrs.
Avian myeloblastosis. virus		(Information not available)		glucose and NH_4^+	5 yrs.
Epstein-Barr virus	tissue culture	lymphoblasts	renewable	glucose and NH_4^+	5 yrs.
Hepatitis........ viruses		(Information not available)		glucose and NH_4^+	5 yrs.
Herpes .... viruses		(Information not available)		glucose and NH_4^+	5 yrs.
Hoof and mouth.. disease virus		(Information not available)		glucose and NH_4^+	5 yrs.
Influenza........ viruses		(Information not available)		glucose and NH_4^+	10 yrs.
Reoviruses		(Information not available)		glucose and NH_4^+	15 yrs.
Rous sarcoma ... virus		(Information not available)		glucose and NH_4^+	5 yrs.
Rubella .... virus	tissue culture	duck embryonic cells	renewable	glucose and NH_4^+	5 yrs.
Varicella .... virus		(Information not available)		glucose NH_4^+	5 yrs.

[a]Ammonium ion.

Chart 21: Short Peptides (17, 49), Nucleotides (17, 36) and Miscellaneous Proteins (17)—Market Information

	Current market data		
Product category	Market volume 1,000 lb	Bulk cost $/lb	Market value ($ millions)
Short peptides			
Aspartame	40	110.00	4.4
Glycine-histidine-lysine..........	(Information not available)		
Nucleotides			
5'-IMP	4,000	12.00	48.0
5'-GMP	2,000	12.00	24.0
Miscellaneous proteins			
Interferon	N/A	N/A	50.0
Human serum albumin........	250	1,000.00	250.0
Monoclonal antibodies......	(Information not available)		

Chart 22: Short Peptides (17, 49), Nucleotides (17, 36) and Miscellaneous Proteins (17)—Technical Information

Product category	Typical synthetic process	Typical precursor	Is precursor renewable/non-renewable limited	Alternate precursor by fermentation	Time to implement commercial fermentation by genetically engineered strain
Short peptides					
Aspartame	chemical	phenylalanine & aspartic acid	renewable	glucose and NH_4^+	5 yrs.
Glycine-histidine-lysine...........	extraction	human serum	renewable	glucose and NH_4^+	5 yrs.
Nucleotides					
5'-IMP	extraction	yeast	renewable	glucose and NH_4^+	10 yrs.
5'-GMP	extraction	yeast	renewable	glucose and NH_4^+	10 yrs.
Miscellaneous proteins					
Interferon	extraction or tissue culture	leukocytes, lymphoblasts, or fibroblasts	renewable	glucose and NH_4^+	5 yrs.
Human serum albumin.........	extraction	human serum	renewable	glucose and NH_4^+	5 yrs.
Monoclonal antibodies.....	somatic cell hybridization	various cells	renewable	glucose and NH_4^+	10 yrs.

Chart 23: Antibiotics (17, 37, 47), Gene Preparations (17, 50) and Pesticides (17, 27)—Market Information

	Current market data		
Product category	Market volume 1,000 lb	Bulk cost $/lb	Market value ($ millions)
Antibiotics			
Penicillins.	49,300	22.11	1,080.0
Tetracyclines	29,300	34.13	1,000.0
Cephalosporins . .	4,210	114.00	480.0
Erythromycins . . .	(Information not available)		
Gene preparations			
Sickle cell anemia	0.0	0.0	0.0
Hemophilias.	0.0	0.0	0.0
Thallasemias	0.0	0.0	0.0
Pesticides			
Microbial.	N/A	N/A	25.0
Aromatics.	N/A	N/A	75.0

Chart 24: Antibiotics (17, 20, 37, 43, 51), Gene Preparations (17) and Pesticides (17, 27)—Technical Information

Product category	Typical synthetic process	Typical precursor	Is precursor renewable/non-renewable limited	Alternate precursor by fermentation	Time to implement commercial fermentation by genetically engineered strain
Antibiotics					
Penicillins.	fermentation semisynthetic	lactose & nitrogenous oils	limited	—	10 yrs.
Tetracyclines	fermentation	lactose & nitrogenous oils	limited	—	10 yrs.
Cephalosporins . .	fermentation	lactose & nitrogenous oils	limited	—	10 yrs.
Erythromycins . . .	fermentation	lactose & nitrogenous oils	limited	—	10 yrs.
Gene preprations					
Sickle cell anemia		(No process exists currently)		glucose and NH_4^+	15 yrs.
Hemophilias.		(No process exists currently)		glucose and NH_4^+	20 yrs.
Thallasemias		(No process exists currently)		glucose and NH_4^+	20 yrs.
Pesticides					
Microbial.	fermentation	molasses & fishmeal	renewable	—	5 yrs.
Aromatics.	semisynthetic	naphthalene	nonrenewable	—	10 yrs.

Chart 25: Aliphatics—Market Information (9, 17, 34, 47)

	Current market data		
Compound	Market volume 1,000 lb	Bulk cost $/lb	Market value ($ millions)
Acetic acid	823,274	0.23	189.4
Acrylic acid.	46,503	0.43	20.0
Adipic acid	181,097	0.50	90.5
Bis (2-ethylehexyl) adipate	43,015	0.49	21.1
Citronellal.	394	3.90	1.5
Citronellol.	1,443	4.50	6.5
Ethanol	1,048,000	0.24	251.5
Ethanolamine. . . .	320,236	0.46	147.3
Ethylene glycol . .	3,137,000	0.31	972.5
Ethylene oxide . . .	525,113	0.36	189.0
Geraniol	2,307	3.25	7.5
Glycerol	116,612	0.54	63.0
Isobutylene.	597,712	0.95	567.2
Itaconic acid.	200	0.83	0.2
Linalool.	3,341	2.60	8.7
Linalyl acetate . . .	1,535	3.50	5.4
Methane	878,000,000	0.013	11,573.0
Nerol	462	4.20	1.9
Pentaerythritol. . .	117,085	0.62	72.6
Propionic acid . . .	62,848	0.21	13.2
Propylene glycol .	525,527	0.73	173.4
Sorbic acid	20,000	2.15	43.0
Sorbitol.	160,267	0.36	57.7
α-terpineol	2,416	1.28	3.0
α-terpinyl acetate.	1,066	1.30	1.4

Chart 26: Aliphatics—Technical Information (17, 34, 40, 52)

Compound	Typical synthetic process	Typical precursor	Is precursor renewable/non-renewable limited	Alternate precursor by fermentation[a]	Time to implement commercial fermentation by genetically engineered strain[b]
Acetic acid	chemical	methanol or ethanol	nonrenewable	glucose	10 yrs.
Acrylic acid	chemical	ethylene	nonrenewable	glucose	10 yrs.
Adipic acid	chemical	phenol	nonrenewable	glucose	10 yrs.
Bis (2-ethylhexyl) adipate	chemical	phenol	nonrenewable	glucose	20 yrs.
Citronellal	chemical	isobutylene	nonrenewable	glucose	20 yrs.
Citronellol	chemical	isobutylene	nonrenewable	glucose	20 yrs.
Ethanol	chemical	ethylene	nonrenewable	glucose	5 yrs.
Ethanolamine	chemical	ethylene	nonrenewable	glucose	10 yrs.
Ethylene glycol	chemical	ethylene	nonrenewable	glucose	5 yrs.
Ethylene oxide	chemical	ethylene	nonrenewable	glucose	5 yrs.
Geraniol	chemical	isobutylene	nonrenewable	glucose	20 yrs.
Glycerol	chemical	soap manuf.	nonrenewable	glucose	5 yrs.
Isobutylene	chemical	petroleum	nonrenewable	glucose	10 yrs.
Itaconic acid	fermentation	molasses	renewable	----	5 yrs.
Linalool	chemical	isobutylene	nonrenewable	glucose	20 yrs.
Linalyl acetate	chemical	isobutylene	nonrenewable	glucose	20 yrs.
Methane	chemical	natural gas	nonrenewable	sewage	10 yrs.
Nerol	chemical	isobutylene	nonrenewable	glucose	20 yrs.
Pentaerythritol	chemical	acetaldehyde & formaldehyde	nonrenewable	glucose	10 yrs.
Propionic acid	chemical	ethanol & carbon monoxide	limited	glucose	10 yrs.
Propylene glycol	chemical	propylene	nonrenewable	glucose	10 yrs.
Sorbic acid	chemical	crotonaldehyde & malonic acid	nonrenewable	glucose	15 yrs.
Sorbitol	chemical	glucose	renewable	----	10 yrs.
α-terpineol	chemical	isobutylene	nonrenewable	glucose	20 yrs.
α-terpinyl acetate	chemical	isobutylene	nonrenewable	glucose	20 yrs.

[a]Wherever glucose is mentioned, other carbohydrates may be substituted, including starch and cellulose.
[b]In many cases these times are based on more readily developed fermentations using nonrenewable or limited hydrocarbons as precursors.

Chart 27: Aromatics—Market Information (9, 47)

	Current market data		
Compound	Market volume 1,000 lb	Bulk cost $/lb	Market value ($ millions)
Aniline	187,767	0.42	78.9
Aspirin	32,247	1.41	45.5
Benzoic acid	36,822	0.47	17.3
Cinnamaldehyde	1,098	2.10	3.4
Cresols	94,932	0.54	51.2
Diisodecyl phthalate	151,319	0.42	63.6
Dioctyl phthalate	391,131	0.42	164.3
p-acetaminophenol	20,000	2.65	53.0
Phenol	1,431,000	0.36	515.2
Phthalic anhydride	646,289	0.40	258.5

Chart 28: Aromatics—Technical Information (17, 52)

Compound	Typical synthetic process	Typical precursor	Is precursor renewable/non-renewable limited	Alternate precursor by fermentation	Time to implement commercial fermentation by genetically engineered strain
Aniline	chemical	benzene	nonrenewable	aromatic[a]	10 yrs.
Aspirin	chemical	phenol	nonrenewable	aromatic	5 yrs.
Benzoic acid	chemical	tar oil	nonrenewable	aromatic	10 yrs.
Cinnamaldehyde	chemical	benzaldehyde acetaldehyde	nonrenewable	aromatic	20 yrs.
Cresols	chemical	phthalic anhydride	nonrenewable	aromatic	10 yrs.
Diisodecyl phthalate	chemical	coal tar	nonrenewable	aromatic	20 yrs.
Dioctyl phthalate	chemical	coal tar	nonrenewable	aromatic	20 yrs.
p-acetaminophenol	chemical	nitrobenzene	nonrenewable	aromatic	5 yrs.
Phenol	chemical	coal tar	nonrenewable	aromatic	10 yrs.
Phthalic anhydride	chemical	coal tar	nonrenewable	aromatic	15 yrs.

[a]Aromatic refers to benzene or benzene derivative. Eventually it is anticipated that lignin, a renewable resource, would serve as a precursor.

Chart 29: Inorganics and Mineral Leaching (17)—Market Information

	Current market data		
Product category	Market volume 1,000 lb	Bulk cost $/lb	Market value ($ millions)
Inorganics			
Ammonia	33,400,000	0.06	2,004.0
Hydrogen	451,000	0.15	677.0
Mineral leaching			
Uranium	(Information not available)		
Transition metals. (cobalt, nickel, manganese, iron)	(Information not available)		

Chart 30: Inorganics and Mineral Leaching (17, 52)—Technical Information

Product category	Typical synthetic process	Typical precursor	Is precursor renewable/non-renewable limited	Alternate precursor by fermentation	Time to implement commercial fermentation by genetically engineered strain
Inorganics					
Ammonia	chemical	water and coke	nonrenewable	nitrogen(air)	15 yrs.
Hydrogen	catalytic reforming	petroleum	nonrenewable	water and air	15 yrs.
Mineral leaching					
Uranium		(Information not available)			
Transition metals. (cobalt, nickel, manganese, iron)		(Information not available)			

Chart 31: Projected Growth of Selected Markets Involving Applications of Genetic Engineering

Product category	Current market $ millions	Projected market in 20 yrs. $ millions
Amino acids[a].	300	900
Miscellaneous proteins .	300	1,000
Gene preparations	0	100
Short peptides	5	2,100
Peptide hormones	260	1,000
Totals.	865	5,100

[a]Only four amino acids are considered here.

6.0 Appendix

Although genetic engineering will develop new techniques for synthesizing many substances, the direct displacement of any current industry seems doubtful. Biotechnology should be considered more as a software tool that is becoming available to industry. As such, the response of the current chemical and applied products industry will be to use this tool to maintain their positions in their respective markets. The fact that a variety of companies in the pharmaceutical, chemical, and energy industries have invested in or contracted with genetic engineering firms illustrates this point. Some large companies are already developing in-house genetic engineering research capabilities.

Current areas that might be reduced by introduction of genetic engineering technology include engineering and design, blood fractionation, pollution control, and monoclonal anti-body production. Some equipment manufacturers might be expected to profit by the fact that they produce equipment associated with fermentation. A breakdown of typical fermentation processing equipment and suppliers is presented below.

FERMENTATION EQUIPMENT AND SUPPLIERS

1. Agitators - Drives

   Galigher, Salt Lake City, Utah
   Mixing Equip. Co., Rochester, N.Y.,
   Pfaudler, Rochester, N.Y.,
   Philadelphia Mixers, King of Prussia, Pa.

2. Centrifuges/Filters

   De Vaval Separator Co., Poughkeepsie, N.Y.
   Artisan Industries, Waltham, Mass.
   Chemetron, Louisville, Ky.
   Hytrex, Inc., Hilliard, Ohio,
   Bird Machine Co., S. Walpole, Mass.

3. Evaporators

   APV Co., Tornawanda, N.Y.
   Kontro Co., Orange, Mass.
   Nitro Atomizer, Columbia, Md.
   Pfaudler, Rochester, N.Y.
   Luwa, Charlotte, N.C.
   Swenson Evapco, Harvey, Ill.

4. Fermentors

   Hamilton Kettles, Brighton, England
   Pfaudler, Rochester, N.Y.
   Chemapec, Woodbury, N.Y.
   New Brunswick Scientific, New Brunswick, N.J.
   Vogelbusch, Vienna, Austria
   (U.S.) ABEC, Allentown, Pa.

5. Dryers

   Fitzpatrick, Elmhurst, Ill.
   Krauss Maffai, Wichata, Kansas
   Wyssmont, Fort Lee, N.J.
   Pfaudler, Rochester, N.Y.

6. Storage Tanks - Process Vessels

   De Dietrick, Inc., Union, N.J.
   Dean Products Inc., Brooklyn, N.Y.
   ARMCO, Middletown, Ohio
   Pfaudler, Rochester, N.Y.

7. Control Systems/Instrumentation

   Texas Instr. Johnson City, Tenn
   Beckman, Irvine, Ca.
   Foxboro, Foxboro, Mass.
   Bailey Controls, Wickliffe, Ohio
   Taylor Instrument, Rochester, N.Y.
   Rosemount, Minneapolis, MN.
   Leeds & Northrup, N. Wales, Pa.
   Fisher Controls, Marshalltown, Iowa
   Honeywell, Phoenix, Az.
   Analog Devices, Westwood, Mass.

Part II

Chemical Industry

The information in Part II is from *Chemical Industry* (second draft), prepared for the Office of Technology Assessment of the Congress of the United States. This material was prepared at the Massachusetts Institute of Technology.

Overview of the Chemical Industry: Introduction

The chemical industry is one of the largest in the United States, with sales in 1978 of over 3 billion dollars and with 1 million employees, representing nearly 5% of all manufacturing jobs in the United States (C&E News, Nov. 1979). Today's chemical industry traces its roots back to the German coal tar industry of the late 19th century. The gradual rationalization of chemical science from the 1890's through World War II coupled with improvements in chemical technology and stimulated by the cheap availability of petroleum caused the industry to expand dramatically from the 1940's through the 1960's. Some observers have predicted that the chemical industry is now reaching maturity, will cease rapid development, and finally decline like such industries as steel. Others believe that new technical developments in relationship with continued demand for its final products will stimulate its continued expansion in the decades ahead. Applied genetics may lead to such new technical developments.

The chemical industry transforms low cost mineral raw materials into end-use products of greater value. The most important raw materials are petroleum, coal, minerals (phosphate, carbonate), and air (oxygen, nitrogen). Roughly two-thirds of the chemical industry is devoted to producing inorganic chemicals such as minerals (lime, phosphate rock, potash, salt, sodium carbonate and sulfur) and aluminum sulfate, ammonia, ammonium nitrate, ammonium sulfate, carbon dioxide, chlorine gas, bromine, iodine, hydrochloric acid, hydrogen, nitric acid, nitrogen gas, oxygen, phosphoric acid, phosphorus, sodium hydroxide, sodium phosphate, sodium silicate, sodium sulfate, sulfuric acid, titanium dioxide and urea.

The organic chemical industry transforms inexpensive and highly reduced raw materials, petroleum and coal, into more valuable products that are in a higher oxidation state. The transformation is done in a series of physically separate steps, each of which normally increases the value of the material. Each step in the process can be done by a separate reaction and is often done by a separate company. A profit can be made at each step because the value of the product is increased.

Several types of chemical operations are performed in the chemical industry, for example,

- Catalytic Reforming -- A chemical can be processed at high temperature and pressure in the presence of a metal catalyst to make and break carbon bonds. This is the fundamental way by which a carbon skeleton may be rearranged. After these chemical transformations have taken place the products are separated by extraction techniques, particularly distillation. These steps require a large amount of energy and typically require elaborate separation methods because a variety of products are produced in a high-temperature reaction with a non-specific catalyst. They are usually done by oil refiners.
- Reductions Using Hydrogen -- Because of the expense of hydrogen and of the ready availability of fully reduced raw materials, hydrogen is only used for certain specialized reactions. Over one-third of the hydrogen produced in the U.S. is used for the reduction of nitrogen gas to ammonia. This is economical because nitrogen is not available in the reduced form from mineral sources,

and ammonia is needed for fertilizers and for various chemical reactions.

- o <u>Oxidations</u> -- Hydrocarbons can be oxidized with air, with gaseous or liquid oxygen, and sometimes a halogen such as chlorine. In general these oxidations make the products more polar and more water soluable because carbon-oxygen bonds are more polar and more hydrophilic than carbon-hydrogen bonds.
- o <u>Water-Elimination-Condensations</u> -- Alcohols can react with organic acids to generate esters, and amines can react with organic acids to make amides. These reactions are particularly important building blocks for polymerization reactions.
- o <u>Separations</u> -- After any of the above reactions have been performed, the desired products must be separated from the reaction mixture. Often a variety of other compounds are separated out as well. The reaction conditions may be altered to generate more of a desired compound. In other cases where this is difficult or impossible, new uses for the other products will be found. It is not uncommon that economics change with time so that former by-products become major products of reaction, and the reaction may then be redesigned to produce more of them.

The chemical industry is a capital intensive and raw material intensive industry. Chemical plants use large amounts of energy and can cost hundreds of millions of dollars to build. Most chemical companies amortize plant in 11 years of straight-line depreciation; Dow and DuPont use 16 year accelerated depreciation. This means that evaluating the competitiveness of a process and the market for the product are very important issues for the chemical industry. Often the largest cost in the production of the chemical is the cost of the raw material. The value added ratio or the difference between the selling price and the materials cost divided by the materials cost, ranges from 0 to 1 for typical basic chemical transformations. Therefore, a 20% price reduction by process improvement can be considered a major price breakthrough in this industry. This is not true, for instance, in the electronics industry where the cost of the silicon in the chips is a trivial part of the final price of a microprocessor.

Three main factors press for the continuing development of more expedient processes by which more valuable end products are generated: (i) the raw material cost; (ii) the capital intensive nature of the business; and (iii) the additive value of serial chemical transformations. By integrating more chemical reactions into a plant, a chemical company can reduce the cost of production and consequently achieve a higher return on investment.

The chemical industry is highly competitive and new processes are introduced steadily. New processes develop in response to changing technology, changing raw material costs, and changing market demands. Demand for organic chemicals increases, for instance, in the automobile industry because of the high weight and the increasing expense of metals. The raw material costs of petroleum have increased dramatically in recent years and new technologies are constantly being developed. This phase of the report is concerned with the impact of applied genetics and the resultant effects on the chemical industry and how, perhaps, biotechnology may interact with the chemical industry.

Overview of the Chemical Industry: Chemistry, Thermodynamics and Kinetics

Carbon dioxide is the most stable form of carbon in an oxidizing atmosphere. Figure 1 shows the possible compounds to be made from one carbon and hydrogen and oxygen and their relation to multicarbon compounds. The first compound, methane, is the main component of natural gas. It is available from underground deposits that are the end result of a process that began millions of years ago with the deposition of dead plant matter. Oxidation of methane produces methanol, an alcohol known as wood alcohol. Further addition of oxygen and loss of hydrogen yields formaldehyde, formic acid, and finally carbon dioxide. Carbon dioxide is the end product of the combustion of methane. Carbon dioxide is also a product of organisms that use methane as food. Carbon dioxide can also be the end product of decay from any compound made from carbon.

Many processes in the chemical industry start with long chains of hydrocarbons, which are in effect strings of methane. Oxidations, condensations, and reductions are then performed to generate products. Energetically speaking, the addition of oxygen to carbon is favorable and proceeds on its own given enough time.

Normally there are two ways to speed chemical reactions. The first is to increase temperature. As a rough rule, the rate of reaction doubles for every 10 degree increase in temperature.

The second way of speeding reaction is by catalysis. A catalyst provides a surface upon which a certain reaction is promoted. The advantage of a catalyst is that it can promote one of the possible reactions to happen at a faster rate than all the other possible ones. For instance, a certain catalyst can promote the production of ethylene oxide from a mixture of ethylene and oxygen gases. Simply heating the two gases would result in the production of a mixture of carbon dioxide, carbon monoxide, ethylene oxide and many other compounds.

Figure 1.* Relations of oxidation states among single and multi-carbon compounds. (all numbers are electron difference from carbohydrate $(CH_2O)_N$)

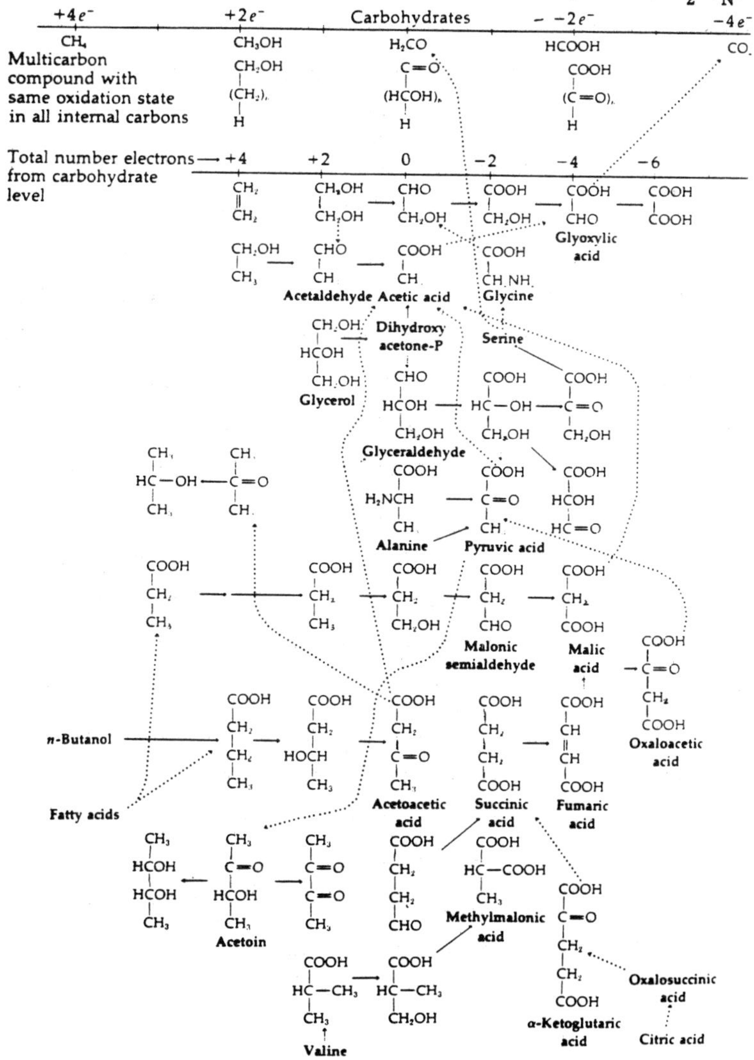

* Metzler, D.E. (1977) Biochemistry- The Chemical Reactions of Living Cells.

Most of the catalysts presently used in the chemical industry are metals or metal complexes. A reason that compounds may not react is the force of the mutual repulsion of the electron shells around the molecules. In order for the compounds to react they must actually be put in close physical proximity. The large and somewhat diffuse electron cloud surrounding a metal atom reduces the repulsive forces between molecules and allows them to react. A sometime disadvantage of the metal catalyst is that it may be non-specific in its action: that is, it may promote a certain type of reaction but it ordinarilly is not limited to the one specific reaction that gives the desired products.

Protein bio-catalysts (enzymes) function somewhat differently and will be discussed in the next section.

Industrial chemical reactions often yield a variety of products that must be separated before they can be used in further chemical transformations. If the products are liquids they can be separated in distillation columns because of different physical properties, normally different boiling points. If the products are in solution they can be separated by extraction into a second, immiscible solvent. For instance, a non-polar product will separate into a non-polar solvent leaving behind a polar product in polar solvent. Energy is required to separate mixtures of chemicals. The more similar the chemicals in a mixture are to each other, the more difficult and expensive separation is.

In summary, the combination of hydrocarbon feed stocks with oxygen to form products is favored by basic thermodynamics. The chemical path from hydrocarbons to carbon dioxide has many branches. Catalysts promote the reaction along either one of or a certain set of these pathways. Products from a reaction must be separated from each other, and this separation requires energy.

Overview of the Chemical Industry: Biochemistry and Enzymes

The oxidation of organic chemicals yields energy which, in an industrial process, is given off in the form of heat. Enzymes in living cells capture the energy from oxidation in chemical energy (usually ATP) and use it to perform the many tasks required of the cell: synthesis of new compounds, transporting materials across membranes, cell motion, and other biological functions.

Of course, if energy is conserved and most life is busily burning sugar to assure its vitality, somewhere that sugar is being made. Green plants capture sunlight and use the energy to reduce carbon dioxide in the atmosphere to sugars. The plants then use some of those sugars to provide for their own energy needs, but the rest is accumulated in what is now commonly called biomass. Therefore, biomass includes starches, cellulose, lignins and other materials. It is this biomass that is the foundation of all renewable resources, that is, being produced by carbon dioxide fixation.

Enzymes promote certain chemical reactions and are

catalysts in the same way that many of the metals used in industrial chemistry are catalysts. Enzymes are proteins and are much less resistant to chemical and physical attack than metal catalysts. But more importantly they are functionally different: (i) they can be extremely efficient catalysts, and (ii) they are extremely specific, often catalyzing only a single transformation. It is this efficiency and specificity that makes enzymes attractive as catalysts for industrial processes.

Although enzymes are catalysts, cells are sometimes called "living catalysts". This means that the cells can take a compound, oxidize it, and usefully recover some of the energy released. This report considers how cells can be used as catalysts in chemical transformations and then how the enzymes themselves are used in chemical transformations. The former is sometimes termed fermentation, the latter is sometimes known as biotransformation.

Fermentations

Many substances are part of the common metabolic pathways shared by almost all life from bacteria to humans. These biochemicals are called primary metabolites and include most sugars, fatty acids, amino acids and nucleic acids. Generally, primary metabolites are the least expensive substances to make by fermentation. They are constantly being produced and consumed within the organism. All that is necessary is to produce a strain that overproduces the primary metabolite and consumes much less of the material.

Other products of cells are known as secondary metabolites. Cells use enzymes that are not part of intermediary metabolism to produce substances for special purposes, examples, the alkaloids and antibiotics. The carbon used to make the secondary metabolites can always be traced back to primary metabolites such as acetyl-CoA or amino acids. Secondary metabolites are also products of fermentation; but generally the yields are lower and genetic manipulation is more difficult.

Therefore potential products of biotechnology may be classed in three groups.

<u>Primary metabolites</u> -- these are already made in substantial quantities by the cell and isolation of strains to hyperproduce them is often possible.

<u>Secondary metabolites</u> -- these substances are already produced by cells, but are only produced under very tight genetic control. Substantial improvements in yields may be developed through applied genetics techniques, but the improvements are often considerably more difficult to achieve than in the case of primary metabolites.

<u>Non-metabolites</u> -- compounds not present in cells but can in principle be made by enzymatic transformations of primary and secondary metabolites, thus these are potential candidates for fermentation. Not only do the problems associated with improving yields in both primary and secondary metabolites have to be addressed, but finding the necessary enzymes in other organisms, isolating the DNA that codes for them, cloning it into the vector and achieving expression must be accomplished. Although this approach holds promise, it has not yet been achieved in any significant commercial scale.

In summary, any compound that is part of intermediary metabolism is a secondary metabolite, or is producible by enzymatic transformation of a metabolite is a potential candidate for production by fermentation.

Bioconversion biotechnology will have impacts on certain aspects of the chemical industry other than fermentation. Enzymes that can be isolated and used as catalysts to perform unique conversions will become more important. This is already done on an economically significant scale for certain hydrolases such as the amylases that are used to cleave corn starch into fermentable sugars. A more exotic use of enzymes is the oxidation of steroids for the synthesis or modification of certain hormones. The difficulties surrounding bio-conversions are related less to producing the required enzymes, and more to technical problems associated with the immobilization of the enzymes and the protection of the enzymes from degradation.

To comprehend the potential scope of biotechnology it is necessary to review how enzymes catalyze reactions. Consider the substance S in Figure 3. It can be transformed into products A, B and C. A metal catalyst catalyzes the transformation from S to A, B and C. While the energy necessary for the transformation is lower than the uncatalyzed reaction, a mixture of all three products will be produced and separation might be difficult. Consider the enzyme catalyzed reaction. It catalyzes only the reaction of S to B. The reaction to A and C are unaffected. The barrier to transformation to B is lower than in the metal catalyzed case and, therefore, the transformation can be conducted at a lower temperature. In addition, the barrier to transformation A and C is so high that the competing side reactions become unimportant.

Fig. 3

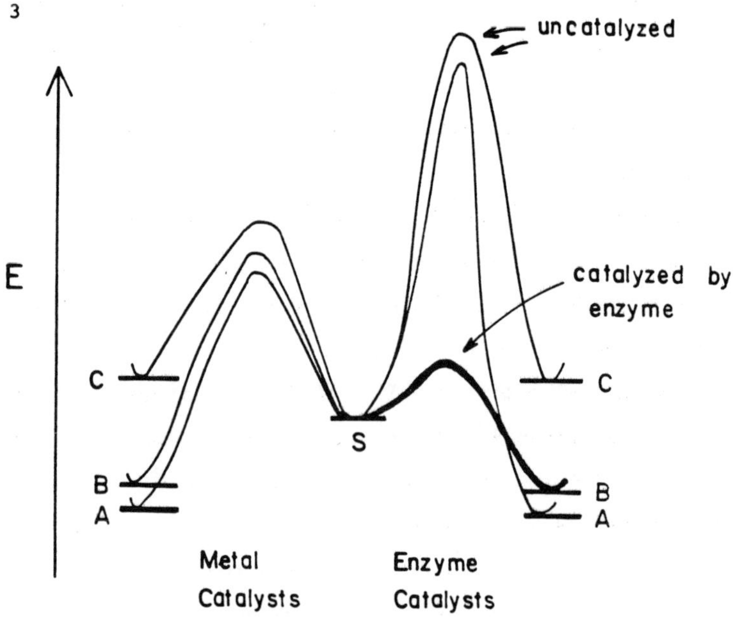

How is it that enzymes achieve such extraordinary specificity?

Enzymes bind their substrate tightly. The substrate becomes a part of the surface of the protein and its rotational and vibrational degrees of freedom are limited. This renders the substrates vulnerable for the reaction that is desired.

Enzyme catalyzed reactions are of four types (Walsh, 1979): (i) group transfers, (ii) oxidations and reductions, (iii) eliminations, isomerations, and rearrangements, and (iv) those that make and break carbon-carbon bonds.

Enzyme Catalyzed Group Transfers

The common feature of these reactions is that some electrophilic group of a substrate molecule is transferred to an acceptor that is plentiful in the cellular environment. Most common is transfer to water which is a nearly ubiquitous nucleophile. These enzymes include esterases, nucleases, and peptidases. Hydrolytic cleavage enzymes are by far the most important commercially in that they include both peptidases and amylases.

A second class is transfer to ammonia rather than water. Ammonia is too reactive to be allowed to remain in the cells; normally ammonia is donated by the amide group of glutamine.

The third group is the transfer of phosphate groups. This is the main reaction by which energy is transferred from one molecule to another.

Enzymatic Oxidations and Reductions

When one molecule is oxidized by an enzyme catalyst, another molecule must be reduced; the electrons removed from the molecule undergoing oxidation must be transferred to some acceptor. Molecular oxygen is a common electron acceptor in biological systems, undergoing a two electron reduction to H_2O_2 and also a four electron reduction to H_2O. When a substrate is oxidized, electrons are removed from it and an electron sink is required.

Enzymatic reduction shows obligate requirements either for co-enzymes to act as electron acceptors or a transition metal to act as a conduit for passage of electrons out of substrate to some acceptor molecule. The most important of these co-enzymes or co-factors are nicotinamide, flavin and pterin.

Enzymes that incorporate molecular oxygen directly into organic molecules are ubiquitous and of high metabolic significance (May, 1979). Such enzymes may become important to the chemical industry. May has clearly summarized the basics of enzymatic epoxidation reactions and points out that from the standpoint of enzyme technology, the ability of these enzymes to incorporate molecular oxygen into organic substrates efficiently and selectively is highly enticing, since these reactions are often poorly accomplished using conventional chemistry. The use of such enzymes to carry out biotransformations is not without significant technical and economic problems. Co-factor recycling and enzyme stability are significant technical problems limiting both the economic and technical applications of such enzymes to the chemical industry. Only when the enzymes can be produced cheaply, stabilized, and the co-factors either are supplied cheaply or regenerated via coupled reactions will such enzymatic epoxidations reactions become economically important to the chemical industry.

Enzymatic Catalyzed Eliminations, Isomerizations and Rearrangements

These reactions are generally related to acid-base chemistry. The common reaction is the elimination of a good leaving group to create a double bond.

These reactions include addition of water to carbonyl groups, the hydrolysis of carboxylic and phosphate esters, the elimination of water from double bonds, many types of rearrangements, and substitution reactions.

Enzymatic Reactions that Make and Break Carbon-Carbon Bonds

These reactions typically involve carbanion chemistry. They include decarboxylation of beta keto acids and decarboxylation of alpha keto acids. Another set of reactions is carboxylations. Condensations of the acetyl group from acetyl-CoA are used to add two carbon units, isoprene is used for 5 carbon units, and folic acid is used to transfer 1 carbon units.

Thousands of enzymes are known and more are discovered each year. Despite this, fundamentally new reactions are now discovered only rarely. It would appear that in principle any functionalized hydrocarbon can be selectively oxidized and reduced, cyclized, rearranged, isomerized or functionalized with nitrogen, sulfur, or phosphorus groups. Within that framework, the potential of biocatalysis appears to many to be limited only by the imagination of the biotechnologist.

Overview of the Chemical Industry: Economics of Chemical Transformations

Chemical transformation, fermentation and bioconversion add value to the raw material so that the producer can make a profit. Table I shows examples of transformations that have been, are, or may become economical. The column headed value-added ratio is the quotient of the price of the product minus the price of raw material over the price of the raw material. Because the basic chemical industry is raw material intensive, one might expect that in order to make a profit all these ratios would have to be greater than zero; however, it is clear that this is not the case. A definition of value more subtle than price per unit mass must be developed.

The increase in value documented in Table I can be divided into two components: increase in the degree of oxidation and increase in the complexity. The column headed hydrogen deficiency is a quantitative measure of the degree of oxidation of the compound. It is the number of hydrogen molecules that can be added per molecule of compound if all double bonds and all rings are fully saturated. It is a more reasonable measure of the degree to which a compound is more valuable than the value of the carbon from which it was derived. The oxygen in these compounds was derived from the oxygen in the air, the cost of which includes only process costs. This is to be contrasted with the carbon which was derived originally from a fossil fuel or from biomass. The second price column gives the price of each compound in terms of dollars per kilogram of carbon. This is a better index to

Table I: Value Adding in Production of Major Organic Chemicals

PRODUCT	FORMULA	HYDROGEN DEFICIENCY	CHIRAL CENTERS	PRICE 3/80 (per kg)	PRICE 3/80 (per kg carbon)	PRESENT SOURCE	YIELD	VALUE ADDED RATIO	CARBON VALUE ADDED RATIO	DOMESTIC PRODUCTION 1978 (metric tons)	POTENTIAL FOR APPLICATION OF BIOTECHNOLOGY (de novo synthesis or bioconversions/organism & enzymes)
acetaldehyde	C_2H_4O	1	0	$0.54	$0.99	Oxidation of ethylene	95%	1.23	1.94		
acetic acid	$C_2H_4O_2$	1	0	$0.51	$1.28	Oxidation of acetaldehyde	90%	0.94	2.03	1,250,000	de novo:
						Fermentation	95%	2.37	2.33		
acetoin	$C_4H_8O_2$	1	0		(1.83x)	Reduction of diacetyl					de novo: Aspergillus, Penicillium; Mycoderma; from 2, 3-butaneodiol: Mycoderma aceti: from diacetyl: yeast
acetone	C_3H_6O	1	0	$0.57	$0.92	Oxidation of cumene (80%)	–	0.98	1.44	1,130,000	de novo: Clostridium acetobutyliticum
						Oxidation of Isopropanol (20%)	85%	1.26	1.31		
						Fermentation	95%	2.86	1.67		
acetylene	C_2H_2	2	0	$1.52	$1.65	natural gas	30%			110,000	
						$CaC + H_2O$	95%				
acrolein	C_3H_4O	2	0	$0.68	$1.06	Oxidation of allyl alcohol	–	0.48	0.47		
acrylic acid	$C_3H_4O_2$	2	0	$0.94	$1.88	Oxidation of propylene	95%	2.58	2.69	146,000	de novo: Clostridium
						CO+ acetylene	80%	0.62	1.14		
adipic acid	$C_6H_{10}O_4$	2	0	$1.10	$2.23	Oxidation of cyclohexane	90%	1.90	1.28	730,000	perhaps
						Oxidation of cyclohexanone	–	1.01	1.51		
butadiene	C_4C_6	2	0	$0.66	$0.74	Petroleum	80%	–	–	1,580,000	low
butanol	$C_4H_{10}O$	0	0	$0.64	$0.77	Reduction of butanal	95%	1.19	0.95	340,000	de novo: Clostridium acetobutyliticum
						Fermentation (6%)	~	2.19	1.40		
butyl acetate	$C_6H_{12}O_2$	1	0	$0.81	$1.17	Butanol + acetic acid	95%	1.41	1.14	55,000	

(continued)

Table I: (continued)

PRODUCT	FORMULA	HYDROGEN DEFICIENCY	CHIRAL CENTERS	PRICE 3/80 (per kg)	PRICE 3/80 (per kg carbon)	PRESENT SOURCE	YIELD	VALUE ADDED RATIO	CARBON VALUE-ADDED RATIO	DOMESTIC PRODUCTION 1978 (metric tons)	POTENTIAL FOR APPLICATION OF BIOTECHNOLOGY (*de novo* synthesis or bioconversions/organism & enzymes)
butyl acrylate	$C_7H_{12}O_2$	2	0		(1.52)	Butanol, acrylic acid	95%	—	--	125,000	yes
butyraldehyde	C_4H_8O	1	0	\$0.54	\$0.81	Oxonation of propylene	—	1.46	1.88	352,000	
cyclohexane	C_6H_{12}	1	0	\$0.58	\$0.68	Butanol + H_2 (80) Hydrocarbon	99%	1.16 / 1.17	1.26 / 1.27		
cyclohexanol	$C_6H_{12}O$	1	0	\$1.08	\$1.50	Cyclohex-anone + O_2	90%	1.86 / 2.07	2.21 / 2.4		
cyclohexanone	$C_6H_{10}O$	2	0	\$1.09	\$1.48	Cyclo-hexanone + O_2	90%	1.88 / 2.09	2.18 / 2.42	552,000	
dihydroxyacetone	$C_3H_6O_3$	1	0	\$17	\$42.50	Fermentation		77	77		f. glycerol: *Acetobacter*
ethanol	C_2H_6O	0	0	\$0.56	\$1.07	Hydration of ethylene	95%	1.27 / 1.34	2.10 / 2.21	570,000 (syn)	*de novo*: *Sacchromyces*
						Ferm. (40%)	95%	.53 / .58	1.95 / 2.05	380,000 (ferm)	
ethyl acetate	$C_4H_8O_2$	1	0	\$0.68	\$1.20	Ethanol + acetic acid	99%	1.27 / 1.28	1.02 / 1.03		
ethyl acrylate	$C_5H_8O_2$	2	0	\$0.90	\$1.65	Ethanol + acetric acid	99%	1.18 / 1.19	1.40 / 1.41	135,000	yes
ethylene	C_2H_4	1	0	\$0.44	\$0.51	Petroleum				11,700,000	
ethylene glycol	$C_2H_6O_2$	0	0	\$0.68	\$1.76	Hydration of ethylene oxide	85%	0.86 / 1.01	1.21 / 1.42	1,760,000	yes
ethylene oxide	C_2H_4O	1	0	\$0.79	\$1.45	Ethylene + O_2	65%	1.80 / 2.77	0.96 / 1.48	2,250,000	yes
formaldehyde	CH_2O	1	0	\$0.16	\$0.43	Oxidation of methanol	90%	0.76 / .84	1.54 / 1.71	2,890,000	yes

(continued)

Table I: (continued)

PRODUCT	FORMULA	HYDROGEN DEFICIENCY	CHIRAL CENTERS	PRICE 3/80 (per kg)	PRICE 3/80 (per kg carbon)	PRESENT SOURCE	YIELD	VALUE-ADDED RATIO	CARBON VALUE-ADDED RATIO	DOMESTIC PRODUCTION 1978 (metric tons)	POTENTIAL FOR APPLICATION OF BIOTECHNOLOGY (de novo synthesis or bioconversions/organism & enzymes)
glycerol	$C_3H_8O_3$	0	0	$1.19	$3.04	Oxidation of propylene	50%	3.22	7.07	59,000	de novo: "Protol" fermentation
						Cl_2,OH propylene	80%	1.32			
						fermentation	—	5.41	5.53		
isoprene	C_5H_8	2	0	$0.34	$0.39					83,000	
isopropanol	C_3H_8O	0	0	$0.42	$0.70	Hydration of propylene	70%	1.14	1.63	780,000	yes
methanol	CH_4O	0	0	$0.21	$0.28	CO + H_2	60%			2,900,000	$CH_4 \rightarrow CH_3OH$
methyl ethyl ketone	C_4H_8O	1	0	$0.62	$1.11	Oxidation of NGL Oxidation of butenes	85%			300,000	
methyl methacrylate	$C_5H_8O_2$	2	0	$1.01	$1.81	Acetone + HCN	85%				yes
propylene	C_3H_6	1	0	$0.37	$0.43					5,900,000	
propylene glycol	$C_3H_8O_2$	0	1	$0.80	$1.69	Hydration of propylene oxide	—	1.21	1.59	250,000	yes
propylene oxide	C_3H_6O	1	1	$0.66	$1.06	Oxidation of propylene	93%	1.78	2.47	920,000	yes
styrene	C_8H_8	5	0	$0.84	$0.91	Benzene + ethylene	90%			3,200,000	
vinyl acetate	$C_4H_6O_2$	2	0	$0.68	$1.22	Ethylene + acetic acid	90%	1.43	1.36	760,000	

use when comparing compounds. The carbon value added ratio was determined by analogy with the value added ratio column but prices per kilogram of carbon were used instead.

The source of value in chemical transformations is addition of complexity and/or functionality. A compound with a variety of functional groups (for instance one that contains both a ketone and a hydroxyl group) is more difficult to make than one that contains only one type of functional group. Chemically speaking the transformation necessary to make one functional group will often affect the other. This suggests why dihydroxyacetone has a much greater price than acetone itself even though they are rather close chemically, differing only by the fact that dihydroxyacetone has two additional hydroxyl groups. Also note that dihydroxyacetone is normally made by fermentation, thus taking advantage of the ease with which living cells make multifunctional products. The resulting high price may also reflect the small volume of the market. A second type of complexity is the presence of rings in compounds. A third type of complexity, and the most important to biology, is a chiral center. A carbon that has four different groups attached to it can exist in one of two forms: the left handed and right handed forms (See Fig. 4).

Fig. 4

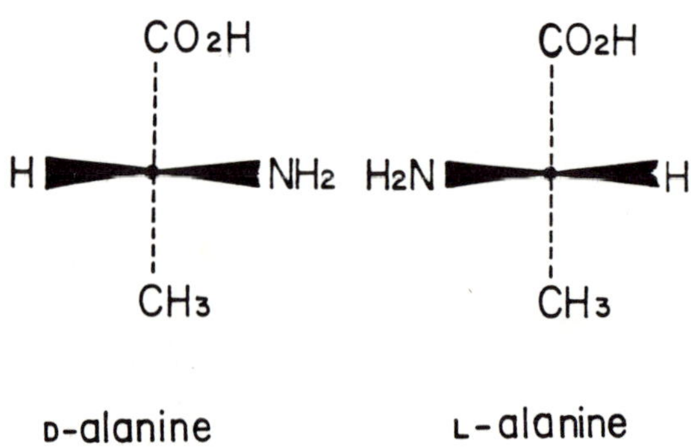

These are chemically identical, but differ only in the fact that one is a mirror image of the other. Compounds that have chiral centers and for which a demand exists for optically pure products are prime candidates for biological production.

Consider transformations one through eight in Table II. Ethylene, a basic petrochemical feedstock, is air oxidized to acetaldehyde. The unsaturation between the carbons in ethylene is eliminated but an unsaturation between carbon and the added oxygen is added so that the hydrogen deficiency remains the same. The molar increase in mass of the acetaldehyde as a result of the oxygen addition explains why the carbon value added ratio is significantly higher than the value added ratio. Transformation 2 shows a further oxidation along the same pathway yielding acetic acid. The hydrogen efficiency again remains the same but another oxygen has been added so that the carbon value added ratio is greater than the value added ratio. This is a good example of why the carbon value added ratio is a better guide to the real value. The cost per unit mass of acetic acid is actually lower than the cost per unit mass of acetaldehyde, so it might appear that the manufacturer is selling at a loss. Actually the oxygen costs virtually nothing and the carbon value added ratio reflects the fact that a profit can be made at these prices for this transformation.

One and two in Table II are very similar transformations: oxidation of organic compounds containing two carbons. Why is the carbon value added ratio so much higher for transformation one? The answer is that acetaldehyde is much more sensitive to further oxidation than acetic acid. The reaction conditions must be more tightly controlled, and the greater expense involved is reflected in the price of acetaldehyde.

Example 3 shows one of the few bioconversions currently being used on a large scale. Sorbitol, made by catalytic hydrogenation of glucose, is oxidized by *Acetobacter suboxydans* to

Table II: Adding Value in Selected Chemical Transformations, Fermentations and Bioconversions

Transformation	Substrate	Hydrogen Deficiency	Chiral Centers	Product	Hydrogen Deficiency	Chiral Centers	Value-° Added Ratio*	Carbon° Value-Added Ratio*
(1) chemical	ethylene	1	0	acetaldehyde	1	0	1.23	1.94
(2) chemical	acetaldehyde	1	0	acetic acid	1	0	0.94	1.30
(3) bioconversion	sorbitol	0	4	ascorbate	2	2	11.4	11.5
(4) fermentation	hexoses	-	-	acetic acid	1	0	2.32	2.33
(5) fermentation	hexoses	-	-	glutamic acid	2	1	18	17.6
(6) fermentation	hexoses	-	-	penicillin V	7	4	770	560
(7) chemical	benzaldehyde	5	0	phenylalanine	5	1	20	24
(8) fermentation	hexoses	-	-	phenylalanine	5	1	250	152

* Market price for chemicals, March 1980; fermentable sugars at $0.22 per kg.

o All values should have one subtracted from them.

sorbose and then chemically converted to ascorbic acid. This reaction involves oxidation at a specific hydroxide out of four and creation of a new chiral center. It is, therefore, not surprising that bioconversion is most efficient.

Examples 4, 5, and 6 show three commercially important fermentations. Hexoses (fermentable sugars such as glucose) are the raw material in each case. Acetic acid is a primary metabolite and is derived from a single reaction from the main pathway of anaerobic glycolysis. Glutamic acid is also a primary metabolite and is available in one step from the tricarboxylic acid cycle by a simple transamination. Penicillin is a secondary metabolite and is produced in huge quantities by the chemical industry. It has two rings including a strained beta-lactam ring and four chiral centers. Note that even for an antibiotic fermentation that has been constantly improved for 40 years, the value-added ratio is several hundred.

Examples 7 and 8 show a compound where chemical and fermentation processes are currently competitive. Phenylalanine is one of the more expensive amino acids currently, costing about $55 per kilogram. The demand is not large now but is expected to increase. The details of the two processes are discussed below, but suffice now to say that improvements are possible in both chemical technology and biochemical technology for production of phenylalanine and it is not yet clear which will be ultimately more economical.

In summary, the chemical industry will be changed by biotechnology if strains of microorganisms, either as whole cells or a source of enzyme catalysts, can carry out chemical transformations at competitive costs. Indeed, it is not merely a single chemical transformation which requires consideration but an entire process. Like any other current fermentation process, a number of conditions must be satisfied:

a) the starting material, or fermentation substrate, must be relatively plentiful and inexpensive;

b) the product must be sufficiently valuable to warrant large-scale production;

c) the yield must be sufficiently high so that the process is profitable;

d) total income from production must be large enough to justify research and development costs;

e) environmental and safety concerns have to be satisfied.

Improvements in the manner in which biotechnology could have impact into the chemical industry may be brought about via recombinant DNA technology and could be put to greater use if the following ideas relevant to all fermentation processes, are also used.

a) In any metabolic scheme, reactions cannot run "uphill," that is, the net reaction must proceed with a decrease in free energy. Enough chemical energy must also be produced for the organism to survive. Also, the "electron flow" must balance. An oxidation must accompany a reduction and vice-versa. These basic concepts appear to get lost in many of the recent analysis being put forth by some of the applied genetics firms.

b) A fermentative product itself may not be useful, but it may be easily converted to a useful compound via chemical means. Such coupled bacteriological/chemical processes may prove useful if economically sound. What is needed is insight into the chemical industry to identify the appropriate chemical reactions that may be conducted via biotechnology processes.

c) If an enzyme in some organism is non-specific, a useful product may be obtained from an inexpensive compound which also acts as a substrate for the enzyme. Research in this area is necessary to explore specific examples and applications.

d) Some chemical transformation processes may be operable using the pure enzyme(s) alone as discussed above. This strategy would probably prove more efficient, and genetic engineering could contribute to the production of such enzyme(s).

e) Finally, the price and availability of substrates may change such that a process which was once unprofitable is now economically feasible. This is particularly relevant to world oil and sugar markets, which are both undergoing unsettled periods.

The Prospect for Change: An Overview of New Process Introduction in the Chemical Industry

The development of biotechnology should be viewed not so much as the creation of a new industry, but as the revival and revitalization of an old one. Many basic organic chemicals were produced by fermentation before petroleum became relatively cheap. As recently as the 1950's, the majority of glycerol and a substantial quantity of industrial ethanol were made by fermentation techniques (see ethanol case study). Now, of course, interest in industrial fermentation is greatly revived, particularly for ethanol (under the name "gasohol" when added to gasoline). Many specialty organic chemicals have always been made by biological methods; but now with the rapidly increasing cost of petroleum the potential to produce certain feedstock chemicals (methane, acetone, butanol, acetic acid, etc.) by biological processes is being reconsidered. It should be cautioned that the increase in petroleum alone is not sufficient cause for one to change processes as cost of sugars and other biological raw materials will also increase with rising petroleum costs. To be convincing one needs to demonstrate such a process is currently competitive with petroleum feedstock.

Public awareness and interest in biotechnology has increased dramatically in the past year. It is difficult to turn on a radio or TV, pick up a newspaper, or read a magazine without encountering a discussion of gasohol, interferon, or the promise of recombinant DNA technology. The surge of public concern over the safety of recombinant DNA that followed the <u>Asilomar</u> conference in 1975 has been replaced by great interest in the potential economic, environmental, and health benefits that biotechnology may offer.

Like investigations of the properties of coal for derivatives that gave us the chemical age, or studies of the electrical properties of special silicon crystals that gave us the electronic age, the biological age is based on discoveries that did not seem at the time to be of great industrial importance. It was only in 1953 that key concepts for understanding DNA structure were proposed and only in the 1960's were some of the basic mechanics of transcription, translation and expression elucidated. The discovery in the 1970's of restriction enzymes was quickly followed by techniques to manipulate and splice DNA, and to sequence it. Many feel that this accelerating rate of scientific discovery will promote the development of an industry in the 1980's that could affect large segments of the chemical industry.

Less frequently discussed are the developments in technology necessary to exploit science. The mechanics of life are fascinating, but they do not necessarily let you produce a cheaper fuel or plastic. The technologies which are required to exploit microbes include development of more efficient, and creatively engineered fermentation facilities and product separation processes. Microbes may produce molecules but they will not isolate, purify, concentrate, mix or package them for human use.

The use of enzymes for industrial processes has been discussed since their discovery. Presently, enzymes have found limited application in the pharmaceutical, chemical and food industries. Perhaps the most spectacular success story in recent years has been the use of enzymes to produce high fructose sweetners from corn syrup.

Many problems remain to be solved. Recombinant DNA technology as well as other genetic manipulations allow us to transfer the gene coding for a particular enzyme to an organism which conceivably could produce that enzyme in a fermentation process more efficiently than the native organism. Additionally genetic manipulation may even change substrate specificities as well as the optimal pH and temperature of a particular enzyme. But genetic engineering does not completely circumvent the problems associated with building and operating a large fermentation or enzyme bioconversion facility.

The application of biotechnology in a modern setting as applied to the chemical industry is unproven and underdeveloped. Investment involves high risk and maintaining a long term perspective, because even the most optimistic of the companies realize that returns from this research cannot be realized until, as best, late in the 1980's.

Recently, the difficulty of maintaining profitability in an environment with escalating raw material costs has resulted in concentrated efforts to achieve short-term goals. Wall Street has exacerbated the problem by magnifying quarterly earnings reports. Another major influence is that careers tend to be rewarded on performance over a three or four year period.

The chemical industry is heavily staffed with analyzers and optimizers of the status quo, and forces against change are formidable. The increasing number of tools and techniques for business analysis are generally limited to examining past performance and the short term future (Malpus, 1979).

It is more difficult to justify investment in new technology in periods of slower growth, particularly when it is coincident with high inflation and high discount rates. High discount rates reduce long term benefits to small values, and the risks which usually occur in the early years are correspondingly heightened. High inflation and high discount rates further complicate the already difficult task of evaluating a new process. New technology may seem of questionable superiority at the time of its introduction, but may later be proven to be much better. Few of the processes introduced in the chemical industry since World War II caused the old process to be abandoned soon after the introduction of the new one. Often the new process makes products which are only slightly different from that already supplied to the market. Even if advantages are substantial, high discount rates tend to reduce them to insignificance. To make the evaluation more difficult, it is extremely difficult to attribute to new technology the effects of future "learning."

The effect of the "learning curve" has been substantial in many industries and is responsible for the explosive growth of the electronics industry. Time will prove whether biotechnology will show similar learning curves. In many cases, chemical processes already are close to thermodynamic limits, whereas the electronics industry is only now beginning to approach these limits.

One analysis of investments in new technology over a long period produced the following interesting conclusions (Malpus, 1979).

- The first plant should not necessarily be the largest technologically possible, if, by making it smaller, a better on-stream factor or greater operability is insured. The profit potential of the first plant thus sacrificed is recovered in the subsequent one.
- Advancing or retarding the timing of plant two by a short time such as two years does not make much difference to the discounted cash flow of both projects combined. This suggests that the overall business is affected more by insuring a high return for the second project than by advancing its introduction in an attempt to correct the low return of the first plant. Thus, using your pilot plant to assure that your second plant will be of the highest possible quality may be one of the best investment strategies.
- The analysis suggests also that an older plant can experience "excess de-bottlenecking". It may appear that the best investment of money is to make small improvements in an old plant, thus increasing the return on

investment. However, the longer term view suggests that at some point investment in a new process becomes a better long term investment.

The introduction of any new process biotechnology or chemical technology should be evaluated carefully,including consideration of:

a) change of raw materials and change in their costs;

b) improvement in process and energy costs;

c) improvement in separation costs;

d) changes in market demand;

e) long-term profitability relative to amortized research and development costs.

The chemical industry strives to find and fill new market demands. It strives to serve established markets with higher quality and lower prices. It strives toward greater profitability. Biotechnology may help to some extent in achieving these goals.

Future Products and Processes in the Chemical Industry

The strategy of chemical companies is towards higher valued products, towards specialty and performance chemicals with a high element of utility in their sales price, and away from commodity chemicals with their requirements for high capital intensity, and lower profit margins. The drive towards specialty chemicals is not to simply improve financial performance. The market has an ever increasing need for higher yield of existing products and an even present demand for new products.

The impact of applied genetics in the chemical industry will be primarily on high value products.

Direct Production of Fuels

It seems to many observers that directly making liquids fuels from sunlight would be a more efficient way of using biotechnology than the industrial fermentation of plant matter. One proposal is to use photosynthetic microbes that excrete ethanol as a waste product and to recover the ethanol by vacuum distillation. A different proposal has been made by Melvin Calvin of the University of California. Professor Calvin suggests that plants can be developed from wild species of the genus *Euphorbia* (a relative of the *Hevea* rubber tree) that produce an oil that can be used as chemical feedstock or used directly as fuel. These plants grow wild in semi-arid regions of the wrold including Africa, Japan, Israel and the United States.

Experimental plantings have produced on average 10 barrels of oil per acre per year. Calvin estimates that oil from a 1,000 barrel per day facility would cost $60 per barrel. It is estimated that genetic manipulation could raise the yield of oil from Euphorbia to 50 barrels per acre, and Calvin believes that they can achieve 20 barrels per acre this year. A two thousand acre farm and associated extraction facility may be able to produce oil for $15 per barrel.

Organic Chemicals

Two modes of biotechnology will have an impact on chemical process development: fermentation and enzyme engineering. The first will primarily change resource entry into the chemical industry, and affect the transition from nonrenewable (petroleum) to renewable (biomass) feedstocks (see Fig. 5). The second will have two functions: to allow the entry of fermentation-derived feedstocks into the chemical intermediate chains, and to compete directly with traditional chemical transformations.

Limitations of Fermentation. The source of feedstocks for all fermentation must be biomass, and for reasons discussed above, must be transformed into 5 and 6 carbon sugars by either chemical or enzymatic processes before fermentation can begin. Sugars can be fermented with no pretreatment, starch and cellulose must be saccharified with acid or with enzymes before fermentation.

Problems with biomass utilization include:

1) competition with food needs for starch and sugar;
2) cyclic availability;
3) biodegradability and associated storage problems;
4) high moisture content for cellulosics, and high collection and storage costs associated;
5) requirement of mechanical processing for cellulosics;
6) heterogenous nature of cellulosics (mixtures of cellulose, hemi-cellulose and lignin);
7) need for disposal of non-fermentable portions of biomass.

Figure 5. Flow of industrial organic chemicals from raw materials to consumption. Values of shipments in US$billion are the estimated 1977 values. Data sources: US Industrial Outlook United States Department of Commerce, 1978; Kline Guide to Chemical Industry Fairfield, N.J. Adapted from Tong (1979). • Billions of dollars.

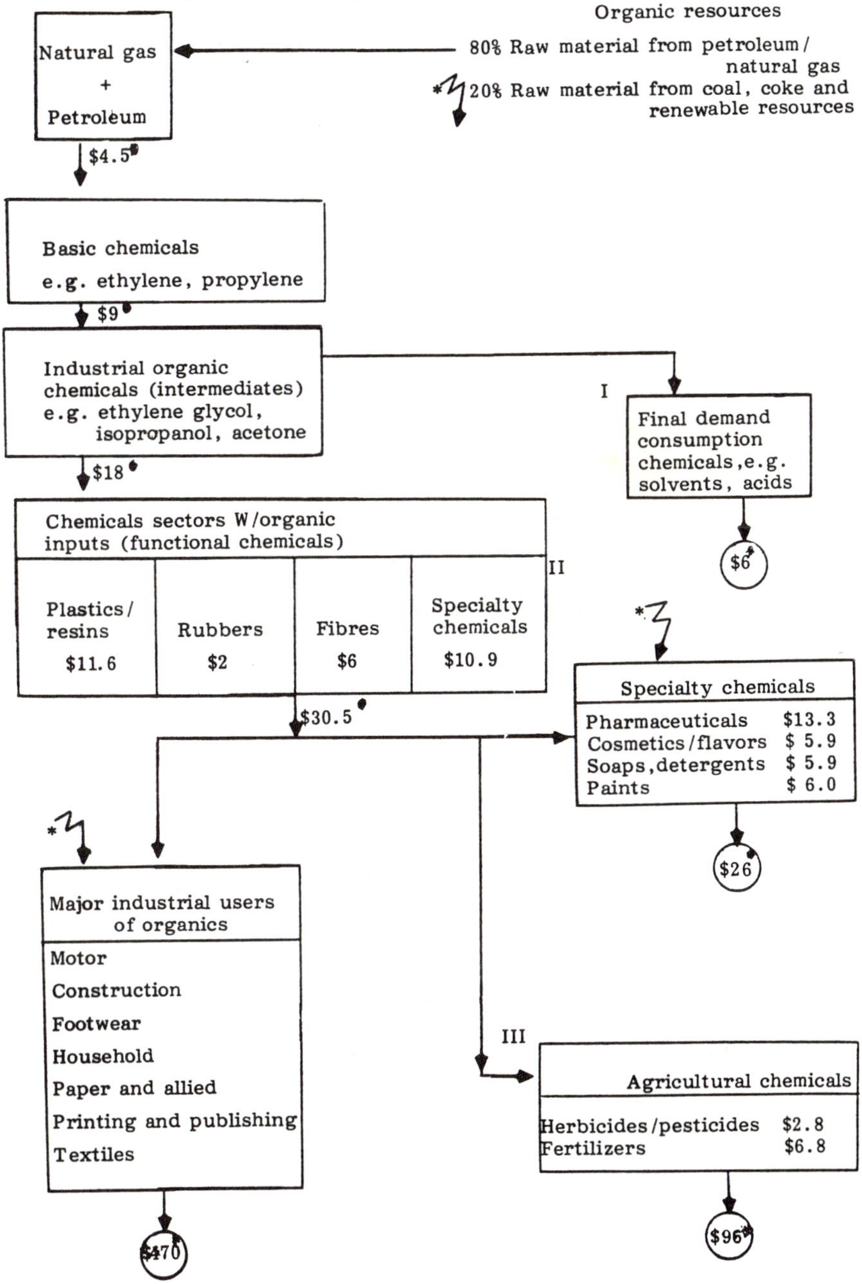

Two generalizations can be drawn about the problems associated with obtaining solutions of sugars suitable for fermentation. For food-related biomass sources such as sugar, corn and sorghum, few technological barriers exist for conversion to fermentable sugars, but domestic economics require subsidies to make fermentation to fuels and basic chemicals profitable. For cellulosic biomass sources such as agricultural wastes, municipal wastes and wood, the economics still do not favor fermentation to fuels and basic chemicals, and technological barriers exist in collection, storage, pretreatment, fermentation and waste disposal.

The nature of biochemical processes places further limits on fermentation. The starting material is half-reduced carbon (carbohydrate, $C_nH_{2n}O_n$). Since the cell is living off the energy that is available from the oxidation of the sugar, the products are inherently oxygenated carbon compounds ranging from simple molecules such as ethanol and acetic acid to larger molecules such as lactic, fumaric, and citric acids. Ethanol is in a lower oxidation state than sugar, but the fermentation produces a mole of CO_2 for each mole of ethanol, so the net change is the oxidation the cell needs to live on. In general the fermentation of a compound that is in a lower oxidation state than sugar requires a compensating production of CO_2. This generalization of thermodynamics is formalized for the theoretical yields for major fermentation, shown in Table III and graphically summarized in Fig. 6.

Expansion of Fermentation Chemical Industry. A schematic representation of fermentation processes in the chemical industry. both at present and projected is shown in Figure 7 . The impact can be divided into three parts. In the order that they have been and probably will be experienced:

1) Aerobic fermentation to produce enzymes, vitamins, pesticides, growth regulators, amino acids, nucleic acids, and other specialty chemicals.

Incomplete

Table III. Theoretical yield for chemicals in major fermentations.

Chemical	% Overall Weight Yield
Solvents	
n-Butanol, acetone, ethanol	41.2
n-Butanol, formaldehyde	67.0
n-Butanol, isopropanol	37.2
Glycerol, acetaldehyde	75.5
Ethanol	51.1
Acids	
Acrylic	
Acetic, propionic	50- 74.4
Acetic	100.00 *
Butyric	
Caproic	
Lactic	
Citric	
Methane	27.00
Amino Acids	
Glutamic Acid	
Lysine	40 - 70 ~~g lysine/g substrate~~
Aromatic	

* Theoretical from C. *thermoaceticum*.

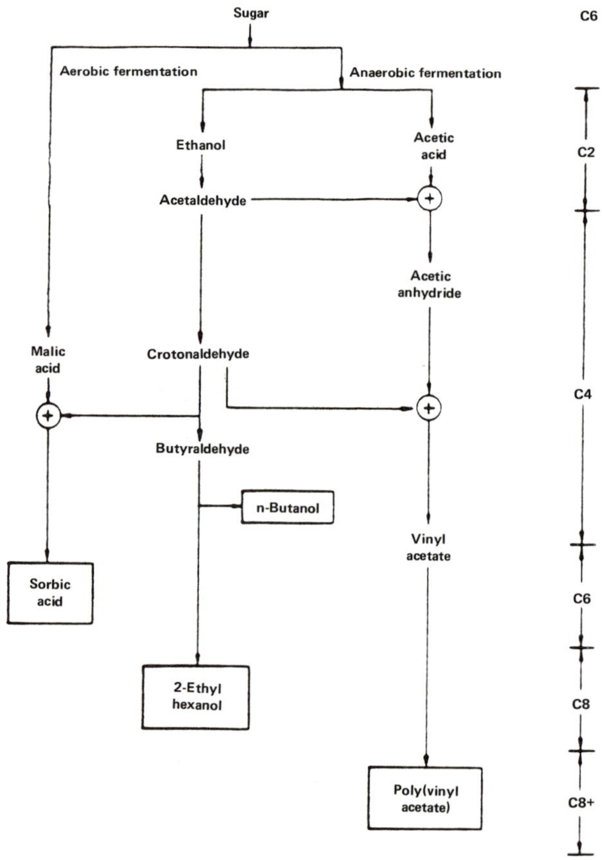

Figure 6. Oxygenated organic chemicals produced from fermentation chemical derivatives. From Tong (1979)

2) Anaerobic fermentation to produce organic acids and solvents.

3) Chemical modification of fermentation products of both aerobic and anaerobic fermentations.

These are summarized in Table IV.

Fermentation is already well-established in the first category and should grow. Very complex biochemicals are now made almost invariably by fermentation (antibiotics and growth factors).

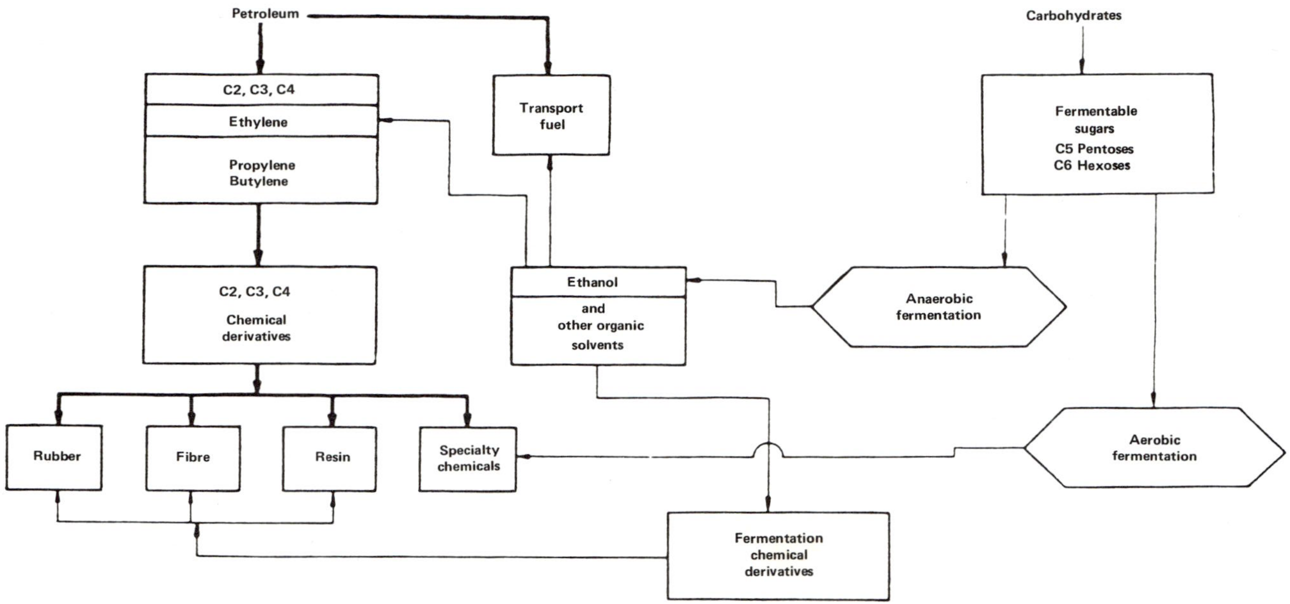

Fig. 7: Diagram of alternative routes to organic chemicals. From Tong (1979)

Table IV: Expansion of Fermentation into the Chemical Industry

I. Aerobic Fermentation	examples
Enzymes	Amylases, proteases
Vitamins	Riboflavin, B_{12}
Pesticides	*Bacillus thuringiensis*
Growth regulators	Gibberellin
Amino Acids	glutamine, lysine
Nucleic Acids	
Acids	Malic Acid, citric acid
II. Anaerobic Fermentation	**examples**
Solvents	Ethanol, Acetone, n-Butanol, Iso-
Acids	propanol, Acetic, Propionic, Acrylic
III. Chemical Modification of Fermentation Products	**examples**
Basic Chemicals	Ethylene
Pesticides	Propionates
Other: plastizers, additives, gums	Ethyl hexanol, sorbic acid, xanthan, pullulan

The same is true for enzymes. Amino acids and nucleotides, being somewhat less complicated molecules, are only in some cases fermentation products, but in some cases fermentation production may be expected to increase. Separate sections are devoted to these compounds.

If the first category is the realm of established fermentation, the second is its area of greatest current growth. Already 40% of the ethanol produced in the U.S. is from fermentation. The barriers, as discussed above, are related primarily to the need for a cheaper technology for converting cellulosics to fermentable sugars. New fermentations for acetic, propionic and acrylic acids will help this area to expand.

The longest development time or wait can be expected for compounds in the third category. A scheme for making certain valuable products from fermentation chemicals developed by Tong (1979) is shown in Figure 7 . In addition, dehydration of fermentation ethanol provides access to the bulk of the chemical indsutry's products through established chemical technology. The development of this area will be slow for the following reasons: (i) under the present economic structure, petroleum is much cheaper as a source of chemicals than biomass processes. The relative price of oil over biomass would have to increase many fold to make it competitive. (ii) The limited availability of biomass and the huge demand for motor fuel will require that virtually all fermentation ethanol be used as a solvent or motor fuel.

Fermentation Routes to Chemicals and Previous Impact of Applied Genetics

Two types of fermentation processes have evolved. One type has been for the production of simple chemical products such as isopropanol, n-butanol, acetone, ethanol, 2,3 butylene glycol, glycerol and the acids such as acetic, fumaric, and citric acid. The other type process has been for the production of more complex chemicals such as antibiotics, enzymes and hormones.

Traditional processes have evolved, to date, for the production of the small molecular weight chemicals. This section attempts to assess the role that genetic research has played in the development of such fermentation processes. As one will see the role of genetic research has mainly been one of strain isolation. Very little genetic research has been employed to improve the industrial processes for small molecular weight chemicals because only a few are presently produced by fermentation. Therefore, one can safely assume that little of the "new genetics" has had an impact on the production of these small molecules.

Acetic acid

Acetic acid is a chemical which can be produced via biological means. For example, ethanol may be oxidized to acetic acid by *Acetobacter suboxydans*; this is the procedure by which vinegar or food grade acetic acid is made. The theoretical yield is 0.67 gm acetate per gram/glucose.

Annual production of vinegar was reported to be 125 million gallons in 1976; and according to Chemical Week, January 12, 1977, vinegar will be one of the fastest growing markets for ethanol (10% per year).

Other microbial processes are possible. One of the most attractive involves *Clostridium thermoaceticum* wherein quantitative conversion of glucose to acetic acid is theoretically possible, i.e.,

$$1 \text{ glucose} \xrightarrow[\textit{thermoaceticum}]{\textit{Clostridium}} 3 \text{ acetic acid} + H_2O$$

The reason that the conversion is quantitative is that the organism is able to fix carbon dioxide. *C. thermoaceticum* is one of the few bacteria that can use carbon dioxide as an electron acceptor and thus can quantitatively convert glucose to acetate. In this case the theoretical maximum yield is one gram of acetate per gm of glucose. For most chemical feedstock fermentations, the most important cost factors are the raw material cost, which is usually 50-60% of the cost, and actual productivity (gm acetic acid per unit volume per day). It is in the latter aspect that difficulties arise with the utilization of *C. thermoaceticum* for acetic acid production. The present productivities are extremely low because of two factors. One is that the cell mass in the fermentation broth is low. The final concentration of acetate is also low, a maximum of 60 gm/l, due to the fact that growth and production of acetate is inhibited by acetate.

The question then becomes can biological processes compete with chemical processes? At present acetic acid is produced by seven companies in the U.S. Domestic capacity for production of synthetic acetic acid (1,605 thousand metric tons) - according to the Chemical Economics Handbook, SRI, Menlo Park, California - is split almost equally between three processes:

a) n-butane oxidation (33%)

b) methanol carbonylation (32%)

c) acetaldehyde oxidation (32%)

From 1968 to 1978, production of acetic acid increased at the average rate of 2.7% per year, with most growth occurring from 1968 to 1973 (6.9% average/year). The carbonylation process will increase in capacity and will probably become the major source of acetic acid in the U.S. by the mid-1980's.

The major cost components for two of the three major synthetic processes were summarized by SRI for 1978 as:

	Cents/lb. of Acetic Acid	
	Acetaldehyde Oxidation * With Air	Low Pressure * Carbonylation of Methane
Net variable costs	13.0	6.5
Labor	0.3	0.3
Other production costs	2.2	1.8
Total production costs	15.5	8.6

*Assumes 63¢ lb/ of methanol and 15.4¢ lb of acetaldehyde.

It is obvious why the methanol carbonylation process will dominate the market for acetic acid. In view of the fact that methanol will be a major product of large-scale coal gasification processes, the carbonylation process begins to seem even more significant. Obviously biological processes for production of acetate will have to compete with the above process technologies, and the raw materials for biological processes will have to be competitive with methanol as a feedstock.

There has been little genetic research of any kind with regards to the organism C. thermoaceticum for the production of acetic acid and no commercial processes have ever been developed.

Acetobacter suboxydans can also be used for oxidation of other chemicals besides ethanol as listed in the table below and described by Greenshields (1978).

List of Substrates Oxidized, and Products Formed, by *Acetobacter Suboxydans*

Substrate	Product
2,3-Butylene Glycol	Acetylmethylcarbinol
Erythritol	L-Erythrulose
Ethylene glycol	Glycollic acid
D-Gluconic acid	D-2-Oxogluconic acid
D-Gluconic acid	D-5-Oxogluconic acid
D-Glucose	D-Gluconic acid
Glycerol	Dihydroxyacetone
D-Mannitol	D-Fructose
Perseitol	Perseulose
α-Propylene glycol	Acetol
D-Sorbitol	L-Sorbose

As described by Greenshields (1978)

Greenshields, R. N. 1978. In: Economic Microbiology, Vol. 2. Primary Products of Metabolism, ed. by A. H. Rose, Chap. 4, 121-186.

Oxidation of mannitol to fructose was conducted on an industrial scale during the 1950's, but the process was discontinued because of the high cost of the raw material (mannitol) and other economic considerations.

A modern industrial fermentation utilizing *Acetobacter suboxydans* for preparing dihydroxyacetone from glycerol has only recently been mounted on a commercial scale, despite the fact that this conversion was known in the late nineteenth century and that the usefulness of the product as a pharmaceutical became known in the early part of the twentieth century. Although manufacture of dihydroxy-acetone depends on efficient chemical purification with carbon treatment, resin treatment and evaporation, the fermentation process has a high conversion efficiency (95-97% yield) and a pure product (0.2% ash).

The sorbose fermentation is of special interest since L-sorbose is used in synthesis of L-ascorbic acid (vitamin C). The substrate (sorbitol) is a cheap raw material produced by electrolytic or catalytic hydrogenation of glucose. Again A. suboxydans is used, and 20 to 20% solutions of sorbitol plus accessory nutrients with vigorous aeration and agitation allow almost quantitative conversion and subsequent 70% yield after chemical purification to a crystalline product.

Perseulose, a rare ketoheptose, gluconic acid and 5-oxygluconic acid, acetylmethylcarbinol or acetoin, L-erythrulose and D-tartaric acid have all been prepared industrially by Acetobacter species.

Except for some strain isolation procedures no major applied genetic studies have been reported with respect to A. suboxydans. The organism is not listed in R. C. King's book entitled "Word Books of Genetics", (Vol. 1, Bacteria, Bacteriophages and Fungi. Plenum Press, New York), and thus any strain improvement work may have been in industrial laboratories. Conner and Allgeier (1976) reviewed the microbiology of vinegar production. The bacterium is described as having an abnormally high mutability by some investigators but others have disagreed. One investigator found that with 39 strains of Acetobacter there were no obvious changes in properties and a controversy still exists. It appears that one of the reasons that genetic research has not had any impact with Acetobacter suboxydans is that it is still somewhat difficult to enumerate and handle the organism in the industrial laboratory. (Conner, H.A. and Allgeier, R.J., 1976. Vinegar; Its History and Development. Advances in Applied Microbiology. Vol. 20, 82-133, Academic Press, Inc.).

Citric Acid

Citric acid is the most important food acidulant and historically has held over 55-65% of the total food market. The two other important acidulants are phosphoric acid (used almost exclusively in cola-flavored beverages) with 20-25% of the total acidulant market, and malic acid (used primarily in newly marketed food and beverage products) with 5% of the market.[1]

As of June 1977, only two companies were producing commercial quantities of citric acid. They are Miles Laboratories, Inc. and Pfizer, Inc. SRI estimates annual production capacity in 1977 as follows:

Company and Plant Location	Annual Capacity June 1977 (Millions of Pounds)
Miles Laboratories, Inc. Dayton, Ohio	25
Elkhart, IN	75
Pfizer, Inc. Chemicals Division Brooklyn, NY Groton, CT Groton, CT	165
Southport, NC[x]	50
	315

[x]Takeda n-paraffin process or fermentation of beet molasses

[1]From SRI Chemical Economics Handbook.

U.S. Citric Acid Consumption*

	Millions of lbs.	% of Total	Millions of lbs.	% of Total	Millions of lbs.	% of Total
Food & beverage uses	120	71	150	70	190	66
Pharmaceutical uses	30	18	24	11	30	10
Industrial & miscellaneous applications	20	11	40	19	70	24
Totals	170	100	214	100	290	100

From SRI, Chemical Economics Handbook

*Total citric acid consumption is expected to continue growing at a rate of approximately 6-7% per year.

For twenty years prior to 1970 the list price of citric acid remained fairly stable at 26 to 30¢ per lb. From July 1970 to March 1977 the list price increased from 32 to 59¢ per lb.

Citric acid is produced commercially by _Aspergillus niger_ using one of three processes:

a) the Koji fermentation process (mainly used in Japan);

b) liquid surface culture process;

c) submerged culture process.

Miall[1] has reviewed in detail the historical developments of these citric acid processes. He discusses the significant developments concerning trace metal requirements and elucidates the biosynthesis route of these industrially important fermentations. Much of what follows is quoted directly from Miall.[1]

The development of the citric acid fermentation owes much to the work of Johnson and his colleagues at the Uni-

[1]Miall, L.M. 1978. In: Economic Microbiology, Vol. 2, Primary Products of Metabolism, ed. by A.H. Rose, Chapter 3, pp. 48-111.

versity of Wisconsin, as does early development of penicillin fermentation. Relevant to this study is the fact that surprisingly little work has been published on improving citric acid producing strains of *A. niger* by mutation.[1] In 1951 (British Patent No. 653,808) Miles Laboratories mentions the use of a mutant strain of *A. niger*. Weight yields of 110% have recently been reported using *A. niger* mutants and solutions containing 160 g sucrose per liter with added salts in surface culture.[1] The mutants were obtained by irradiation of a strain for which a maximum yield of 29% was reported.

Commercial processes for production of citric acid by other organisms have recently been developed. The most important is the Takeda n-paraffin process. The most significant development is formation of citric acid from hydrocarbons by *Candida lipolytica*. Although Kyowa Hakko Kogyo in 1970 patented the use of *Arthrobacter paraffineus* (British Patent 1,187,610) and the Takeda Chemical Industries in 1971 (British Patent 1,257,900) the use of strains of corynebacteria, most of the patents refer to the use of various strains of *Candida*. (British Patents are listed since Miall is with Pfizer Central Research, England.) The earliest patents all claim production of citric acid by submerged fermentation from C_9 to C_{20} normal paraffin mixtures using *C. lipolytica* and other *Candida* strains. One patent (Takeda Chemical Industries, 1970, British Patent 1,211,246) mentions a 114% yield of anhydrous citric acid. From later patents, it is obvious that these organisms can also make substantial amounts of citric acid.

[1]Miall, L.M. 1978. In: Economic Microbiology, Vol. 2, Primary Products of Metabolism, ed. by A.H. Rose, Chapter 3, pp. 48-111.

Takeda Chemical Industries in 1972 patented (British Patent 1,297,243) a procedure for isolating and testing mutant strains of microorganisms that will give improved yields of citric acid with less isocitric and other contaminating acids. In the patent the mutants are primarily tested for growth on n-paraffins and on citrate, and those selected for further study grow on the former but not the latter compound. An example is given of a mutant giving a 138% yield of citric acid from a C_{13} to C_{15} n-paraffin mixture with 17% of isocitric acid, whereas the parent strain gives a 61% yield of citric acid and 65% isocitric acid. Other examples using procedures to obtain mutants low in aconitate hydratase activity are also presented.

It is educational to note that most of the recent patents concerning the acid fermentation are granted to Japanese firms and that Pfizer Corp., Inc., has licensed the technology from Takeda Chemical Industries.

Lactic Acid

Lactic acid is produced both synthetically and by fermentation. Over the last 10 to 20 years manufacture by fermentation has been under heavy pressure from chemical processes. At one time there was the possibility that only the L-acid would be allowed for food use, which would have been a boon to fermentation processes.

The organisms used for production of lactic acid are all species of *Lactobacillus* most commonly *L. delbruckii*. The organism of choice depends on the substrate to be fermented and the type of acid to be formed. Starting materials may be glucose, sucrose, lactose (whey) and the final product may be DL-acid or L-lactic acid. The fermentation *per se* is very efficient and can result in yields of 90%, depending on the original carbohydrate. Since most of the problems in the manufacture of lactic acid lie in recovery and not in the fermentation step, few genetic studies have been employed for improvement of the industrial processes. Most of the genetic studies have been related to the fermentation of milk. Here one can find genetic work that has had important ramifications. Rates of acid production are critical and have resulted in improvements.[1]

Genetic approaches to control lysis of culture bacteria - probably the single most important factor in retarding milk fermentation - have had a significant economic impact in the dairy industry. Additional discussions of the impact of genetics on lactobacilli are to be found in the food section.

Additional Organic Acids

Itaconic acid. The industrial processes for itaconic acid are derived mainly from studies done at the Northern Regional Research Laboratory of the USDA in Peoria, Illinois. A screening program resulted in isolation of several species of *Aspergillus* that were able to form itaconic acid. A most promising species was *A. terreus* and additional work at NRRL screened 30 cultures of *A. terreus* for production of itaconic acid in promising yields. Only one was found. Lockwood and Reeves (1945) screened 308 strains isolated from soil samples. Eleven of these strains made itaconic acid in yields greater than 45% of theoretical (which was accepted as one molecule of itaconic acid from one molecule of glucose consumed, or a 72% weight yield).[2]

[1]Lawrence & Thomas, 1979. The Fermentation of Milk by Lactic Acid Bacteria. In: "Microbial Technology - Current State, Future Prospects", ed. A.T. Bull, D.C. Ellwood, C. Ratledge, Cambridge University Press.

[2]Miall, L.M. 1978. Organic Acids, Chapter 3. In: Economic Microbiology, Vol. 2, Primary Products of Metabolism, ed. A.H. Rose, Academic Press, New York.

Little additional work on strain improvement for the production of itaconic acid has been reported. The overall metabolism of itaconic has been elucidated by Cooper and Kornberg, using a *Pseudomonas* sp.[1] However, mutation experiments with *A. terreus* must have been conducted and evaluated since Stodola et al. (1945)[2] report that *A. terreus* mutants make not only itaconic acid, but also itatartaric acid.

Pfizer is reported by Miall (1978) to be the only U.S. manufacturer of itaconic acid; 75% of itaconic acid is used in styrene butadiene copolymers and for lattices and emulsions in general.

Fumaric Acid

Fumaric acid was at one time made on a commerical scale by fermentation by Pfizer with a strain of *Rhizopus*, but production was stopped when the economically more attractive synthesis by oxidation of benzene was developed (Miall, 1978; and SRI Chemical Economics Handbook).

In the 1940's, *R. nigricans* was patented as the production organism in submerged fermentation by Merck & Co. and Pfizer & Co. Again NRRL evaluated strains for production and found a unique strain *R. nigricans* 2582, which not only produced fumaric acid but also was able to synthesize invertase. Thus, sucrose could be used as a fermentation substrate without prior hydrolysis.

As more substrates are evaluated for fermentation processes, one finds reports on soil isolates capable of forming fumaric acid from different substrates. For example, Yamada

[1]Cooper, R.A. and Kornberg, H.L. 1964. Biochem. J. 91, 82.

[2]Stodola, F.H., Friedkin, M., Mayer, A.J., and Coghill, R.D. 1945. J. Biolog. Chem. 161, 739.

et al. (1970)[1] isolated a species of *Candida* from soil called *C. hydrocarbofumarica*, which made fumaric acid from n-paraffins.

In 1960 about 23 x 10^6 pounds of fumaric acid was produced and in 1975 5.0 x 10^6 pounds (SRI Chemical Economic Handbook). The price per pound in 1975 was 43¢ and in 1960 it was 23¢ per pound. According to SRI, all fumaric acid is currently produced by isomerizing maleic acid, which is obtained directly from benzene oxidation or from maleic anhydride produced through oxidation of benzene. Tong (1978)[2] reports that 15% of U.S. production of fumaric acid is by fermentation. Asymmetric addition of ammonia to fumaric acid catalysed by aspartase containing microorganisms, giving optically pure L-aspartic acid, is an industrially applied process.[3]

Malic Acid

Fermentation processes are not currently economical for production of L-malic acid. L-malic acid can readily be made enzymatically from fumaric acid with whole cells or the isolated enzyme from *Brevibacterium ammoniagenes*.

Several species of microorganisms, mainly yeasts, have been shown to convert fumerate to l-malate, including *Candida utilis*, *Pechia membranaefaciens*, *Pullularia pullulans*, as well as a species of *Brevibacterium*. In 1973 Tachibana and Murakami (1973)[4] report that *Schizophyllum commune* can produce L-malate in high yield from ethanol.

[1] Yamada, K., Furukawa, T. and Nakahara, T. 1970. Agricul. Biolog. Chem. 34, 670.

[3] Leuenberger, H.G.W. (1978). Microbial Transformations - Some Applications in Natural Product Chemistry. In: Antibiotics and Other Secondary Metabolites: Biosynthesis and Production, ed. R. Hutter, R. Leisinger, J. Nuesch, and W. Wehrli, Academic Press, pp. 87-100.

[4] Tachibana, S. and Murakami, T. 1973. J. Fermentation Technol. 51, 858.

Other tricarboxylic acid-cycle acids could readily be made by fermentation if required but no commercial use has yet been found (for example, α-oxoglutaric acid).

Epoxy Succinic Acid

This acid has been isolated as a metabolic product of *Aspergillus fumigatus* (Miall, 1978). Apparently, epoxy succinic acid is formed by a direct oxidation of fumaric acid. The epoxy succinic acid can be further converted to mesotartaric acid by a variety of microorganisms including *Pseudomonas putida*. The enzyme catalyzing the reaction has been isolated from *P. putida*.

Kojic Acid

Kojic acid can be made in good yields by a number of aspergilli, particularly those of *A. flavus-oxyzae* group. Kojic acid has a number of potential uses including production of the flavoring agent maltol via kojic acid via comenic acid and 6-methyl comenic acid.

Gluconic Acid

A variety of microorganisms have been investigated for production of gluconic acid including *Aspergillus niger*, *Acetobacter suboxydans*, various species of *Pseudomonas*, *Pullularia pullulans* and a species of *Moroxella*. *A. niger* and *Acetobacter* sp. are the organisms with which some industrial processes have been pursued.

Strain development work again was undertaken at NRRL and the research activity concentrated on *Pseudomonas* with 16 strains being examined. They found that *P. ovalis* NRRL B-8 made only gluconic aicd from glucose and a 99% yield was eventually reached.

Recovery of gluconic acid is troublesome in this fermentation.

Oxogluconic Acids

Further bacterial oxidation of gluconate leads to 2- or 5-oxogluconate with *Acetobacter melanogenium*. Employing *Pseudomonas albosesamae*, 2-5 dioxogluconate can be formed. The type of organism used primarily controls whether 2 or 5-oxogluconate is made. 5-Oxogluconate has never been made on a commercial scale while 2-oxogluconic acid is made on a commerical scale as an intermediate in the production of erythorbic acid.

Usually it is made by oxidation of glucose with strains of *Pseudomonas*. However, *Serriatia marcescens* NRRL B-486 has been used in pilot plant runs, where yields of 95-100% in 16 hrs are reported. Ten strains of *S. marcescens* all gave over 80% yields and it is claimed that both yield and productivity are greater than when using *Pseudomonas* sp.

Erythorbic Acid

Penicillium notatum can be used for the direct production of erythorbic acid from glucose. A strain was obtained following a mutation and screening program which made erythorbic acid in a medium containing 80g glucose per liter in about 40% yield in five days.

Tartaric Acid

Besides conversion of 5-oxogluconate to tartaric acid by chemical catalysts, production of tartaric acid by fermentation with *A. suboxydans* or *P. fluorescens* have been reported.

Yamada et al. (1969)[1] screened about 3600 strains of fungi and 4000 strains of bacteria and yeasts for tartaric acid production.

[1] Yamada, K. Minoda, Y., Kodama, T. and Kotera, U. 1969. In; Fermentation Advances (D. Perlman, ed.), Academic Press, New York, p. 541.

Six strains of bacteria, all isolated from Japanese persimmon, using a mannitol-based medium, were shown to produce L(+) tartaric acid. Four strains were identified as *Flucono-bacter suboxydans*. Kotera et al. (1972) report on a mutation program with *S. suboxydans* and were able to obtain concentrations of 11 g/l tartaric acid, while the highest concentration reported by others was 6.0 g/l. Subsequent mutation program resulted in 14.68/liter from 50 g glucose, 35% of theoretical yield, in six days in shake flasks.[1]

Conclusions

Only a few organic acids are presently made by fermentation, the main ones being acetic, citric and lactic. The role that genetics has played in the industrial process development has been small. Most of the processes have been developed on an empirical basis. With some of the more specialized fermentation acids, classic microbial isolation and screening program resulted in production strains. Much of the recent research on strain screening and mutation programs for the production of organic acids appears to be originating from Japan. Most notable accomplishment is the citric acid process from n-paraffins employing yeast.

Industrial Gases

Carbon Dioxide

Most carbon dioxide is produced as a byproduct of industrial manufacture of ammonia. The source of the hydrogen to reduce the nitrogen in air to ammonia is via steam reforming:

$$CH_4 + 2H_2O \longrightarrow 4H_2 + CO_2$$

[1]Kotera, U., Kodama, T., Minoda, Y. and Yamada, K. 1972. Agricul. Biolog. Chem. 36, 1315.

In addition, carbon dioxide is produced as a by-product in fermentations. About 1% of the U.S. carbon dioxide was from fermentation in 1978. Production and removal of carbon dioxide can be a significant technical problem in many types of fermentation processes.

Over one third of carbon dioxide used in the U.S. is used in refrigeration (solid CO_2, or dry ice). Other important uses are carbonating water for beverages, chemical synthesis, and inert atmospheres. The major cost of carbon dioxide is in the post production steps: compression, purification, liquifaction, and in the case of dry ice, snowing. The price of CO_2 is a function of locality. Because of the great abundance of carbon dioxide, it is unlikely to be little more than a minor byproduct of fermentation where local demand is sufficiently large. Table V
duction in 1978.

Ammonia

Ammonia is made industrially by the reaction of methane, air, and water. The methane and water react to form hydrogen (see above); then the hydrogen and nitrogen react:

$$N_4 + 3H_2 \longrightarrow 2\ NH_3$$

About 15 billion metric tons of ammonia were produced this way in 1978.

Nitrogen can be "fixed" to ammonia by the nitrogenase enzymes using reducing potential (electrons) derived from the tricarboxylic acid cycle. It is estimated that 175 billion metric tons of ammonia are fixed by cells per year. One square meter of land planted with nodulated legumes (such as soybeans) can fix 1-30 gram of nitrogen (Metzler, 1978).

Several prokaryotes are known to fix nitrogen; the most important of these are Rhizobium and blue green algae. The infected roots of legumes develop nodules in which the Rhizobium bacteria degenerate into "bacteroids". Blue-green algae are also quantitatively

Table V

Industrial Gases

Product	Formula	Hydrogen Deficiency	Chiral Centers	Price 1978 (per kg)	Present source	Domestic Production 1978 (metric tons)	Potential For Application of Biotechnology
Ammonia	NH_3	0	0	$0.13	hydrogen+ nitrogen	15,250	Enzyme reduction of N_2
Carbon Dioxide	CO_2	2	0	--	$CH_4 + H_2O$	2,060,000	By-product of ethanol and butanol/acetone fermentation

significant nitrogen fixers; a square meter of rice paddy can contain algae that fix from 2-10g of nitrogen per year.

The full 6 electron reduction is performed by a complex of enzymes and iron/molybdenum/sulfur groups that are inactivated by oxygen and requires 12 ATP molecules to be hydrolyzed per N_2 reduced. The enzyme system is remarkable in the amount of energy it consumes and in the total lack of detectable partial reduction products. It is the subject of intense scientific inquiry (Walsh, 1979).

Microbial production of ammonia from nitrogen appears to be uncompetitive. Aside from the difficulties associated with the enzyme's sensitivity to oxygen and our near total lack of understanding of its mechanism, it consumes the equivalent of the energy in 4kg of sugar to make one kg of ammonia. Since ammonia is presently $0.13/kg and sugar is $0.22/kg, the chemical Haber process does not appear to be threatened.

Biochemicals: Amino Acids

Amino acids are the building blocks of proteins. Twenty have the distinction of being incorporated into proteins manufactured on the ribosomes; others serve specialized structural roles (hydroxyproline), are important metabolic intermediates (homoserine), or are hormones and neurotransmitters (dihydroxyphenylalanine (DOPA)).

Many bacteria can manufacture all the amino acids they need; higher animals cannot. In general the more complicated amino acids are required in the mammalian diet. In fact, if one of the essential amino acids is missing, protein synthesis halts, and sufficient quantities of the other 19 will be wasted. In human beings the essential amino acids are threonine, methionine, lysine, valine, isoleucine, leucine, arginine, histidine, phenylalanine, tyrosine, and tryptophan. The requirements vary in other animal species.

The alpha carbon of amino acids is chiral; it can exist

in the D or L form (see figure 4). The vital form is L. In some cases, cells can convert the D to L, but it is the L that is used in protein synthesis.

All of the amino acids are used in research and in nutritional preparations. Most are used in one way or another in the preparation of pharmaceuticals. Three are used in large quantities for two purposes: glutamate is a flavor enhancer (MSG); lysine and methionine are feed additives.

Amino acids are primary metabolites and are prime candidates for production via biotechnology. About half of the amino acids sold are now produced by fermentation. The leading country in acid fermentation processes is Japan. The food industry there now produces 14 amino acids at least partly by biotechnology processes.

Table VI gives production data for the amino acids. "Enzymes needed" enumerates the number of enzymes to convert an intermediary metabolite on one of the central pathways to the amino acid concerned -- a rough measure of the difficulty of engineering a microbe for fermentation. Production data are given for Japan because in most cases domestic production is very small.

Glutamic Acid -- MSG is used in huge amounts as a flavor enhancer, particularly in oriental cooking. In 1978, the U.S. manufactured about 18,000 metric tons and imported about 11,000 metric tons. 97% was consumed by the food industry. The entire U.S. production is accounted for by the Stauffer fermentation plant in San Jose, California.

The microbes used in glutamic acid fermentation (*Corynebacterium glutamicum*, *C. lileum*, and *Brevibacterium flavum*) produce glutamic acid in 60% of theoretic yield. Therefore the potential for use of applied genetics to improve yield is slim.

Lysine -- As explained above, the lack of a single amino acid can retard protein synthesis, and therefore growth, in a

Table VI: Amino Acids

Amino Acid	Formula	Hydrogen Deficiency	Chiral Centers	Price March 1980, Pure L ($/kg)	Present Source	Value-Added Ratio	Enzyme Needed	Domestic Production 1978 (metric tons)	Potential for Application of Biotechnology (*de novo* synthesis or bioconversion; organisms and enzymes)
Alanine	$C_3H_7NO_2$	1	1	80	hydrolysis of protein chemical synthesis	–		10-50 (J)	
Arginine	$C_6H_{14}N_4O_2$	2	1	28	gelatin hydrolysis	–	–	200-300 (J)	fermentation in Japan
Asparagine	$C_4H_8N_2O_3$	2	1	50	extraction			10-50 (J)	
Aspartic acid	$C_4H_7NO_4$	2	1	12	bioconversion of fumaric acid	–	–	500-1,000 (J)	bioconversion
Citrulline	$C_6H_{13}N_3O_3$	2	1	250				10-90 (J)	fermentation in Japan
Cysteine	$C_3H_7NO_2S$	1	1	50	extraction			100-200 (J)	
Cystine	$C_6H_{12}N_2O_4S_2$	–	–	60	extraction			100-200 (J)	
Dopa (dihydroxy-phenylalanine)	$C_9H_{11}NO_4$	5	1	750	chemical			80-200 (J)	
Glutamic	$C_5H_9NO_4$	2	1	4	fermentation			18,000-100,000 (J)	*de novo: Micrococcus glutamicus*
Glutamine	$C_5H_{10}N_2O_3$	2	1	55	extraction			200-300 (J)	fermentation in Japan
Histidine	$C_6H_9N_3O_2$	4	1	160				100-200	fermentation in Japan
Hydroxyproline	$C_5H_9NO_3$	2	2	280	extraction from collagen			10-50	
Isoleucine	$C_6H_{13}NO_2$	1	2	350	extraction			10-50 (J)	fermentation in Japan
Leucine	$C_6H_{13}NO_2$	1	1	55				50-100 (J)	fermentation in Japan
Lysine (HA)	$C_6H_{14}N_2O_2$	1	1	350	fermentation (80%) chem: f. cyche-chemical (20%) 4.7			10,000 (J)	(80% by fermentation) *de novo: Corynebacterium glutamicum* and *Brevibacterium flavum*
Methionine	$C_5H_{11}NO_2S$	1	1	265	chemical acrolein	2.7		17,000 (D, L) 20,000 (D, L)(J)	
Ornithine	$C_5H_{12}N_2O_2$	1	1	60				10-50 (J)	fermentation in Japan
Phenylalanine	$C_9H_{11}NO_2$	5	1	55	chemical from benzaldehyde	20		50-100 (J)	fermentation in Japan
Proline	$C_5H_9NO_2$	2	1	125	hydrolysis of gelatin			10-50 (J)	fermentation in Japan
Serine	$C_3H_7NO_3$	1	1	320				10-50 (J)	bioconversion in Japan
Threonine	$C_4H_9NO_3$	1	2	150				50-10 (J)	fermentation in Japan
Tryptophan	$C_{11}H_{12}N_2O_2$	7	1	110	chemical from-iadole	1.02		55 (J)	
Tyrosine	$C_9H_{11}NO_3$	5	1	13	extraction			50-100 (J)	
Valine	$C_5H_{11}NO_2$	1	1	60				50-100 (J)	fermentation in Japan

mammal. The limiting amino acid is a function of the animal and its feed. The major source of animal feed in the United States is soybean meal. The limiting amino acid for feeding swine is lysine; the limiting amino acid for feeding poultry is methionine.

L-lysine is produced both by fermentation and by chemical synthesis. Fermentation has been gradually replacing chemical synthesis, and in 1980 80% of lysine produced world-wide is expected to be made by microbes. No lysine is produced in the U.S. Imports were about 7 thousand metric tons in 1979, mostly from Japan and S. Korea.

Because of increased poultry demand, world demand for lysine is climbing. Eurolysine is spending 27 million dollars to double its production capacity in Amiens, France, to 10 thousand metric tons. The Asian and mideast markets are estimated to increase to 3 thousand metric tons in 1985. Again, the microbe produces lysine at over 90% of theoretical efficiency, and no genetic improvement is likely.

Recently Shvinka *et al.*, (1980), reported on the yield regulation of lysine biosynthesis in *Brevibacterium flavum*. Experimental data are presented showing how the increase in yield (Yp (g product/ g substrate) can be controlled by enzyme induction and elimination of by-product synthesis. Consequently, they report close to theoretical yields for lysine production and concentrations of 78 g/liter lysine-HCl with *B. flavum* in molasses medium in pilot-plant fermentors.

A summary of the cost differences for chemical and fermentation production is shown in Table VII.

Methionine -- Methionine is the limiting amino amino acid for feeding soybean meal to poultry. However, there is an important difference from lysine: the fowl has an enzyme that can convert D-lysine to L-lysine, so a D,L mixture can be used. D,L methionine

Table VII *

The following table summarizes recent estimates of primary U.S. cost factors in the products of L-lysine monohydrochloride by fermentation and chemical synthesis.

	Cost factors in production of 98% L-lysine monohydrochloride					
	By fermentation[a]			By chemical synthesis[b]		
	Requirement (units per unit product)	Estimated 1976 cost per unit product		Requirement (units per unit product)	Estimated 1976 cost per unit product	
		Cents/lb	Cents/kg		Cents/lb	Cents/lb
Total labor[c]	—	8	18	—	9	20
Materials						
Molasses	44	7	16	—	—	—
Soybeanmeal, hydrolized	0.462	4	9	—	—	—
Cyclohexanol	—	—	—	0.595	17	37
Anhydrous ammonia	—	—	—	0.645	6	14
Other chemicals[d]	—	7	15	—	4	10
Nutrients and solvents	—	—	—	—	4	8
Packaging, operating, and maintenance materials	—	10	22	—	9	21
Total materials	—	28	62	—	45	90
Total utilities[e]	—	6	12	—	7	16
Total direct operating cost	—	42	92	—	56	126
Plant overhead, taxes, and insurance	—	10	21	—	10	21
Total cash cost	—	52	11	—	66	147
Depreciation[f]	—	16	35	—	13	28
Interest on working capital	—	1	3	—	1	3
Total cost[g]	—	69	151	—	80	178

[a]Assumes a 23-percent yield on molasses.
[b]Assumes a 65-percent yield on cyclohexanol.
[c]Includes operating, maintenance, and control laboratory labor.
[d]For both the process of fermentation and chemical synthesis, assumed use of hydrochloric acid (36 percent) and ammonia (29 percent). For fermentation includes also potassium diphosphate, urea, ammonium sulfate, calcium carbonate, and magnesium sulfate. For chemical synthesis also includes nitrosyl chloride, sulfuric acid, and a credit for ammonium sulfate byproduct.
[e]Total utilities for both processes include cooling water, steam process water, and electricity. For chemical synthesis, natural gas is also included.
[f]Ten percent per year of fixed capital costs for a new 20 million lb per year U.S. plant built in 1975 at assumed capital cost of $\$38.6 \times 10^6$ for fermentation and $\$32.5 \times 10^6$ for chemical synthesis exclusive of land costs.

* From SRI

is produced by chemical synthesis. The process uses acrolein, methyl mercaptan, and hydrogen cyanide as raw materials.

Domestic consumption of D,L methionine is 28,000 metric tons, of which about 17,000 metric tons is produced domestically. The major producers are Degussa, DuPont and Monsanto. Monsanto recently announced plans to build a 40,000 metric ton plant, which would be the largest in the world, to be completed in 1982.

Microbial production of methionine appears to be unlikely for the following reasons:

(1) pure L-methionine is not required for poultry feed addition;

(2) chemical production is cheap; and

(3) microbial production is difficult because fixation of sulfur is energy intensive and the metabolic pathway to methionine is long and complicated.

Other Amino Acids -- Figure 8 shows the central metabolic pathways with the branches leading to amino acids. The most likely use of recombinant DNA to increase yields of an amino acid would be to clone the gene(s) coding for the enzymes needed to synthesize the amino acid. Each pathway of Fig. 8 is labelled with a number indicating the number of enzymes on the pathway, and thus gives a rough indication of the difficulty of the microbiology involved.

The aromatic amino acids represent a special problem. Figure 9 shows the feedback inhibition pathways for aromatic amino acid production. Because aromatization is metabolically expensive, the pathways are tightly controlled. Further controls exist at later steps. Genetic manipulation to improve yields of aromatic acids will require manipulation of these complicated control mechanisms.

Aspartic acid will be required in the coming decade in large amounts for the production of the sweetner aspartame (L-aspartyl-L-phenylalanine methyl ester). It is presently made by a very efficient bioconversion of fumaric acid.

Phenylalanine also will be required for the production of aspartame. The pure L enantiomer will be required. Figure 10 shows

Fig. 8

Biosynthesis of Amino Acids

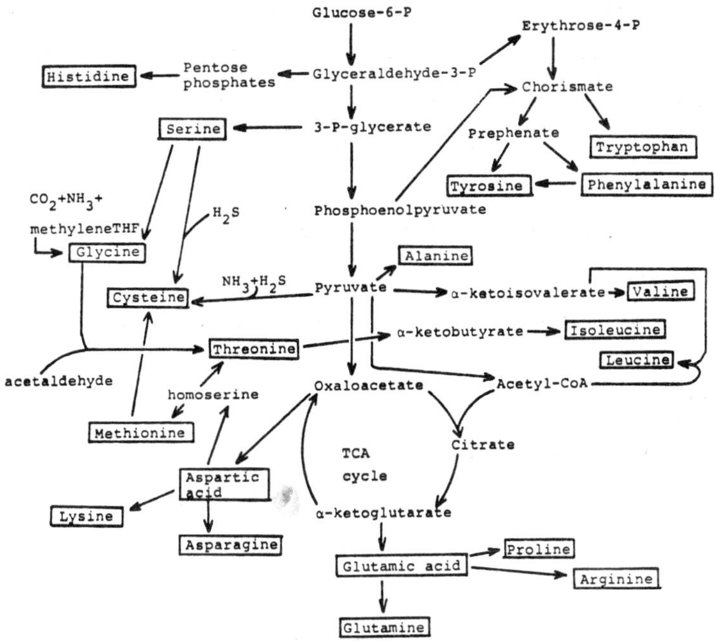

Fig. 9

Feedback Controls on Aromatic Amino Acids Biosynthesis

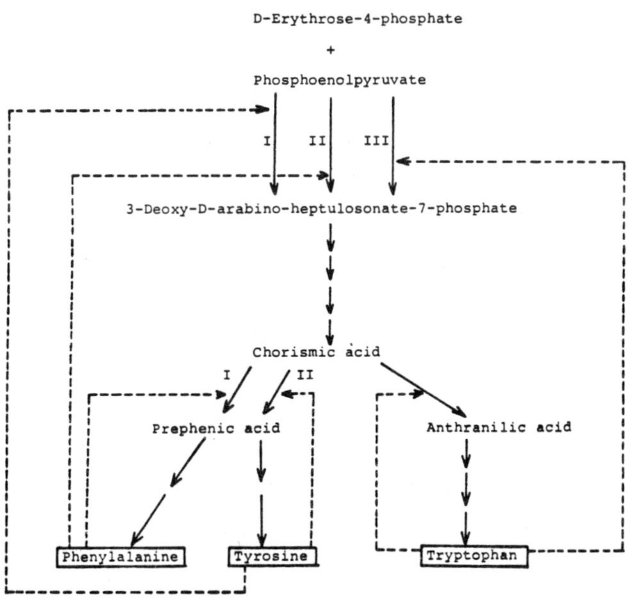

the two ways in which L-phenylalanine can be made: by chemical production from benzaldehyde and acetylamino acetic acid or by fermentation from sucrose. The difficulty in making L-phenylalanine chemically is that an ordinary reduction catalyst for reaction produces D- and L-phenylalanine in equal amounts. The two optical isomers must then be separated (reaction 3). However, an optically active reduction catalyst can give a higher yield of one optical isomer over the other, and in some cases over 90%. Compound III, because it is highly functionalized around its reduction center, is a good candidate for making the L-isomer exclusively.

Fig. 10

Chemical and Fermentation Production of Phenylalanine

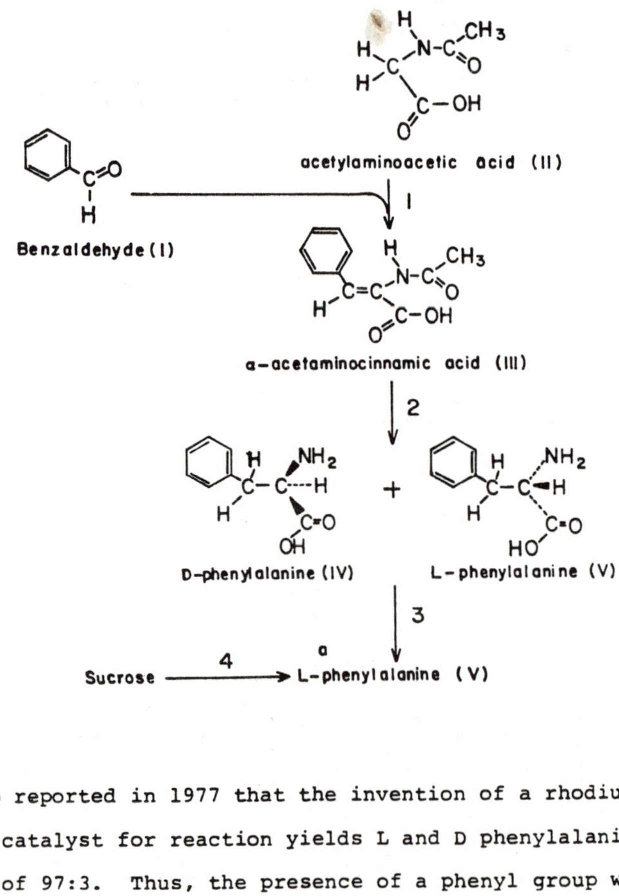

Monsanto reported in 1977 that the invention of a rhodium reduction catalyst for reaction yields L and D phenylalanine in a ratio of 97:3. Thus, the presence of a phenyl group which is energetically expensive to produce microbially, and the availability of an optically active catalyst to produce the L enantiomer exclusively in a chemical process, makes it unlikely that fermentative production of phenylalanine will be competitive.

Proline is an amino acid for which improvements in production may be brought about by genetic engineering. Bethesda Research Laboratory announced in April, 1980, that it had cloned one enzyme necessary to increase proline production.

Tryptophan is the third limiting amino acid for most feed uses. It is not used as a feed additive now because of its expense (it is made by chemical synthesis from indole). An inexpensive source of tryptophan may have a large market. Considerable research has been conducted on genetic means to increase production of tryptophan.

Monomers and Polymers

A very large segment of the chemical industry is devoted to the production of plastics, fibers, and rubber; 4.3 million metric tons of fibers, 12 million metric tons of plastics, and 1.1 million metric tons of synthetic rubber were produced in the United States in 1978. With the exception of cellulose fibers, this entire mass was derived from petroleum. A breakdown of sales is shown in Table IX.

Microbial production of monomers was mentioned in the previous section on organic chemicals. Because the bulk of monomers are chemically simple and are presently available in high yield from petroleum, microbial production of most monomers is unlikely. The most likely exceptions are polyamides (chemically related to proteins), acrylics, and isoprene type rubber.

The fact that present technology appears not to be replacable by biotechnology should not lead one to the conclusion that the eventual impact on polymer production will be slight.

Table IX. Monomers and Polymers

PRODUCT	PRICE 3/80 (per kg)	DOMESTIC PRODUCTION 1978 (thousand metric tons)	POTENTIAL FOR BIOTECHNOLOGY
I. Plastics			
THERMOSETTING RESINS			
Epoxy		135	fair
Polyester		544	fair
Urea		504	fair
Melamine		90	
Phenolic		727	low
THERMOPLASTIC RESINS			
Polyethylene			
Low density		3200	low
High density		1890	low
Polypropylene		1380	low
Polystyrene		2680	low
Polyamide, nylon type		124	good
Polyvinyl alcohol		57	fair
Polyvinyl chloride		2575	low
Other vinyl resins		88	
II. Fibers			
Cellulosic Fibers			
Acetate		139	fair
Rayon		269	fair
Noncellulosic Fibers			
Acrylic		327	good
Nylon		1148	good
Olefins		311	low
Polyester		1710	fair
Textile glass		418	low
Other		7	
III. Rubber			
Styrene-butadiene		628	low
Polybutadiene		170	low
Butyl		69	low
Nitrile		33	low
Polychlorophene		72	low
Ethylene-Propylene		78	low
Polyisoprene		62	good

Biopolymers represent a new way of thinking. Most of the important constituents of cells are polymers: proteins (polypeptides, from amino acid monomers), polysaccharides (from sugar monomers), polynucleotides (from nucleotide monomers).

Cells normally assemble polymers with extreme specificity: the monomers occur in a well-defined order (e.g., the AGTC sequences of DNA) and in addition each monomer is added to the chain in a chemically specific way (e.g., the head-to-tail structure of natural rubber from isoprene units). In contrast, man-made polymers are random ordered (e.g., the styrene and butadiene monomers of stryrenebutadiene rubber are randomly ordered, the numbers reflecting only the proportions in the monomer mix) and each monomer is added to the chain in a haphazard way (e.g., the non-specificity of synthetic isoprenoid rubber). The higher quality of biological polymers is achieved by the specific order of monomers.

It appears that the ideal situation would be to imitate the biological production of polymers in all possible respects: a single biomachine to convert a raw material (for instance, sugar) into the monomer, polymerize it, then form the final product. Commerical dextrans are in fact produced by converting sucrose in polymeric glucose. Much experimental work must be done on the methods cells use to make structural polymers.

A more likely application is the development of new monomers for specialized applications. Polymer chemistry has largely been the study of modification of properties through the regulation of co-polymerization reactions, addition of plasticizers, and regulation of temperature and pressure to yield a polymer with desired properties. Conceivably polymer chemists could be able to modify functionalization and stereochemistry of the monomers. As with any new approach to refining a new technology, it is difficult to guess the impact now; but perhaps a valid analogy is the transition in the 30's from exclusively cellulose-based polymers to chemical synthetic polymers, i.e., at least 30 years.

Pesticides

Mankind will always need some means of controlling pests.

Under primitive conditions crop plants such as cereals and cabbages grow interspaced with other species as part of a complex community of plants and animals. Because of the dispersed distributions of wild plants, pests shown adaptation that enable them to survive long enough or in sufficient numbers to get from host to host. They often produce large numbers of progeny or exhibit parthenogenesis (in which a female reproduces without fertilization), so that they stand a chance of survival in spite of the difficulties of finding a host on which to feed. Once agriculture arrived, this propensity showed that it could result in devastation. A pest that infected one plant in the field found it had thousands of identical plants all around it -- and so the ability to survive under difficult circumstances became, once man had discovered how to farm, the origin of widespread outbreaks of plant disease. Insecticides that worked by inhibiting insect respiration, such as DDT, were invented and used in large amounts; their use was responsible for great improvements in disease control and crop protection, but the undesirable secondary effects are well known. The problem with DDT and related substances is due principally to the fact that they are concentrated as one moves higher in the food chain: birds ingest large amounts of DDT that have been absorbed by the insects that they eat, it becomes concentrated in their bodies, and causes effects such as the weakening of egg shells. Another major problem with DDT-like pesticides is the ability of the insects to rapidly gain resistance through mutation natural selection.

Microbes could provide us with a range of new pesticide preparations that are harmless to man, ecologically compatible, highly efficient, and selective. Microbes have been used in the past to control pests, and it is surprising that microbial pesticides have not been more systematically explored. The potential market for pesticides is indicated in Table X.

Table X

Pesticides

Product	Domestic Market 1978 (Millions $)	Domestic Production 1978 (metric tons)
Fungicides	$220	66,000
Herbicides	1,760	295,000
Plant Growth Regulators	223	3,300
Insecticides and Raderticides	1,040	272,000

The feeling of workers in the field seems to be one of enthusiasm combined with pessimism. Norris, 1978, has published an excellent review on the microbial control of pest insects. The most general complaints are the unattainability of patents and the strict standards of environmental agencies more familiar with toxic pesticides than behavior modifying chemicals and biodegradable pesticides. Both of these questions should be addressed.

The impact of applied genetics on pesticide development can be divided into four areas:

1. natural toxins;
2. viruses;
3. bacteria, fungi, and protozoa; and
4. pheromones.

Natural Toxins

Bacillus thuringiensis toxin has grown in the 15 year period from 1962 to 1977 to sales of $13 million. The crystalline toxic compounds present on the outer coating of the bacterial spore destroy the larvae of a number of strains of insects. The World Health Organization has recently shown great interest in the development of the new *B. thuringiensis* strain (H.14). It is known to be extremely lethal for the larvae of certain diptera, including some mosquitos. WHO sees a possibility of using it to control the black fly, simuliumdanosum, the vector of onchocerciasis (river blindness).

The high density of the crystalline toxic compounds is a problem because they tend to sink to the bottom in water. A formulation that will keep the toxic compounds afloat long enough for them to reach the target larvae downstream at the point of application is needed. It is therefore necessary to put the maximum amount of crystal toxin in the primary powder used as a basis for an actual pesticide, and then to formulate the latter in such a way as to give the desired residual effect under normal field operation conditions. From the commercial point of view, the development of such a formulation could probably enable the manufacturer to patent his material, a matter of great importance in the highly competitive pesticide industry.

As summarized by Norris (1978) the present effectiveness and future potential of microbiological insecticides depend on a subtle interaction between cost/effectiveness as compared with established chemical agents, safety in the sense of absence of toxic effects on man and other animal and plant species and efficacy in the field against the target insect.

A major factor which complicates the development of a microbiological control agent is the confused economics which are almost totally dependent upon each situation being evaluated (Norris, 1978).

Viruses

Many viral diseases of insects are known; over 300 insect viruses are known to exist and many of these cause rapidly epizootics amongst insect populations in nature. An infestation of sawflies in Canady during the 1930's was fought with a polyhedrosis virus grown in the larvae of sawflies and dispersed in water to the young spruce trees. The effort was considered very successful (Norris, 1978).

The use of myxomatosis virus to control rabbits is an example of an ill-considered application of virus to pest control. The virus was discovered in Uruguay in 1896 and introduced in 1926 in Australia where rabbits imported from Europe had become a major pest in the country. The attempt was a failure. Several further attempts failed until one in 1950 finally succeeded probably because of favorable weather conditions. More than 80% of the rabbits of Southeast Australia were wiped out in the first wave of the epidemic and in some areas millions of them were found piled high against rabbit proof fencing. Two years later the virus was introduced into France. Two years after that it crossed into England. By the end of 1955 it is estimated that 90% of the rabbits in Britain had been killed. Since that time the disease has remained present in those rabbit populations in epidemic form. The effect of natural selection has been to produce rabbits that are more resistant to the disease, and animals blinded by it may still survive and feed normally. The population of rabbits now is increasingly immune to the disease and may be developing tolerance.

Viral agents of pest control should be more carefully selected in the future, and the ecology of their use should be thoroughly understood. Although most viruses are characteristically host specific, the others are not.

Bacteria, Fungi and Protozoa

Bacteria, true microbes, pose different issues. There are fewer bacterial viruses known to cause diseases in insects and which might have the ability to act as insecticidal agents. Approximately 100 species of bacteria are known to infect insects.

Popillia japonica is a Japanese beetle which damages crops and lawn grasses and which produces wide-spread outbreaks. The natural pathogen of this beetle, Bacillus popilliae, has been used in a successful demonstration of microbial pest control. Spores of Bacillus popilliae are eaten by the larva along with its normal diet of vegetation. The spores can resist the normal digestion process of the larva and in due course germinate and release the rod shaped bacteria which produce the disease.

In cases where bacterial infection can be identified in a pest species, similar applications of biological control can be used.

Two fungi are of special interest in the search for microbial control agents. They are Beauvaria bassina and Metarrhizium anisopliae, and both of which were used on a somewhat experimental basis at the beginning of the 20th century. Their spores germinate on the body wall of the hast. The growing fungus colony penetrates the insect larva through a breathing pore or spear pore. Thre are other fungi that could be used to control pests by producing a specific infection.

Pheromones

Insect pheromones have long been believed to be sex attractants, produced by females to attract males. Twenty years of research with pheromones has made it clear that they are not nearly so easy to categorize: in some cases males release pheromones to attract females and in many cases the communication is not related to sex, and in fact, may be inter-special. The low toxicity of pheromones, high specificity, and fast degradation time have long made them attractive as pesticides.

The EPA recently approved the use of a synthetically produced pheromones to control the pink ball worm caterpillar. The pheromone is sprayed on an infected cotton field to prevent the male insects from mating. It has been shown that resistance to this chemical through evolution is almost nonexistent.

Three applications of pheromones have been found in pest control. The first is mass trapping. This is a large scale capture in pest traps baited with pheromones. Israeli farmers spray their cotton fields with pesticides less often since they began the Egyptian cotton lead worm beginning in 1975. Mass trapping is also the basis of the Norwegian campaign against the spruce bark beetle. Workers at the State University of New York have reported progress in trials using the aggregation pheromones with the Europena elm bark beetle, which carries the fungus that carries Dutch Elm disease. Rhodesian workers have painted cattle with a mixture of tick pheromone and acaricide. Small areas of the mixture attract unfed nymphs and adults, killing them before they bite the cattle, thus reducing the chance that they would spread disease.

A second approach is to fill the atmosphere with the pheromone to confuse the insects and prevent the males from finding the females for mating. Tests in the Philippines on the use of mating disruption by pheromone inhibitors of the striped rice stem bore have been very encouraging. The next step will probably be to microencapsulate the inhibitors for slow continuous release and set up trials in the open fields.

The third application is to use pheromones to attract pests for population monitoring.

Although pheromones have been known for 20 years, the complexity and subtlety of their action is only now being appreciated. Pheromones are needed in such small amounts that they will probably be manufactured chemically rather than biogically. However, biological production may be practical if the insect cells that secrete the pheromones can be cultured to produce the pheromones _in vita_.

Microbial Hormones

Related to the above chemicals are microbial hormones. Little effort has been put forth on possible application of such compounds. Gooday (1978) has reviewed the recent scientific literature and defines microbial hormones as powerful metabolites that control fundamental processes in microorganisms. The mode of action of the few chemicals characterized thus far is concerned with bringing cells together and with diverting cells from activities concerned with growth to those concerned with reproduction. There are now a handful of microbial sex hormones that have been fully characterized similar to the insect pheromones described above. One of the most studied microbial hormones is trisporic acid which controls sexual differentiation in the *Mucorales*.

Although such chemicals have been mainly the subject of basic studies, it is possible that novel applications of such chemicals could be developed for control of microbial growth and reproduction in unwanted situations.

Difficulties

The inability of regulatory agencies to comprehend the difference between biological pest control and use of chemically-toxic materials for pest control will have to be addressed. Many of the regulatory issues are discussed elsewhere. However, for emphasis it should be realized that the safety issues are also related to the disruption of the natural ecological web caused by the sudden reduction in the number of pest animals, rather than solely by the toxicity caused by a compound that is applied in minute amounts, decays rapidly and has high specific activity.

Another problem is the limited markets for these products. Since each one by its nature will be specific for a single pest or organism, the market will be small and the development cost for each product will be very high.

Patent protection is the final problem. Formulations and methods of application will surely be patentable. Novel and reliable delivery systems are required for this technology. In the case of natural toxins and pheromones this will probably provide a fair incentive on the part of industry to sponsor research. However, the patentability of viral and bacterial pest control agents is less clear. The protection of the commercial interests of the companies investing large amounts of money in developing biological pest control agents should be considered carefully.

Dyes and Pigments

Domestic production of dyes and pigments is shown in Table XI.

Table XI. Dyes and Pigments

Products	Market 1978 (Millions $)	Present Source	Domestic Production 1978 (metric tons)	Potential for Application of Biotechnology (*de novo* synthesis or bioconversion; organisms and enzymes)
Dyes, Total	$734	cyclic intermediates (aromatic)	18,000	*de novo*: very unlikely. Bioconversion: specialized applications possible
Pigments, Total	$319	cyclic intermediates (aromatics)	34,500	*de novo*: very unlikely. Bioconversion: specialized applications possible

The pigment and dye industries are an outgrowth of the coal tar chemistry of the 19th century. Most of the compounds are not referred to by their chemical identity, but by a scheme related to their origin, application, pH, and color characteristics.

The principle desirable features of these compounds are their fastness and stability. Since biological pigments are usually sensitive to physical treatments and are biodegradable, it is unlikely, therefore, that they will find significant markets among those currently dominated by coal tar dyes and pigments. However, many of the natural microbial pigments have never been tested and characterized for such applications.

Part III

Medical Impacts of Molecular Genetics

The information in Part III is from *Implications of Molecular Genetics for Medicine,* prepared by W.P. O'Neill, Ph.D., M.B.A., of Poly-Planning Services for the Office of Technology Assessment of the Congress of the United States, June 1980 (revised June 1981).

Summary and Conclusions

PRINCIPAL OPPORTUNITIES

Nowhere is the interrelationship between prógress in fundamental bioscience and improvements in modern medicine better illustrated than in the field of molecular genetics. Contemporary advances in bioscience are simultaneously creating powerful tools to understand life in molecular detail and also providing the means to synthesize extraordinarily complex biomolecules in sufficient quantity and purity for clinical testing and use. An impressive variety of enterprises are probing this potential and elaborating specific combinations of production and utility, for both their own sake and as models of what is widely expected to come. In this summary, we can do little more than name the specific objectives and catagorize them with respect to a few parameters.

No doubt, the application area which is the subject of the most attention at this time is the interferon system. Important in themselves, perhaps, on the basis of their apparent utility against certain viral diseases and cancers, interferons exemplify the unnumbered rare molecules in the body which are involved in regulation, communication, and defense to maintain health. Now scientists have the capacity to recognize, isolate, characterize, and produce in quantity many times more of these molecular constituents of human physiology than was possible five years ago. Identifying those agents that could be beneficial will be the pivotal activity which will determine the directions and rates of progress in exploitation of physiological compounds, such as those involved in the manifold processes that are encompassed by the term *immune system*. Besides interferon, many messenger (e.g. lymphokines) and effector (e.g. antibodies) molecules can (or will, probably) be produced by one or another genetic intervention. It is reasonable to anticipate that they will be employed in cancer therapy, to prevent or combat infections, to facilitate transplantation of organs and of skin, to treat allergies and for other diseases in which the immune system harms the organism of which it is an element.

The furthest advanced applications are in the field of *hormones*. The new capacity to synthesize proteins has been pioneered in large part with the immediate objective of preparing peptide hormones - with the amino acid sequence peculiar to humans - that have already been proven to be therapeutically useful, i.e. *insulin* and *growth hormone*. The merits of synthesizing other peptide hormones by this route depend upon clarification of the pharmacology of these agents, their derivatives (fragments) and analogues. Bits of evidence that such hormones might be used to counteract the atrophy of bones that often accompanies aging (osteoporosis), improve the treatment of diabetes, promote wound healing, or stimulate the regrowth of nerves, will stimulate scientific investigations that were really impossible until a way of producing quantities of specific polypeptides became available. Should the promise of certain relatively small polypeptides with respect to analgesia, appetite suppression, or cognition/memory enhancement prove out, molecular clones will certainly be evaluated for production. When it becomes possible to create clones that will synthesize glycopeptides - in which complex carbohydrate substituents are attached to specific sites in the peptide sequence - additional hormones will become accessible. One of the most interesting glycopeptide candidates, which has never been available in sufficient quantity for clinical evaluation, is the hormone, erythropoietin, that stimulates the production of red blood cells. It might be used for treating persons with diseased kidneys, for burns, along with blood transfusions, or for some anemic conditions.

Antibiotics were gifts of the genetic diversity of microorganisms and, in addition to relief of much infectious disease, their development/production provided the training ground for many who will participate in the utilization of various gene manipulation techniques to obtain other desirable agents. Recently developed methods for fusing microbial cells in a sort of forced mating makes it possible to combine the benfits that have been achieved in separate lines of conventional genetic selection and obtain hybrids with heretofore unattainable pedigrees. This, and other techniques in which plasmids or viruses carry genes from one line (strain, species or genus) to another, will be used to improve the costs of antibiotic production and to create new antibiotics. Furthermore, screening of the compounds produced by such organisms seems likely to identify agents with useful pharmacological effects on human physiologic processes - based on the side effects manifested by many antibiotics.

Synthesis of enzymes is not possible by conventional chemistry. Hence, almost all the enzymes presently employed in medicine are extracted from human blood or urine, or fortuitously produced by microorganisms. One is produced by culture of human kidney cells. Immediately, the possibility of substituting molecular clones as the source of those enzymes that have demonstrated clinical value - primarily to reduce the cost of production - is being explored. Creation of bacteria that synthesize the human enzyme, urokinase, which initiates blood clot disintegration, was announced in April 1980. Most enzymes are either too complex for the state of the art (e.g. blood clotting factors) or not reasonable candidates for therapeutic use on the basis of what is known today. The capacity to produce large quantities of pure enzymes by cloning and fermentation practices readily accessible to any biological researcher - either directly or through service organizations - will inevitably give rise to therapeutic concepts worthy of clinical evaluation. Deficiency diseases are possibilities, but the small number of patients with any particular condition, as well as the technical difficulties will probably deter commercial interests. Success of one or more enzyme products - such as urokinase or a novel entity made practical for clinical evaluation by molecular cloning - would do a great deal to stimulate interest in enzyme therapy, which has never generated much enthusiasm in the USA to date.

Novel vaccines, based on antigens synthesized by molecular clones, have been proposed to address communicable disease problems that have resisted prior efforts, especially those involving organisms that are difficult or impossible to grow outside a human host. Pure immunogens have always been scarce, and the potential availability of such proteins offers great promise of reduced adverse effects and suggests reexamination of the routes and dosage forms of vaccine administration. The creation of functional organisms that will colonize specific body spaces, without pathogenic effect, and present one or more antigens in order to immunize the appropriate tissues is forseeable, should that prove necessary to establish or maintain a particular immune defense. There is reason to be hopeful that the combination of molecular cloning with other modern biomedical techniques will make possible the development of vaccines against the parasitic diseases that burden the people and economies of most of the under-developed areas of the world.

Table 1, below, illustrates the billion-fold range of values, based on the apparent current cost of production by other means, of equivalent masses of exemplary pharmaceutical products that could, at least in principle, be produced via genetic engineering.

Table 1. Author's Estimate of Manufacturing Costs ($/Gram), 1980

Interferons from Leucocytes	$25,000,000
Blood Clotting Factor VIII from Plasma	200,000
Urokinase from Cell Culture	75,000
Synthetic Calcitonin and Beta-endorphin	10,000
Human Growth Hormone from Donated Glands	1,500
Synthetic Peptides of <20 Amino Acids	<1,000
Synthetic ACTH Fragment (24 Amino Acids)	500
Asparaginase from *E. coli* Fermentation	100
Insulin from Slaughter House Glands	<50
Gentamicin Antibiotic from Fermentation	<10
Human Serum Albumin from Blood	1.50
Penicillin from Large Scale Fermentation	0.02

PRINCIPAL CHALLENGES

The identification of desirable biomolecules is the primary challenge. Meaningful screens, cellular, animal and clinical models must guide the exploitation of the almost infinite synthetic capacity that has been opened. Pharmaceutical industry R & D establishments and the National Institutes of Health are the principal definers of such screens. The degree of real benefit derived from this technology will depend upon the quality of those screening efforts. Multi-disciplinary cooperation is essential and the successful pharmaceutical houses maintain a diverse R & D staff, usually with screening capability focused around three to six kinds of pharmacological activity. Historically, synthetic organic chemistry has been the primary source of new candidates, except in the antibiotic realm. Increasingly, protein biochemists, enzymologists, endocrinologists, immunologists, microbiologists and cellular physiologists will be submitting purified extracts from tissues and microorganisms. Pilot plants will be faced with adapting state-of-the-art laboratory separations of complex, unstable mixtures to scale up preparations for expanded animal, preclinical and clinical tests. In number, if not in nature, those evaluations of biomaterials will be unprecendented. Adequate characterization of tested entities to facilitate inference will be a substantial challenge.

To bring molecular cloning into play it becomes necessary to define assays that amplify rare phenomena from the sub-cellular level to the perceptible. Methodological innovation influences not only the rate but the scope of application.

Rapid transfer of insights and techniques from the universities and research institutes into the application-oriented labs will continue to be a challenge. The fact that new companies were the first to attempt applications of *in vitro* recombination, suggests that established firms were out-of-touch or could not appreciate what their recently expanded activities indicate to be their present high expectations. A few with on-campus outposts (or on-plant campuses) seemingly have been better prepared, and the uncertainty associated with the Asilomar Conference (1975) and Guidelines made investment in (or even association with) recombinant-DNA more risky than alternative R & D programs, but many refrained because they simply did not know what was going on - or rather what it implied. Talking with industry scientists one still hears of much bootlegging, dabbling, and getting acquainted. Thus, it is a continuing challenge to maintain communication between American industry and fundamental bioscientists.

Furthermore, it is a challenge to devise mechanisms for replenishing the store of basic knowledge. Basic research must be recognized as an asset that requires maintenance. There ought to be some way for a portion of the value generated from the pool of knowledge to be directly recycled, so that basic research can participate in the benefits derived from earlier contributions. Lewis Thomas, Clifford Grobstein and others have urged that, in Thomas' words, "partnership be worked out between the corporate world and the world of academic science, specifically designed for the very long term." (Thomas, 1980). Direct industrial support of academic research could redress the imbalance that has developed as the federal government has assummed an increasing share and absolute magnitude of basic research spending. In an era of federal budget cutting, enlightened self interest should encourage industry to devise means to maintain the vitality of its sources.

Among the technical challenges, the most immediate, other than screening, has to do with glycoprotein synthesis. So many important molecules - including interferons and erythropoietin, and blood clotting factors - are in this class, that publicly funded multifaceted programs related to the biology and chemistry of glycoconjugates are definitely in order. However, in order not to disturb the commercial incentives to build on the basic findings, it is urged that academic and non-profit institute research efforts center on such important physiologic species as the gonadotropins, with the expectation that profit incentives will lead to parallel work on the more marketable entities. (Blood clotting factors may also be appropriate topics for Federal emphasis for reasons elaborated on in the text.)

Another set of technical challenges lies in the realm of macromolecular separation and purification for testing and, especially, for mass production. Never has it been necessary to purify large quantities of proteins to the degree that is likely to be required for the products from molecular clones. Existing techniques for separating similar proteins involve costly, time and energy consuming, batch processes for the most part. Assuring maintenance of the specified structural integrity - free, for example, from aberrant amino acid sequences resulting from mutation - will be challenging from both the microbiological (strain stability) and analytical (sequence determination) perspectives.

FACTORS MOST LIKELY TO GOVERN PROGRESS

The most important influence on progress is the expectation that these efforts will lead to a multiplicity of products. This confidence will ensure perseverance through the inevitable set-backs. Consider the consequences of the finding that interferons exert some heretofore unrecognized toxicity when trials are expanded or - as has happened from time to time with new drugs - after market introduction. To the extent that the investments are recognized in advance to be broadly useful, the flow will be unperturbed, but there is a danger that many investors who are not experienced in pharmaceutical development, or well positioned to perceive and exploit alternative applications, may loose heart. Most newcomers to the biotechnology sphere are attracted by anticipated applications outside of medicine, and their accomplishments or disappointments with respect to those other applications will influence the emphasis placed on medicinal agents.

Another aspect of this entry of new participants into the pharmaceutical realm is the delay and cost imposed by the Food and Drug Administration product approval procedure. Even veteran pharmaceutical houses chafe under the burden of the alleged vicissitudes of the regulatory bureaucracy . Newcomers may be shocked by the extent and depth of probing involved in the clearing of a pharmaceutical product. Since both FDA and the sponsors need to learn what are the appropriate questions to ask and how to go about answering them, the first products of a novel type are likely to move slower than later ones. The fact that Eli Lilly, which has a great deal of experience both generally and with insulin specifically, probably will be the sponsor of the first NDA on a peptide pharmaceutical made by in vitro recombination could facilitate the process a great deal.

Frequency of administration, complexity and novelty of the type of agent proposed for pharmaceutical use will influence the time period and cost of testing required to obtain approval for marketing. Severe conditions, for which satisfactory therapeutic alternatives are not available, are likely to be the basis for much more rapid approval than prophylactic applications, for example, but the initial response of the FDA to all submissions for test approval have been very encouraging.

Doubt about the patentability of inventions in this realm may impede investment somewhat, but even more it will discourage the publication and transfer of findings which, in the presence of adequate patent protection would stimulate the overall development of this field. Extension of the duration of patent protection, e.g. for a period to compensate for the time consumed satisfying the FDA, could be expected to stimulate development, especially of a very long range nature, such as the development of processes to produce serum albumin or blood clotting factors, for example.

The availability of production scale technology to economically separate and purify proteins, as well as grow and contain novel production organisms, could limit application of this technology, especially for the lower unit value products. Since insulin derived from animals has low unit value (~$50/gram) relative to interferon or urokinase, for example, cost criteria will be prominent in the priorities of products likely to be ready for commercialization most quickly. The early development of proprietary solutions to general problems with respect to production economics is a major incentive for vigorous participation in this field even before companies recognize which specific products made possible by the technology best suit their own business position and plans.

Progress in the utilization of the technology for certain applications - especially vaccines - is likely to be deterred by the <u>scale of clinical testing</u> required to demonstrate safety and efficacy relative to the sales volume and profit potential of the markets. Vaccines, which are administered only once and often through public sector agencies, are particularly sensitive. Government support of clinical trials will probably determine whether certain vaccines are developed at all. Government stimulus, e.g. Dept. of Defense support of the development of synthetic serum albumin to assure availability of plasma expander for military use, could transform a marginally attractive commercial prospect into a bellwether with far-reaching implications.

Acceptance of some therapies employing agents that could be synthesized by means of molecular genetics will require the <u>development of dosage forms</u> that take into account the chemical and pharmacokinetic properties of those molecules. Physiologic compounds, in particular, are rapidly metabolized or excreted for the most part, and if they are to be exploited in other than hospital settings, it will be necessary to make use of emerging technology for controlled drug delivery to facilitate administration of the appropriate concentration patterns to achieve the desired effects.

Agents That May Be Produced by Molecular Genetics

1. INTRODUCTION

The promise of molecular genetics is two-fold. On the one hand, the science of molecular genetics offers the most fruitful approach yet to the unraveling and interrelating of the myriad processes that underlie life, health and disease. On the other hand, the technology of molecular genetics offers a largely unknown and inestimable potential for the synthesis of agents useful to man. With regard to this latter exploitation, it appears that a number of medically valuable products will be among the first generation of inventions to spring from the science through the technology. In this section we shall outline the nature of the most probable applications and attempt to depict their impact upon the practice of medicine. Although the emphasis shall be upon such applications and their practical implications, the farther reaching significance of the scientific insight that will flow simultaneously from the molecular genetic approach must not be missed. By means of molecular genetics, along with the many other valuable ways of viewing living creatures, modern biology will facilitate the development of the information and tools necessary to overcome many age-old problems and satisfy curiosity about the phenomena of life.

In essence, molecular genetics focuses upon those aspects of the complex interlocking processes in living organisms which concern the translation of information encoded in DNA, through messenger RNA intermediates, into polypeptides. Advantage is taken of the recently developed techniques for isolating nucleic acid segments bearing specific messages and inserting them into organisms in which they conveniently can be propagated and their messages translated into abundant copies of the desired product--a polypeptide. In some cases (*in vitro* recombination), small pieces of nucleic acid are separated from adjacent genetic material and manipulated chemically outside the biological realm, while in other cases (cell fusion and viral transformation) entire genetic contents, genomes, of organisms are pooled and allowed to sort themselves out biologically under conditions which favor the propagation of those offspring which express the desired combination of genetic characteristics.

At this time, the immediate product of the genetic translation process--a polypeptide (whether it be a peptide hormone, enzyme, antibody, or other protein) will be the product objective of most efforts. However, it is a matter of time and evolution of technology until the manifestation of a number of enzymes so-engineered into an organism could be a complex non-protein, e.g., an antibiotic.

2. HORMONES

Many of those potent messenger molecules, by which the human body coordinates the actions of various tissues, are simply chains of amino acids (polypeptides). Some are polypeptides that have been modified by the attachment of carbohydrate substituents (glycopeptides). Along with such non-peptides as certain steroids , these hormones trigger events in cells remote from the cells that produced them by interaction with specific receptors. Some act over relatively short distances, such as between segments in the brain and in glands closely linked to the brain, while others act on distant sites or on tissues throughout the body. Many such hormones manifest multiple activities depending both upon their local concentration and the specific cells with which they are interacting. They are arrayed in complex balancing and controlling interactions that seldom permit the consideration of a single effect when one contemplates intervention in the scheme. Nevertheless, certain hormones have attained a place in pharmacology and undoubtedly more will do so. Furthermore, it seems apparent that some pharmacologically useful agents originating from sites altogether outside the body, e.g., plants and synthetic chemistry, manifest their desireable activity through fortuitous interaction with the receptors for physiologic hormones.

To some extent the testing, and therefore utilization of hormones has been impeded by their scarcity and consequent cost. Extraction from animal glands, or rarely from human-cadaver glands, and from urine has been the only source of hormones until recently. However, in the past decade chemical synthesis has become feasible for some members of the class and a few totally synthesized polypeptide hormones are being marketed. Synthesis of glycopeptides exceeds the capabilities of synthetic chemists.

The fact that polypeptides are digested and inactivated

if administered by the oral route means that they have to be taken by injection, which has significantly impeded their adoption. Furthermore, most of them act briefly, because they are rapidly eliminated, and many give rise to multiple effects.

2.1 Polypeptides

It is a fact of nature that polypeptides are the first abundant product of the genetic apparatus, and they are the obvious focus of much of the initial effort to harness genetically-directed processes for the production of useful molecules. Thus, the peptide hormones insulin, growth hormone and somatostatin were the first objectives selected for applying *in vitro* DNA-recombination with practical consequences in mind. One point that complicates the relatively simple matter of applying molecular genetics for this purpose is the fact that many of the peptide hormones arise in nature as products of structural genes that are considerably larger than the secreted hormones. (In the case of many enzymes, this is even further complicated by the fact that zymogens - not active enzymes - are secreted, only to be activated elsewhere through the action of another enzyme.) Furthermore, disulphide (-S-S-) links are often critical to the active structure of polypeptide hormones in addition to the basic polyamide (-CONH-) linkage that is specified by the genetic code in the DNA. Folding of the polyamide backbone into the proper configurations for formation of the correct disulphide bridges is one of the reasons nature synthesizes the prohormones, but their utility rests upon the presence of the appropriate enzymes to form the disulfide links while the segments are properly configured and excise the intervening peptide segments when they have served their purpose. The discussion of insulin will illustrate this point.

A major variable in the family of peptide hormones is the length of the amino acid chain. Tables 2 and 3 list some polypeptides of interest in declining order of number of amino acids (AA). Modern chemical synthesis methods have made the preparation of relatively low molecular weight polypeptides a fairly straight-forward task and chemically synthesized hormones up to at least 32 AA i.e., calcitonin, have become competitive with those derived from current biological sources.

Table 2. Large (>20 Amino Acid) Polypeptides Potentially Useful in Medicine Which May Be Prepared by Genetic Engineering and Fermentation

	Amino Acid Residues	Molecular Weight	No. of Chains	No. of -S-S
Prolactin	198	~ 23,000		
Placental Lactogen	192			
*Growth Hormone	191	22,005	1	2
Calmodulin	148	16,700		
Nerve Growth Factor	118	13,000 (mouse)	2	2
Parathyroid Hormone (PTH)	84	9,562 (bovine)	1	
Proinsulin	82		1	3
Insulin-like Growth Factors (IGFI & IGF-2)	70, 67	7,649, 7471	1	3
Somatomedin A	59		1	
Epidermal Growth Factor		6,100	1	3
*Insulin	51	5,734	2	3
Thymopoietin	49			
Somatomedin B	44	5,000	1	4
Gastric Inhibitory Polypeptide (GIP)	43	5,104 porcine	1	0
*Corticotropin (ACTH)	39	4,567 porcine		
Cholecystokinin (CCK-39)	39			
Big Gastrin (BG)	34	(contains sulfated tyrosine)		
Active Fragment of PTH	34	4,109 bovine	1	0
Cholecystokinin (CCK-33)	33	3,918 porcine (contains sulfated tyrosine)	1	0
*Calcitonin	32	3,421 human 3,435 salmon	1	1
Endorphins	31	3,465		
*Glucagon	29	3,483 porcine	1	0
Thymosin Alpha-1	28	3,108	1	0
Vasoactive Intestinal Peptide (VIP)	28	3,326 porcine	1	0
*Secretin	27	3,052	1	0
*Active Fragment of ACTH	24		1	0
Motilin	22	2,698	1	0

*Currently employed in medical practice.

Table 3. Naturally Occurring Small Peptides of Potential Medical Interest

	No. of Amino Acids	Mol. Weight	No. of -S-S-
Dynorphin	17		
Little Gastrin (LG)	17	2,178	0
Somatostatin	14	1,639	1
Bombesin	14	1,620	0
α-Melanocyte Stimulating Hormone	13	1,655	0
Active Dynorphin Fragment	13		0
Neurotensin	13		0
Mini-Gastrin (G13)	13		0
Substance P	11	1,347 bovine	0
Luteinizing Hormone Releasing Hormone (LHRH)	10	1,183	0
Active Fragment of CCK	10		
Angiotensin I	10	1,297	0
Caerulein	10	1,252 porcine	0
Bradykinin	9	1,060	0
*Vasopressin (ADH)	9		1
*Oxytocin	9	1,007	1
Facteur Thymique Serique (FTH)	9		
Substance P (4-11) Octapeptide	8	966	0
Angiotensin II	8	1,046	0
Angiotensin III	7	931	0
MSH/ACTH 4-10	7		
Enkephalins	5	575	
Active Fragment of Thymopoietin (TP5)	5		
*Thyrotropin Releasing Hormone (TRH)	3	362	

*Currently used in medical practice

A significant advantage of chemical synthesis, for the research purposes to which most of these products are put these days, is that one can prepare readily, analogs which differ slightly from the natural hormone by incorporating different amino acids or even mirror image isomers in place of certain AA subunits. Such analogs may possess desirable pharmacological differences from the hormone. Further complicating the picture is the finding that fragments of peptide hormones express activities comparable to or, for some purposes, superior to the intact hormone.

2.11 Chemical vs. Biological Synthesis

In principle, it will be seen, a genetically-engineered biosynthetic scheme could be devised for most desirable peptide hormones and analogues, although the practicality of doing so must be assessed on a case-by-case basis. Ultimately, the principle factors bearing upon the practicality of the alternatives will be the following:

- The cost of raw materials, which in the case of engineered-biosynthesis is primarily the nutrient broth plus some amortization of the cost of developing the synthetic organism, and in the case of chemical synthesis involves the pure, optically-active AA subunits plus the other chemicals that are employed as activating, protecting, coupling, liberating and supporting agents for incrementally building up the sequence.
- The differential costs of separating the desired product from the cellular debris and culture medium on the one hand and from the supporting resin, by-products and excess reagents on the other.
- Any difference in the cost of purification and attainment of freedom from toxic contaminants, in particular, traces of potent bacterial endotoxins that will be a somewhat greater concern for biologically-produced material than that produced by conventional chemistry, although hormone from any source could pick up pyrogens from adventitious contamination.
- Differences in labor and equipment requirements between the chemical synthesis, which involves a sequence of similar (but different) operations throughout a time period roughly proportional to the length of the AA chain (~3 AA/day) in pilot-plant-scale apparatus sufficient for the production of quantities of 100g-1kg per batch, over a period measured in weeks, in contrast to fermentations in vats, with capacities of several thousand gallons, for a few days regardless of chain length.

- The cost and suitability of comparable materials derived by extraction from organs or fluids obtained from animals or people.

Our impression, reinforced by discussions with people directly involved in the production of peptides by chemical synthesis (e.g., Armour, Beckman, Diosynth, Sigma), is that the cost of preparing polypeptides of less than 25 AA and, probably up to 50 AA, in length by the chemical route is extremely sensitive to volume of off-take. They say that for peptides of the order of 30 AA in length, costs may approach $1 per milligram as the volume approaches the kilogram level - a level of demand rarely existing today but likely to be generated by work in progress. Today, the cost of the 32 AA polypeptide calcitonin, which is synthesized chemically and marketed as a pharmaceutical product by Armour, is probably in the range of $20 per milligram (since the wholesale price in vials containing approximately 0.15 mg is about $85 per milligram). However, that $20 per milligram is purely speculative since such costs are closely guarded secrets, and the pricing of a pharmaceutical product must take into account so many variables that the material cost of the agent is a small consideration, especially early in its lifetime, as is the case with the calcitonin product, Calcimar. Bradykinin (9 AA) can be purchased for $1.60 per milligram and LRH (10 AA) for $2 per milligram in one gram quantities. β-endorphin (31 AA) could be had for about $20 per milligram if one ordered a gram, although its price at the 5 mg scale is $114 per milligram (Beckman).* The conclusion is that we are talking about

---

* A ballpark estimate of $5 per milligram for a 34 AA fragment of parathyroid hormones in kilogram quantities was suggested by Sigma, but Mr. Kovitch there said they have given up predicting such costs.

substances with values measured in millions of dollars per kilogram.# Fortunately, they are also the most potent pharmacologic agents known, with a kilogram representing of the order of million doses.

As to the merits of the alternative sources, each case must be considered individually. However, one can assign them to rough categories which we shall discuss as we go through a number of cases, outlining at the same time how the technology of recombinant-DNA has evolved since its inception in 1973.

2.2 Low Molecular Weight Polypeptides

2.21 Somatostatin

In 1977, a totally synthetic DNA sequence was inserted in an *E.coli* plasmid and demonstrated to lead to the production of the 14 AA polypeptide sequence corresponding to somatostatin fused to a protein corresponding to most of the bacterial enzyme β-galactosidase. The fusion was such that, after separation from the bacterium, the chimeric protein could be cleaved with cyanogen bromide to liberate the 14 AA polypeptide. The fact that the amino acid sequence was known permitted the design and synthesis of DNA oligonucleotide with the appropriate triplet codon sequence, and the existence of sensitive assays for somatostatin enabled the detection of its expression. Although these were the primary motivations for using that particular molecule for the first demonstration of the expression of a foreign gene in a bacterium, Genentech (which funded the work) has stated that it plans to market

The value of gold at recent record-high levels was less than $30 thousand per kilogram.

somatostatin made by this route for research purposes. As a point of reference, somatostatin can presently be purchased for $60 per milligram in milligram quantities, and a number of the companies* investigating it are probably synthesizing their own at much lower costs. Furthermore, much of the interest has shifted to analogs of the natural molecule. Be that as it may, somatostatin is a good example of the low-molecular weight peptide hormones that could be made by molecular cloning.

Obtaining somatostatin as a "fusion product" was convenient in terms of expression, i.e., it was included within the lac operon and therefore read out when that operon was transcribed. Furthermore, its production as a segment of a large protein probably sequestered it from the bacterial proteolytic enzymes which customarily degrade polypeptides that do not belong in the bacterial cytoplasm (exogenous somatostatin is indeed rapidly degraded by extracts from E.coli). This design also permitted demonstration of the fact that expression of a foreign gene could be controlled, through the use of an inducer of the lac operon to increase production. All expression of the somatostatin fused-gene was assessed immunologically (radioimmuno assay, RIA), not by hormonal activity. The yield of the somatostatin fragment corresponded to from 0.001% to 0.03% of the cellular protein.#

---

* In 1975, the Salk Institute granted licenses to four companies (Wyeth, Merck, Ciba-Geigy, and Kabi) on somatostatin which was discovered by Roger Guillemin, et.al., at Salk in 1972.

However, ~3% of the protein produced by these E.coli consisted of the fused somatostatin – galactosidase hybrid on account of the size of β-galactosidase and the number of plasmids (>30) in the average E.coli cell.

Somatostatin is one of about 20 recognized human polypeptides of molecular weight less than about 2000 that can be made without difficulty by the Merrifield solid-state synthetic route and, unless a very sizeable market develops for one of them, it is unlikely that biosynthesis will prove competitive in the forseeable future. Its particular activity includes the inhibition of pituitary secretion of growth hormone and thyrotropin and pancreatic secretion of insulin and glucagon among other effects, and it has been especially investigated as an adjunct to insulin for treatment of some diabetics. However, in view of its two-minute half-life, most workers in the field feel it is unlikely that somatostatin *per se* will become an important factor in therapy. It is of interest that, while the polyamide backbone of somatostatin in the body is cyclized by a disulphide link between its two cysteine residues, the linear molecule appears to be as active.

2.22 Neuroactive Peptides

The past five years has seen an explosion of scientific interest in low molecular weight polypeptides as they have been identified in many tissues and with a great variety of activities. In large part, this is the result of the finding that such peptides are a major group of neurotransmitters with roles distinct from their involvement in neuroendocrine processes as classical hormones. At this time, the presence and activity of particular peptides at certain sites in the central nervous system and their relevance to physiological processes are the subjects of study and conjecture, but the significance is not at all clear. Some of these peptides certainly bind to the same receptor sites as the pain-relieving opiates related to the morphine family produced by plants, and such peptides have been classified as endogenous opiates: the lower molecular weight (5 AA) ones being termed enkephalins and the higher (31 AA) endorphins.*

The pentapeptides, (Met^5) and (Leu^5) enkephalins produce analgesia of very brief duration when instilled directly into the brain (in mice). Synthetic analogs which are less susceptible to enzymatic inactivation give rise to longer lasting analgesia even by conventional intravenous infusion, as does the larger β-endorphin molecule. Very recently a 17 AA polypeptide, dynorphin, has been reported to be the most potent pain killer yet identified.# Determination of its AA sequence is considered a tour de force accomplishment, having been performed with only 2 micrograms of the material, which is very difficult to isolate from the posterior pituitary. The preparation of new analgesic agents** appears an almost certain outcome of this research, but there are problems (tolerance and physical dependence), similar to those

---

* They are all related to sequences in the 91 AA β-lipotropin, which in turn appears to be biogenetically derived along with corticotropins from a common glycoprotein precursor (pro-opiocortin) of molecular weight ~31,000. Curiously, when glycosylation of this pro-opiocortin is prevented through the use of an antibiotic, tunicamycin, the unglycosylated pro-opiocortin peptide is rapidly degraded to different peptides than those normally found in such tissue. (Nature 278, 394-5, 423-7, 3/29/79.)

\# With activity 50 times that of β-endorphin, 200 times that of morphine and 700 times that of (leu^5) enkephalin in a guinea pig intestine model system. The first 5 AA residues are identical to (leu^5) enkephalin.

** Various results suggest that mood altering agents might be an outcome of this line of research, and opiates are commonly used to suppress coughing and to control diarrhea.

associated with classical opiates, that must be overcome, and unnatural analogues - including ones made with amino acids that are not produced or utilized (no transfer RNA) by microorganisms - seem likely to turn out most useful. Thus, the relevance of microbial biosynthesis is highly doubtful at this time. The importance of genetic technologies,(especially for the production of specific antibodies and elucidation of the underlying mechanisms in which the physiologic molecules are involved),to the progress of such research should not be underestimated, however.

A related area is the brain activity of certain other small peptides, such as the 7 AA sequence known as MSH/ACTH 4-10, which is reputed to influence memory, concentration and other psychological-behavioral effects. Should such agents prove out in wider testing, the enormous volume potential might well justify development of a biosynthetic route, including the technology for protection of the desired small peptide from microbial peptidases.

2.23 Gastro-intestinal Hormones

Neurotensin, Substance P, and several other relatively small peptides are found in certain cells in both the central nervous system and in the intestine (gastroenterohepatic and related endocrine systems). Along with certain somewhat larger peptides - e.g., cholecystokinin (33 AA), secretin (27 AA) and vasoactive intestinal peptide (28 AA) - they are categorized as the APUD* series, on the basis of a common embryologic ancestry

* APUD stands for "amine precursor uptake and decarboxylation," which characterizes the cells referred to.

shared by the specific cells in which they appear. None of these is currently employed therapeutically, but there is considerable interest in the activity of gastrin (a fragment or an analog) with respect to repair and increased integrity of the gastric mucosa and for reducing esophageal reflux by increasing lower esophageal sphincter pressure. Similarly, ways of utilizing the effects of cholecystokinin (CCK) on gall-bladder emptying and pancreatic secretion or to encourage motility in pseudo-obstruction of the intestine are being investigated.

Incorporation of polypeptides into liposomes to protect against digestion in the G-I tract is being studied by a number of investigators, but non-peptide agents remain of greater interest to pharmaceutical companies. Gastrin and CCK are used experimentally for diagnosis of gastric and pancreatic secretory funtionality, but FDA approval has not yet been obtained. Both CCK and bombesin (10 AA) have been shown to suppress appetite, presumably as a satiety signal from stomach to brain. The market for anti-obesity agents is great - approximately $85 million per year at the manufacturers' level.

2.24 Hypothalamic Hormones

Actually somatostatin is named for its action as an hypothalamic hormone, i.e., those hormones which are released by nerves in the hypothalamus into the portal vein to stimulate or inhibit release of the tropic hormones by cells of the anterior and intermediate lobes of the pituitary. Only two others (TRH and LHRH) have been fully characterized yet, but extracts appear to contain factors which stimulate the release of corticotropin (CRF), growth hormone (GHRF), melanocyte stimulating hormone (MSH-RF) and perhaps prolactin (PRF). There are also factors which inhibit the releasing of prolactin (PIF) and melanocyte stimulating hormone (MSH-IF), as well as somatostatin (GHIF). Through the action of these releasing and inhibiting factors, the hypothalamus translates neural signals into endocrine signals.

To date, hypothalamic hormones have been used for diagnosis of pituitary function, but not for therapeutic purposes. Some analogues of LHRH are being investigated as contraceptive agents, and preliminary clinical results are encouraging.

2.25 Pituitary Hormones

The posterior lobe of the pituitary secretes two hormones, both of which are cyclic octapeptides, and both of which are medically useful.* Vasopressin, or as it is more descriptively termed, antidiuretic hormone (ADH), acts on the kidney to control water loss. Its potency is amazing; less than 0.1 microgram per hour produces maximal antidiuresis in man. It is used to treat diabetes insipidus. Annual sales are about $3 million at manufacturers' level. Depot injections that last three to four days illustrate one manner in which short-lived peptides can be given somewhat more conveniently than daily injections. However, another dosage form stimulates the imagination even more. This version, a nasal spray, takes advantage of the fact that polypeptides are readily absorbed through mucosal membranes into the blood stream. Absorption varies a great deal, however, from dose-to-dose and person-to-person. For vasopressin this route is quite

* Actually vasopressin and oxytocin are synthesized in the hypothalmus, and are only stored in the posterior pituitary.

satisfactory since the consequences of diuresis are readily perceptible and not acutely harmful. (A dose is reported to last three to eight hours.)

Should therapeutic research lead to significant new indications for polypeptides to be used chronically, (e.g. anti-obesity, anti-senility, contraception, memory or learning enhancement), use and improvement of the mucosal absorption route will probably prove important. That could also be true for agents to be discussed later, such as interferon or antibodies.

The second octapeptide from the posterior pituitary is oxytocin. Its two principal actions (simulating contractions of the pregnant uterus and causing milk ejection from the lactating breast) are medically useful, but they do not have a big market because they are one-time uses, i.e., only at parturition. The total sales of oxytocic drugs is about $5 million (manufacturer level) and 30% of that is taken by ergot alkaloid preparations that exert similar effects on the uterus.*

All of the oxcytocin and some of the vasopressin pharmaceutical products sold today are made synthetically. The largest selling vasopressin product uses pituitary extract.

We shall save for later discussion (see lymphokines) the pentapeptide (TP5) recently identified as possessing activity comparable to the 49 AA thymopoietin.

---

* Recent use of another ergot alkaloid to treat acromegaly due to hypersecretion of growth hormone suggests a market opportunity of a somatostatin-like compound with fewer sites of action than somatostatin.

2.3 Higher Molecular Weight Polypeptides

It is presently impractical to assemble more than fifty amino acids by chemical synthesis. Among those listed in Table 2, insulin and growth hormones stood out as the obvious targets for production via recombinant technology, and the initial laboratory demonstrations of possibility have been accomplished already. They were important because it was felt desirable (in the case of insulin) or essential (in the case of growth hormone) to have a peptide corresponding to the uniquely human AA sequence rather than the slightly different peptides produced by livestock. Furthermore, each presented substantial technical challenges:

- Insulin is composed of two polypeptide chains joined by disulfide links that are essential to its activity, and in nature, these are assembled through the use of an intermediate piece that must be deleted to render the hormone functional. The literature indicates that spontaneous association of the A and B chains into functional insulin is inefficient.
- Growth hormone is more than ten times as large as somatostatin, the synthesis of which by recombinant-DNA technology (fused to a larger protein) has been described above. It is a single chain with one disulfide link.

Since the market for insulin to treat diabetes mellitus is relatively large, while the present use of growth hormone is quite limited, it is not surprising that the first developments occurred with insulin. We shall defer discussion of the commercial and therapeutic considerations until after a brief description of the technical developments that have been disclosed with respect to the synthesis of human insulin through the use of genes inserted on plasmid-DNA in *E. coli*.

2.31 Insulin

The synthesis and secretion of insulin by the beta cells

of the islets of Langerhans in human pancreas is depicted in Figure 1. We shall not discuss the process but the figure provides a sense of its complexity. The initial natural product, proinsulin, has the structure depicted in Figure 2 and is about 1.5 times the size of the active hormone.*

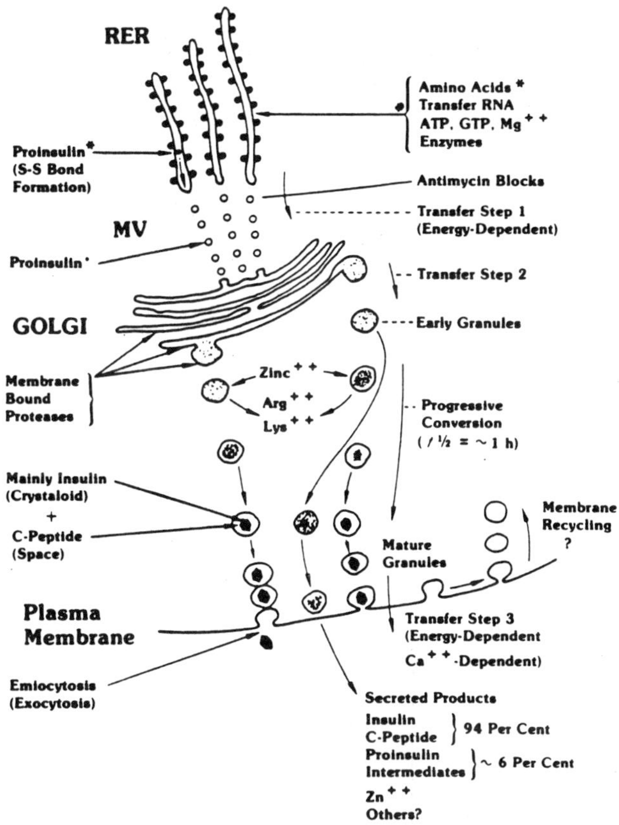

Figure 1. Pancreatic Production of Insulin

---

* "The biosynthesis of proinsulin is selectively modulated by the ambient glucose concentration of the islets, without any immediate requirements for the formation of new nRNA." This translational control is independent of the process of granule release, which is also stimulated by glucose. [Steiner, 1976].

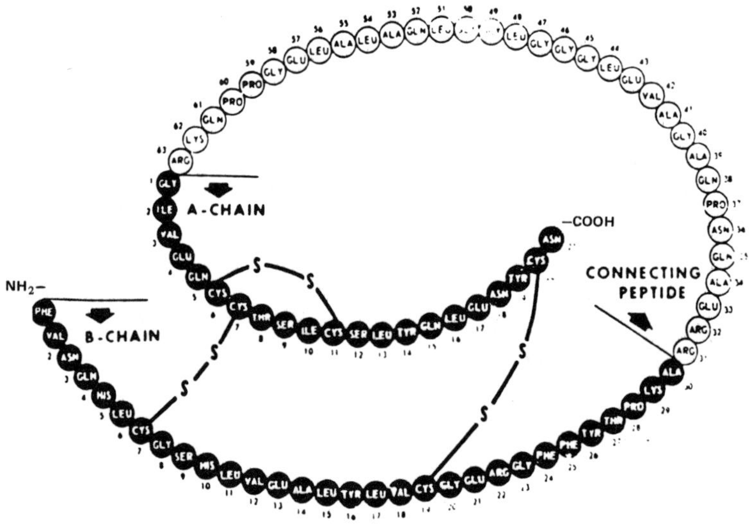

Figure 2. Proinsulin

Most of the extra AAs form a segment, identified as C, connecting the A and B chains. Amino acids are also cleaved from the ends of the A and B chains in the formation of the active hormone. Presumably these cleavages are catalyzed by specific proteases, with activities similar to pancreatic trypsin and carboxypeptidase B.

2.311 <u>rDNA Synthesis of Insulin</u>

The initial announcements with respect to recombinant-DNA (rDNA) synthesis of insulin involved the cloning of the gene sequence encoding for the precursor of proinsulin, pre-proinsulin, of the rat. After one group (UCSF, 5/77) reported incorporation of the gene without expression, another group (Boston, 6/78) managed to get colonies of <u>E.coli</u> that excreted rat proinsulin fused to an enzyme, penicillinase, that the bacterium normally excretes. Actually the gene inserted was not made directly from the rat's DNA, but rather it was made by reverse transcription of the messenger RNA (mRNA) and production of the complement of that DNA to form what is known as cDNA corresponding to the sequence of the mRNA. The yield of fused penicillinase-proinsulin was only about 100 molecules per cell.

Within a couple of months (9/78) after expression of the fused rat proinsulin gene, workers at Genentech, in collaboration with the same group at City of Hope Medical Center who synthesized the somatostatin gene, announced the separate syntheses of the A (21 AA) and B (30 AA) chains of human insulin through the expression of plasmid-borne DNA sequences in E.coli strains. As had been the case for somatostatin, each insulin subunit was attached to the bacterial enzyme β-galactosidase and retained in the bacterium (100,000 molecules/cell). The "genes" were set up so that cyanogen bromide treatment (which cleavespeptides at methionine sites) would liberate the insulin subunits from the cellular protein fraction. According to the initial reports, only the A segment, which is identical in the human and porcine species, was obtained in sufficient quantity and purity for AA sequence verification. Production of the B chain by the other strain was demonstrated only by radioimmune assay. The published experiment indicates the formation of the total insulin molecule by association of the A and B chains derived from the two fermentations to be quite inefficient,* for mature insulin was detectable only by radioammuno assay. [Goeddel, 1979]. For commercial purposes, synthesis of

---

* Less than the 10-15% correct joining normally obtained when purified bovine B chain is reacted with an excess amount of A chain. Another technique with 50-80% efficiency is cited in the Genentech article. [Goeddel, 1979]

proinsulin or something functionally comparable may turn out to be preferable. Genentech disclosed the cloning of the proinsulin gene in March 1980. Proinsulin was produced as a fused protein, cleaved from the precursor *in vitro*, and processed enzymatically to fold and cut out the insulin.

A key aspect of the insulin work was the use of the known AA sequence of the A and B chains of insulin, along with the triplet nucleotide code, to synthesize the corresponding DNA sequences. Rapid accomplishment of that objective was made possible by the development of a new "phosphotriester technique" for DNA synthesis utilizing trimer building blocks (a significant improvement on the phosphodiester approach employed in the pioneering work for which H. G. Khorana was awarded the Nobel Prize) and by recently developed high-performance liquid chromatographic purification and identification techniques. Twenty-nine oligonucleotides of 10-15 bp in length were used to assemble the "genes" corresponding to the A (77 bp) and B (104 bp) chains. Three months were required to synthesize the necessary trimers (in quantity sufficient to make many genes) and another three months to prepare the 29 oligonucleotides. [Crea, 1978]

All of the work was reportedly funded by Genentech and, presumably at least some of the chemistry will be covered by patents.

Each of the "genes" was assembled in segments, e.g., halves of the B-chain "gene" were cloned in separate plasmids (pBR 322) and their sequences verified, before the segments were joined to form a complete "gene" adjacent to the genes for the *lac* control region and most of the β-galactosidase structural gene. Approximately 20% of the total cellular protein produced by the selected strains was the β-galactosidase-insulin A or B chain hybrid expected. From the isotope dilution assays, it appears the yield of insulin chains in the crude extract following cyanogen bromide cleavage was about 0.4 milligram of insulin per gram of wet cells - about 10 times greater than the yield of somatostatin. Under the culture conditions employed, approximately 1.5 mg of crude insulin chain was produced per liter of medium. Of course, the yield might be considerably increased by optimization of the fermentation, but then there are the unsolved problems of coupling and purification.

It is instructive to make a comparison with the traditional source of animal, not human, insulin at this point. If, after all the increases and decreases in yields at various stages it turns out that 0.5 mg of pure insulin can be obtained from a liter of fermentation brew, then 2000 liters (roughly 500 gallons) would yield the equivalent of about 16 pounds of animal pancreas (1 gram of purified insulin). One hundred times that yield would correspond to the level currently achieved in the production of asparaginase by E. coli. (The average diabetic uses the equivalent of about 2 mg of animal insulin per day.)

2.312 Supply and Demand for Insulin

The significance of the potential production of human insulin through the use of microorganisms containing recombinant-DNA is complex to estimate. No one seems to be arguing that there is a shortage today of glands from slaughter houses for the production of animal (principally bovine and porcine) insulin. A study conducted by the National Diabetes Advisory Board (NDAB) concluded that a combination of maximum demand growth and minimum supply growth would lead to shortages in the 1990s. (Figure 3). [DHEW Pub. Nos. (NIH) 78-1588 and 1589] Eli Lilly's projection, presented in that report, also anticipates shortages in the 1990s. (Figure 4). Novo, a major world supplier of insulin, told the NDAB that it estimates that 1976 free-world consumption of insulin of 51 X 10^9 units constituted only 23% of the potential supply, and the 87 X 10^9 units projected for 1996 would only equal 40% of supply, assuming the animal population stays constant.

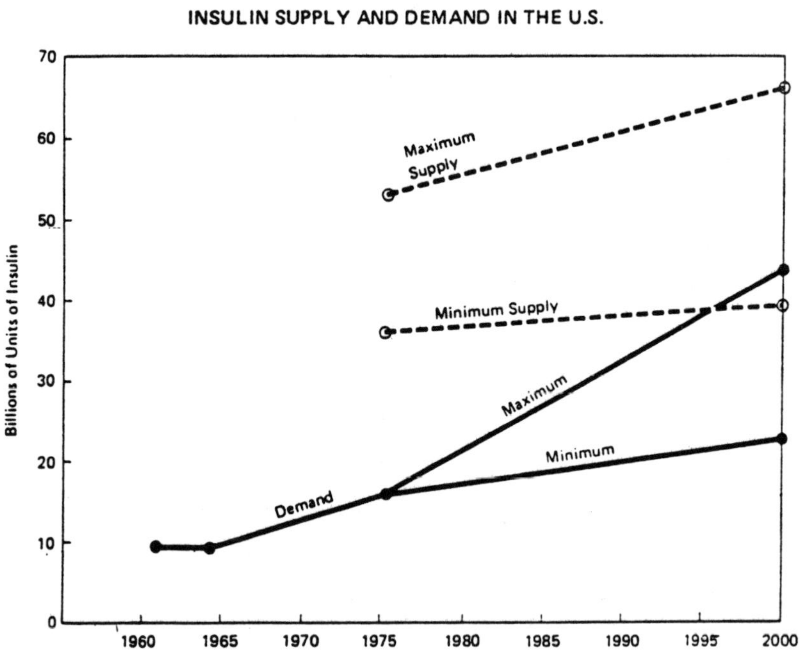

Figure 3. Maximum and Minimum Projections of Insulin Supply and Demand to the Year 2000, as Calculated by the National Diabetes Data Group. (DHEW Pub. No. NIH-78-1588) The lower supply curve, which intersects projected demand in 1996, assumes a 3% increase between 1975 and 2000, in contrast to the 23% increase projected on the basis of projections of animal production by the American Meat Institute.

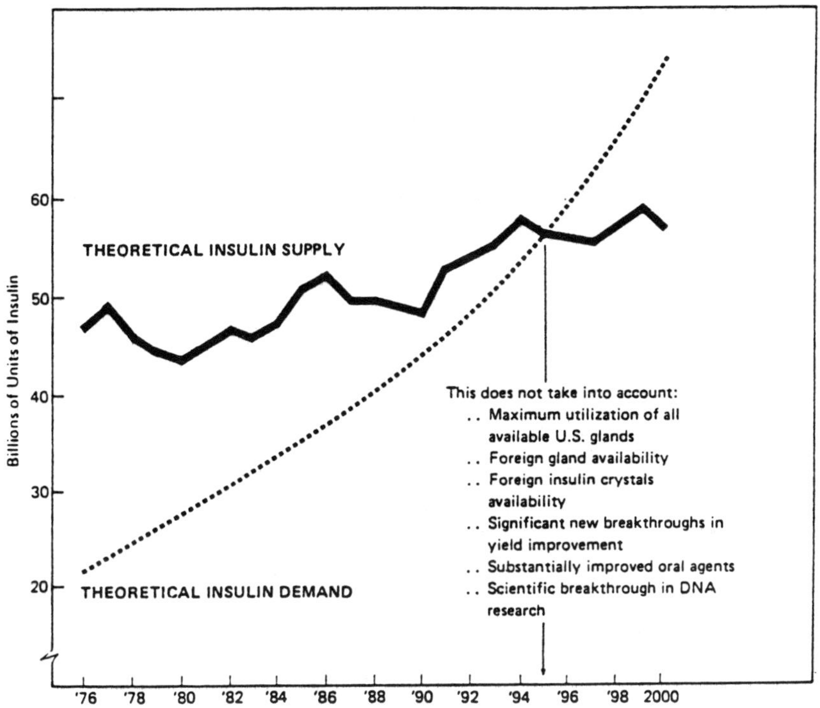

Figure 4. Eli Lilly and Company's Projection of Insulin Supply and Demand as Presented to the Insulin Study Committee of the National Diabetes Advisory Board in January 1978.

Sales of insulin to U.S. pharmacies and hospitals in 1978 amounted to approximately $84 million, for about 2×10^{10} International Units. At highest purity (28 IU/mg) that would be equivalent to roughly 700 kg of pure insulin. We note that the unit value of bulk animal insulin, which must be much less than the $120,000 per kilogram obtained by simple division for obvious reasons, is of the order of one thousand times less than the values attributed to the lower molecular weight (and smaller market) polypeptides described in the preceding section.*

*The market price of bulk insulin (Spring 1980) was about $1.50 per thousand international units for the pharmaceutical grade containing 500 ppm proinsulin. At 28 IU/mg, which is higher than the actual activity of such material, that price corresponds to about $37,500 per kilogram of pure insulin.

2.313 Eli Lilly and Insulin

Ever since it produced the first commercial quantities of insulin more than fifty years ago, Lilly has dominated the market in the United States. The degree and alleged methods of domination led to a Federal Trade Commission antitrust case which was resolved as summarized below:

> FTC alleged that "Lilly monopolized the U.S. finished insulin market through its participation in an international conspiracy to allocate and control the collection and distribution of animal pancreas glands within the U.S." Without admitting guilt, and while stating that the company believes that its activities were responsible and lawful, Lilly agreed to license its insulin related patents and know-how to any U.S. company that asks, at no profit to itself. Not only does this consent agreement apply to existing technology, but to future technology (including, especially, possible advances resulting from Lilly's agreement with Genentech on the development of human insulin through the recombinant-DNA process). FTC's analysis of the settlement explained that Lilly must grant "a no-profit-to-Lilly license to any domestic company under any or all future insulin-related patents and know-how that Lilly acquires from third parties within five years" of the date of the signing of the proposed consent order (May 30, 1979). Futher, it requires Lilly to grant licenses with "reasonable" royalties: (1) to any domestic firm for future patents developed by Lilly within the same five-year period, and (2) to any foreign company for existing Lilly patents. There are a number of additional concessions by Lilly in the agreement with respect to insulin patents, know-how and importation of insulin or insulin precursors.

While that case was being prosecuted, Genentech made the developments described above, and Lilly negotiated an exclusive license to market the rDNA human insulin. It has been vigorously supporting the development, both at Genentech and at Lilly. Lilly is a minor factor in European markets with respect to insulin and expects human insulin to enable its penetration of Europe in competition with Novo.

Besides being the dominant supplier of the insulin market (at least an 81% share of sales),* Eli Lilly is one of the

* E.R. Squibb has at least a 9% share and the supplier of approximately 10% is not stated, but much of it is probably supplied by Squibb and Lilly also.

leading research-based pharmaceutical companies which happens to be extremely strong in the development and use of microbial fermentation technology. Hence, its motivation derives from a broad perspective that envisions many applications for the sort of process of which human insulin is an example. Lilly has reported that its collaboration with Genentech is proceeding satisfactorily, and clinical trials commenced early in 1980.

Permission was granted in April 1980 for Genentech to scale up fermentation production of insulin and other proteins to the 750 liter batch size. At a meeting with the FDA in June, 1980, Lilly revealed that it had developed new chemistry for efficiently combining the A and B chains, and had developed HPLC and other analytical methods for monitoring purification. In July it announced plans to build a $40 million plant to produce human insulin, at the time that the preliminary results of its first clinical trials (in England) were published. No untoward results were observed in these tests involving 17 subjects. Clinical trials in the United States commenced in December 1980. Several hundred patients will be involved in trials worldwide early in 1981, according to J.H.Marsden of Lilly.

2.314 The FDA and rDNA Insulin

Lilly has stated that it anticipates a six-month testing period in humans. No doubt the FDA will examine the evidence presented in the "Notice of Claimed Investigational Exemption for a New Drug" (IND) with special care, for this review will establish criteria that may influence the review of subsequent applications with respect to at least the following considerations.

- o Evidence that the AA sequence of the material is identical to the normal human hormone

- o Freedom from bacterial endotoxin (lipopolysaccharides produced by E.coli, and other gram-negative microorganisms that produce a fever at extremely low concentrations) - an inherent hazard associated with any process utilizing E.coli*
- o Freedom from by-products,# including substances of very similar structure, which may give rise to rare acute or chronic effects of an immune or other nature

Furthermore, as the development progesses, FDA will probably require, in addition to the data and methodology required for all new drug products, stringent methods for monitoring that the molecules produced from batch-to-batch are not subject to some subtle variation resulting from their monoclonal genetic origin. Of course, Lilly has considerable experience meeting the FDA's requirements, and there is a precedent for using E. coli as a production organism for the cancer chemotherapeutic agent asparaginase.

2.315 Human Insulin

The clinical rationale for utilizing human insulin rather than that derived from animals rests upon the differences in structure between insulins produced by different species. Table 4 illustrates the sites of difference in AA sequence for the relevant species. Porcine insulin differs from human only at AA number 30 of the B chain. Bovine and sheep insulins differ from the human hormone at that site and at two or three sites in the A chain.

---

* Testing for endotoxins (pyrogens) is a routine part of the production of all products that are injected or implanted, and there are available exquisitely sensitive assays based on the reation of an extract from the blood of the horseshore crab (Limulus amoebocyte lysate), which far surpasses the ability of currently accepted rabbit tests to detect pyrogens.

In addition to the peptides and other cell constituents produced by the recombinant-DNA-bearing microorganism, there will be the enzymes used to cut and, perhaps, assemble the active insulin molecule and other reagents, some of which may be especially challenging to separate completely from the desired product. This would seem to be an excellent use for polymer-bound enzymes.

TABLE 4. AMINO ACID DIFFERENCES
IN THE INSULIN PRODUCED BY VARIOUS SPECIES

	A Chain Positions			B Chain Position
	8	9	10	30
Human	Thr	Ser	Ileu	Thr
Swine	Thr	Ser	Ileu	Ala
Cattle	Ala	Ser	Val	Ala
Sheep	Ala	Gly	Val	Ala

So far as is known, these differences generally do not impair the functionality of the insulin, but then the relevant knowledge doesn't go all that far. No one has ever been in a position to conduct a significant test of the use of human insulin in a diabetic population. We do know that many consequences of the disease, such as retinopathy and nephropathy, are not prevented by routine injection of animal insulin. Furthermore, we know that patients occasionally react adversely or produce antibodies to the animal insulin with consequent allergic or resistance phenomena.# All preparations of animal insulin contain small amounts of related substances, e.g., proinsulin,* although these substances can be almost

A "patient is said to be in a resistant state when over 200 units of insulin is required daily to afford reasonable control. Frequently the person who develops resistance due to insulin antibodies has had interrupted insulin therapy. The affinity of insulin antibodies is stronger for beef insulin so that many recommend the use of pork insulin in patients who have developed the immune type of resistance. Although pork insulin is available, its use rarely solves the problem." [Bonnar, 1977, p.58]

* While the differences in the AA sequence between man and livestock are minor in insulin itself, they are much more numerous and include 2-4 AA deletions in the C-segment of proinsulin. Historically, commercial "insulin" derived from animals has contained 3-6% proinsulin, along with some glucagon and other pancreatic peptides. For years 92% insulin content was common. The normal human beta-cell secretion contains about 6% intact human proinsulin, plus its C-segment fragment, which are released along with insulin into the blood stream.

completely removed by modern purification procedures. Recent preparations, referred to as monocomponent insulin, that have been more highly purified are less antigenic, and the FDA is currently proposing new standards for proinsulin content (10 ppm for porcine, and 50 ppm for bovine or beef-pork combinations) for products to be labelled "purified."

The real demand for insulin is a controversial issue. Lilly estimates there are 60 million diabetics in the world (35 million in underdeveloped countries, where few are diagnosed or treated). Of the 25 million in the developed countries, perhaps 15 million are diagnosed and, according to Lilly's estimate, 5 million are treated with insulin. Only one-fourth of those diabetics treated with insulin live in the United States but they use 40-50% of the insulin consumed in the world. A number of studies indicate that while the emphasis on diet (alone) and oral antidiabetic drugs* varies, approximately 40% of U.S. patients in large diabetic clinics or practices take insulin injections. Insulin therapy is definitely indicated only for the following (20% of all diabetics):

- Any patient of any age who has had ketoacidosis and coma.
- Juvenile diabetics or growth onset type with onset from infancy throughout adolescence.

Insulin therapy is also prescribed frequently for the following:

- Patients whose diabetes developed prior to the age of 40 (usually require insulin)
- Unstable diabetics; thin, older individuals

* These are agents which stimulate the release of endogenous insulin and are used primarily in connection with maturity-onset diabetes. They constitute a $105 million-dollar market.

- Oral hypoglycemic drug treatment failures
- Diet therapy failure patients
- Pregnant diabetics
- Diabetics with infections and febrile illnesses
- Diabetics undergoing major surgery

It should be recognized that in the USA, diabetes ranks 5th as a cause of death and 2nd as a cause of blindness. Roughly 2 million persons require daily injections of insulin. How many will be better off with human insulin remains to be seen, and the proof of improved therapy will take many years.* Progress on the etiology of the disease, especially the identification (perhaps by immunologic typing) and protection (perhaps by vaccination) of persons at risk of developing juvenile onset diabetes, or improvements in the dosage form and patterns of administration of insulin, may have a far more significant effect on this economically and personally costly disease than new developments in insulin production. Nevertheless, so long as private enterprise sees fit to invest in such developments, and so long as the cost of insulin treatment of diabetics who are properly treated with animal insulin is not increased, then one can view these events as insurance for diabetics

---

* It is interesting to note, although hardly the basis for pessimism, results of limited studies made at the time (1974) Ciba-Geigy scientists synthesized human insulin chemically. These studies indicated that patients who had previously been treated with animal insulin actually showed immune responses to the insulin that was supposed to be identical to human insulin. On the other hand, diabetics who had not yet been exposed to animal insulin did not manifest such a response to the human insulin preparation.

Lilly reports preliminary studies with insulin extracted from human pancreases that show no antigenicity. They also anticipate less local reaction to human insulin and modestly decreased dosage requirements.

in the 21st century and possibly of benefit sooner than that.

2.32 Glucagon

It is appropriate to write of glucagon at this point. Glucagon is a single strand of 29 AA that is secreted by the alpha-cells of the pancreatic islets. It is removed from the circulation by the liver and to a lesser extent by the kidney. Like insulin, it has a half-life in the blood stream of only about 10 minutes. Its primary effect is to increase the activity of adenylate cyclase in the liver by reacting with sites on liver cell membranes. The increased cyclic AMP activates glycogen breakdown, by increasing the activity of the glycogen phosphorylase system and decreasing the activity of the glycogen synthetase system. It also increases gluconeogenesis and lipolysis in the liver.* Its precise physiological role in man is unclear, but it clearly has the opposite effect on blood glucose levels compared to insulin. Excessive glucagon secretion may well be a component of some diabetes mellitus, as experiments with somatostatin have suggested.

At present, the only therapeutic use of glucagon (by injection of 0.5-1 mg) is to counteract severe hypoglycemic reactions in diabetics, and in connection with insulin shock therapy for psychiatric patients. It is a by-product of insulin production and purification, with supply far exceeding demand, although there was a shortage in 1976. It is also used to treat rare insulin secreting tumors.

---

* The release of glucagon is prompted by certain amino acids (e.g., arginine and alanine) and by the gastrointestinal hormones.

2.33 Growth Hormones

Growth hormone (GH) is one of a family of closely related, relatively large peptide hormones, including prolactin and placental lactogen. All are single-chain polypeptides, containing 2-3 disulphide bridges, of 191-198 AA in length (22,000 mw). Human growth hormone (hGH) is present in the pituitary gland in far greater abundance than any of the other hormones, being accumulated rather than immediately secreted, accounting for 10% of the dry weight of the gland. Prolactin is much more rapidly turned over and thus only 1% as much is found in the pituitary.

Furthermore, hGH manifests some of the same activities as prolactin on the latter hormone's target tissues. All three can be shown to possess a wide variety of activites, differing from one another in potency. In fact, the biological actions of these hormones show a range and diversity unparalleled among vertebrate hormones. The major activities are outlined in Table 5. As their names suggest, most of the activities of these hormones can be divided in two broad categories; those of general anabolic and growth promoting nature and those which relate to parenteral nutrition and care of offspring.

2.331 Physiology of Growth

GH is best known for the growth it induces in many soft tissues, cartilage and bone. It is the prime determinant of post-natal growth in man. The growth of an organism is a highly complex process not exclusively mediated by a single hormone, but depending on the correct balance of many variables. For example, the action of GH *in vivo* depends upon the presence of insulin, the secretion of

Table 5: Some Biological Activities of Pituitary Hormones, hGH, hPRL and hPL*

I. General metabolic actions—associated chiefly with hGH

(1) Increased statural growth in children, through stimulation of epiphysial cartilage growth

(2) Increased growth of the following tissues: connective tissue, including bone (due to periosteal growth); muscle; skin and accessory skin structures; viscera, including heart, lungs, liver, kidneys, adrenals, intestines, pancreas

(3) Nitrogen

(a) Increased uptake of amino acids by muscle
(b) Increased protein synthesis
(c) Increased RNA synthesis
(d) Increased DNA synthesis
(e) Decreased protein catabolism; positive nitrogen balance

(4) Fat

(a) Increased lipolysis; increased concentration of free fatty acids in plasma; increased fatty acid oxidation (chronic effect)
(b) Increased lipogenesis (acute 'insulin-like' action, seen only transiently after hGH administration)

(5) Carbohydrate

(a) Decreased glucose utilization
(b) Increased gluconeogenesis
(c) Antagonism of insulin action (plasma glucose may remain normal if pancreas normal because of increased insulin secretion; in diabetics, hGH causes increased plasma glucose and increased ketogenesis)
(d) Transient decrease in plasma glucose after intravenous hGH (acute 'insulin-like' action)

(6) Calcium and phosphorus

(a) Increased absorption of calcium by intestine
(b) Increased urinary excretion of calcium
(c) Positive calcium balance (usually)
(d) Increased renal tubular reabsorption of phosphate
(e) Positive phosphorus balance
(f) Increased serum phosphorus

(7) Sodium and potassium

(a) Decreased sodium excretion; positive balance
(b) Decreased potassium excretion; positive balance

(8) Connective tissue

(a) Increased chondroitin sulfate synthesis
(b) Increased collagen synthesis and degradation
(c) Increased urinary hydroxyproline

(9) Renal function

Increased glomerular filtration rate; decreased blood urea concentration

(10) Hepatic function

Increased conjugating capacity for many substances

II. Mammotropic and other effects—associated chiefly with prolactin, but seen to a considerable extent with hPL and hGH (non-primate growth hormones exhibit little or no prolactin-like activity)

(1) Mammotropic—growth and differentiation of breast tissue
(2) Lactogenic—during later pregnancy and puerperium
(3) Luteotrophic (in rodents)
(4) Crop sac-stimulating (in doves)
(5) Parental behavior-stimulating (birds, some mammals, some fishes; hGH little investigated in this respect)

*hGH = Human Growth Hormone
hPRL = Human Prolactin
hPL = Human Placental Lactogen

which GH stimulates.* Under some circumstances, one or more intermediary polypeptides (mw ~10,000), referred to as somatomedins, that are apparently produced by the liver (and possibly the kidney) under the influence of GH, may actually be the proximate causes of some of the effects attributed to GH. At least some of these resemble insulin,# or more closely proinsulin, in structure and activities at the cellular level. GH is liberated by the pituitary in intermittant bursts, in response to a variety of metabolic or neurologic stimuli, and many have embraced the somatomedin mechanism as providing a systematic rationale for the orderly process of skeletal growth that is observed. What is known is that somatomedins promote the incorporation of amino acids and sulfate into the collagen and polysaccharides, which are built up in the presence of proliferating chondrocytes at the epiphyseal plate in the process of longitudinal skeletal growth, for which GH is regarded as essential. Efforts to isolate and characterize the somatomedins - particularly by researchers associated with the Swedish government-owned firm, Aktiebolaget Kabi - as a potential source of new pharmaceutical agents have not yet yielded practical consequences. However, growth hormone is an established therapeutic modality.

2.332 Growth Hormone Deficiencies

The biological significance of GH is most clearly illustrated by the growth retardation which characterizes

---

* In the absence of insulin, as in a juvenile diabetic, GH is severely catabolic for fat tissue, and its administration may precipitate ketoacidosis.

\# The structures of IGF-1 and IGF-2 have been determined. [See Phillips & Vassilopoulou-Sellin, 1980 for review.)

its absence prior to puberty and by the effects of replacement therapy. Approximately 1.4 milligrams per day GH is secreted by a normal adult, to maintain a blood level around 10^{-10}M, under the control of stimulating and inhibiting factors from the hypothalamus.* The pituitary is able to synthesize GH from about the ninth week of gestation, but the hormone is not essential for fetal growth. At birth GH production is high (~4 mg/day), and it is especially elevated in the first two years, declining progressively thereafter. Growth hormone deficiencies can arise from: (1) cerebral inhibition of GH release at the hypothalamic and/or pituitary levels; (2) releasing, (or inhibiting) factor deficiency (or excess); (3) decrease in production and/or release of GH; (4) release of biologically inactive GH; (5) deficiency or dysfunction at somatomedin level. With the exception of the 5th named malfunctions, each of these conditions that may cause inadequate growth during childhood can be counteracted by the administration of human growth hormone (hGH).

Approximately 50% of children demonstrating hypopituitarism are only deficient in hGH. The largest fraction of these, and others suffering loss of additional pituitary hormones, are in fact victims of a hypothalamic abnormality - although these are generally categorized as having idiopathic hypopituitarism. The most common organic lesion involved in pediatric hypopituitarism is craniopharyngioma. Various causes,(e.g., other tumors, trauma,)account for the remaining cases. From all causes, the incidence of

---

* Hypersecretion, which is usually caused by a tumor, in children produces gigantism and in adults acromegaly. Closure of the skeletal epiphyses determines the manifestations of elevated GH level. In the rat, the epiphyseal plates do not close at maturity, and growth stimulation can be provoked at almost any stage. For humans and domestic livestock, the period of susceptibility does not extend into the adult stages. However, studies of growth promotion in cattle are being launched with rDNA bovine growth hormone made by Genentech for Monsanto.

hypopituitary childhood growth inhibition runs at about 1 per 100,000 live births.

Diagnosis takes advantage of contemporary knowledge of the interrelationships and dynamics of endocrine systems.* Age is critical, in that the majority of those who do not respond to therapy are teenagers, and early diagnosis is a major benefit. The only satisfactory treatment is human growth hormone by injection. A dose as small as 2.5 mg twice weekly may be sufficient, but usually 5 mg is injected three times a week, in addition to the hormonal replacement of any associated deficiences.

2.333 Supply and Demand of Growth Hormone

Growth hormones are readily isolated in substantial quantities from the pituitary glands of slaughtered domestic livestock. However, they are remarkably diverse in structure (witness the differences in mw shown in Table 6), and none of the animal species GH tested have proved satisfactory in man.

Table 6

Molecular Weights of Growth Hormones From Various Species

	Human	Bovine	Porcine	Ovine	Monkey
Molecular Weight	20,500	26,000	22,500	24,000	21,000

In the United States most of the demand for human growth hormone (hGH) is met by the National Pituitary Agency (Baltimore, MD), which was created in the early 1960s by the College of Pathologists and the NIAMDD to collect pituitary glands from coroners and private donors for preparation of hGH and other hormones. Under the programs of the NIAMDD, hGH is provided without charge for the

* Once GH deficiency is demonstrated, the presence or absence of other deficiences of anterior pituitary hormones must be determined. This must be re-checked periodically as the child ages.

treatment of children with hypopituitarism (~1600 patients, each of whom receives therapy for several years) and for research. In addition to those provided for by the National Pituitary Agency, another 100 to 400 patients are treated with hGH from commercial sources. We understand that the commercial price is approximately $15 per unit (roughly $30 per milligram), and that the production cost at the National Pituitary Agency is about $0.75 per unit ($1.50 per milligram).* National Pituitary Agency produces 650,000 IU (~325g) of hGH, along with thyroid stimulating hormone, prolactin, and other hormones, from about 60,000 human pituitaries collected each year. This is sufficient hGH for their current demand plus perhaps another 100 hypopituitary patients.[#] As little as 4-5 years ago, the supply was only sufficient for 500-600 patients, and there was a long waiting list. However, at this moment there is no waiting list. Many physicians feel that there are many undiagnosed cases of pituitary (hGH) deficiences, but the level of treatment abroad (~500 patients per year in UK and France) is consistent with the United States experience. Of course the presence of one or more pharmaceutical firms promoting new products has often been observed to stimulate the apparent incidence of a condition.

While the National Pituitary Agency feels that the demand for hGH to treat diagnosed childhood pituitary deficiences is presently satisfied, it welcomes the promise of

---

* According to Salvatore Raiti, M.D., Director of the National Pituitary Agency.

\# Approximately 5% of the hGH supply is presently used for research on conditions other than hypopituitarism.Dr. Henry Miller, Director of Metabolism and Endocrine Drug Products, Bureau of Drugs, feels thatmost informed observers think "patients currently maintained on hGH may well be treated with sub-optimal levels of hormone, and that there exists a modest shortage of hGH in this country."

additional hGH at relatively low cost ($1/unit), because a number of areas of research appear to warrant larger scale evaluation - individually handicapped by scarcity of funds more than scarcity of hormone - any of which, if successful, would severely strain the present hormone supply. The conditions for which the greatest promise is apparent include:

- Senile osteoporosis
- Turner's syndrome (1 in 3000 live female births) and other non-pituitary growth deficiences
- Intrauterine growth retardation
- Bleeding ulcers that cannot be controlled by other means
- Burn, wound and bone fracture healing.

2.334 rDNA Synthesis of Growth Hormone

Two groups have announced preparation of microorganisms with the capacity for synthesizing growth hormone. Workers at the University of California - San Francisco's Howard Hughes Medical Institute isolated messenger RNA (mRNA) for the GH of the rat and converted it into a DNA sequence from which *E.coli* produced fused peptides containing the rat GH and β-lactamase. Then they repeated the procedure with mRNA from a human pituitary tumor, and they obtained production of about 3% of the bacterial protein in the form of a fusion hybrid (32,000 MW), 70% of which is composed of AA coded for by hGH "gene." Actually, the cDNA turned out to code for a pre-hGH with 26 additional AA in the case of the rat GH. Furthermore, the level rate of synthesis was only about 17% of that expected, for reasons that remain to be seen. [Martial, 1979]. It has been reported that Eli Lilly has obtained a license to the patents relating to this work. Lilly provided some grant money, but NIH and NSF grants were also acknowledged in the publication.

At practically the same time, Genentech, Inc., in conjunction with their associates at City of Hope National Medical Center, disclosed the production of hGH not fused to any other peptide. This was accomplished by obtaining from mRNA about 88% of the DNA sequence corresponding to hGH and attaching to that a synthetic DNA sequence corresponding to the remaining 12% plus a start codon. That assembly was inserted into a plasmid downstream from the lac promoter and a ribosome binding site. Strictly speaking, the product obtained from the bacteria, after cleavage in situ to the fMet-hGH*, is an analog of hGH differing from hGH only in the presence of an additional amino-terminal methionine, the presence of which is not expected to influence the activity of the problem. The work was supported by Kabi Gen AB, and Kabi has the marketing rights.

The level of hGH production (2.4 mg/l) reported in the scientific account of the Genentech work was of the same order as that reported for the insulin fragments. (Goeddel, 1979b) That amounted to approximately 186,000 hGH molecules per cell, prior to efforts to increase yield by design. One aspect of general interest was the observation that the level of hGH, both per cell and overall, declined as cultures matured from log-phase growth (3.6×10^8 cells/ml) to stationary phase (1×10^9 cells/ml), apparently as a result of proteolytic degradation.# Minimization of such proteolysis, along with increased plasmid copy numbers, and improved promoters are

* fMet = formyl methionine

Polypeptides are thought to be especially vulnerable to intracellular proteases if their configuration exposes hydrophobic segments. If hGH were less favorably structured, little accumulation might have been observed.

parameters that can be adjusted to achieve greater yield. The Genentech workers stress the point that design, rather than the classical mutation and selection approach, is the logical way to optimize the sort of system they have developed. One should bear in mind that the "blue-print," so to speak, is incorporated in a plasmid which can be moved from one strain of E.coli to another or between species or even from the prokaryote bacteria into eukaryotes, such as yeast. Once one has the gene, as they do for hGH, it is also a relatively simple series of operations to move it into other vectors to achieve better expression in a given host organism. The combinations are almost infinite, and an optimal combination, it can be anticipated, will probably be regarded a "secret formula." Nevertheless, the general principles and examples of feasibility promulgated in the scientific literature provide stimulation and encouragement to other embryonic activities oriented toward practical applications. Large scale, 750 liter, fermentation production of hGH by Genentech was approved by the RAC of the NIH in April 1980.

2.335 Kabi and Growth Hormone

The world's largest producer of growth hormone, from frozen human pituitaries, is KabiVitrum AB, a firm owned by the Swedish government. It operates internationally, marketing a very small amount of hGH in the United States under an approved NDA.* However, it has never collected human pituitaries in this country, and it incurred the wrath of the National Pituitary Agency (which spoke of "the Kabi Threat") a few years

*Serono, a subsidiary of Applied Science Research Systems NV which is based in Switzerland, also markets hGh in the USA, on the basis of an NDA that it purchased in 1979 from Calbiochem, a Lajolla, California firm that had been acquired by the German giant, Hoechst, in 1977.

ago when it made inquiries concerning the possibility of doing so. In 1968, a joint venture called KabiGen AB was established, with private investors taking a 50% interest, to investigate new sources of raw materials for KabiVitrum. In 1978, as the synthesis of insulin was being accomplished, they contracted with Genentech to have developed a strain of _E. coli_ containing a plasmid bearing the genes for production of hGH via fermentation.

When the objective was achieved in June 1979, they disclosed that the contract granted KabiGen the sole rights to manufacture and market anywhere in the world, except in the U.S. and Canada, where the market would be shared by Genentech, and Genentech has an opportunity to supply a fraction of Kabi's product requirements.

While steps went very smoothly for Genentech in the United States, KabiGen ran into environmentalist opposition to such laboratory work in Sweden. It was December 1980 before they received permission to scale up to 10 liter production.

2.336 Clinical Research

Genentech presented evidence of the activity of bacterial-hGH in promoting the growth of rats at an FDA conference in June 1980, and commenced clinical trials in January 1981, in the UK. Kabi's trials will not begin sooner than mid-1981.

It should be noted that fragments of growth hormone exhibit biological activity _in vitro_ and _in vivo_. A great deal remains to be done in sorting out the potency and selectivity of fragments for the many effects manifested by GH and related hormones. (The somatomedins are definitely not fragments of GH, by the way.) Use of recombinant techniques to prepare fragments of growth hormone and to synthesize those few somatomedins whose structure is known should be relatively straightforward from the present art. No doubt this capacity is of great interest to Kabi. Until the cellular origin of the other somatomedins is clarified, it will be impossible to obtain the mRNA necessary to commence the process of cloning the "genes" and analyzing the peptide products. Bringing to bear the ultra-micro amino acid sequencing capacity of a facility such as that at Cal Tech could provide the basis for a round of analog biosynthesis with significant value in the field of growth and healing factors.* The somatomedins are the first of several examples of rare messenger molecules, poorly characterized or evaluated due to scarcity, the study and exploitation of which stands to benefit enormously because of the techniques based on molecular cloning and new analytical technology.

2.34 Prolactin and Human Placental Lactogen

Neither of the two hormones which are structurally similar to GH, prolactin and human placental lactogen (hPL), appear to have particular clinical promise. Nevertheless, two groups have reported the production of DNA for hPL from mRNA derived from placental troploblasts. This hormone has in the past, and in other cultures, been of interest because of its GH-like effects, but with GH available through the same technology, it is not likely to be of more than scientific interest. For many years, there was no firm evidence of a prolactin distinct from GH in human pituitaries and, since hPL had more effect on lactation than growth, hPL was investigated for induction and prolongation of milk production. Today prolactin,

* The facility referred to is a novel peptide sequenator, developed by L. Hood, et al, which has enabled extended amino terminal sequence analysis on 20 picomoles of protein (e.g., 0.3 mcg apomyoglobin) and 200 picomoles of shorter peptides. [Hunkapiller and Hood, 1980].

prolactin releasing hormone, or an analog thereof, would appear to be more suitable. There have been a number of studies in recent years on the use of polypeptides to prolong lactation for both women and domestic livestock but the initial promising results have not led to practical programs. The potential of increasing the annual productivity of milk cows has obvious economic and nutritional benefits, and, especially in underdeveloped countries where the use of poor synthetic formulas often results in infant malnutrition, the value of mother's milk is not to be underestimated.* The availability of prolactin from molecular cloning would have to be combined with an appropriate dosage form (drug delivery system) in order to have practical consequences.

2.35 ACTH

While we are on the subject of pituitary polypeptides, it is appropriate to consider adrenocorticotropic hormone (ACTH). This linear 39 AA polypeptide, produced by the anterior pituitary, promotes and maintains the normal growth and development of the adrenal cortices and stimulates their secretion of cortisol and other glucocorticoids. ACTH may be derived from a larger polypeptide, as are other hormones synthesized by the same cells, since a so-called "Big ACTH" of 20,000 mw has been identified, from which ACTH can be cleaved with trypsin. The first 24 AA from the N-terminal end have been shown to produce a response indistinguishable from the natural hormone.** Loss of even one AA from the N-terminal end

---

* At the present time, more than 50% of American babies are being breast-fed, a marked increase in the past decade.

** In fact, synthetic ACTH 1-20 has equimolar bioactivity in comparison to intact ACTH, but ACTH 1-19 has only 30% as much activity.

(Ser-1) results in complete loss of activity.# Ovine, porcine and bovine ACTHs differ from human ACTH only at AA positions 25, 31 and 33. There are rare incidents of immune reaction to porcine ACTH - the most commonly used form - which are allegedly less common with synthetic ACTH 1-24, but it has not been determined whether the immune responses relate to the AA sequence differences or trace contaminants. Some experts (Dr. Hansen, Armour) say immune reactions to the gelatin additive mask any improvement. The plasma half-life of ACTH is about 15 minutes. Hence, injectable preparations often include gelatin or zinc hydroxide to prolong activity through gradual desorption of ACTH from the inert substrate.

2.351 Pharmacology of ACTH

The pharmacology texts say that the primary use of ACTH in the USA is as a diagnostic agent for adrenal insufficiency. However, the FDA-approved package insert lists 10 categories with about 50 conditions for which ACTH may be employed therapeutically. A physician's motivation for adopting ACTH therapy generally is to avoid the adrenal suppression and some other side effects of synthetic steriods. In principle, ACTH could be used for at least one-third of the indications listed in Table 7, on account of which about 5 million Americans annually receive corticosteroids. However, being a peptide, ACTH would be degraded by gastrointestinal peptidases if taken orally, and, hence, it must be given by injection. The aversion of American physicians and patients to therapy that requires frequent injection, coupled with the variety of steroid

---

\# Potency and duration of activity can be enhanced by substitutions, e.g., D-for L-serine at position 1, norleucine for Met-4, or valine amide for Asp-25, which increase stability in the presence of proteolytic enzymes.

products available and the confidence placed in them is typified by these quotes from Drugs of Choice: "there is no therapeutic place for ACTH in pituitary insufficiency... it offers few, if any, advantages over steroids in the treatment of allergic disease," and , with regard to rheumatic diseases, "all in all, the evidence to date does not justify the general adoption of self-administered ACTH over an extended period of time."

Table 7: Selected Clinical Conditions for Which Glucocorticoids are Used[a]

Addison's syndrome—replacement therapy
Adrenal hyperplasia due to enzymatic defects (e.g., 11-, 17- and 21-hydroxylase syndromes)
Allergic diseases
- Angioneurotic edema
- Bee stings
- Contact dermatitis
- Drug reactions
- Hay fever
- Serum sickness
- Urticaria

Arthritis, bursitis, and tenosynovitis
- Inflammatory complications of a variety of types of arthritis

Collagen vascular disorders
- Giant cell arteritis
- Lupus erythematosus
- Mixed connective tissue syndromes
- Polymyositis
- Polymyalgia rheumatica
- Rheumatoid arthritis
- Temporal arteritis

Blood dyscrasias
- Acquired hemolytic anemia
- Allergic purpura
- Autoimmune hemolytic anemia
- Idiopathic thrombocytopenic purpura
- Lymphoblastic leukemia
- Multiple myeloma

Eye diseases
- Acute uveitis
- Allergic conjunctivitis

Eye diseases (con't.)
- Choroiditis
- Optic neuritis

Gastrointestinal diseases
- Inflammatory bowel disease
- Nontropical sprue
- Regional enteritis
- Subacute hepatic necrosis
- Ulcerative colitis

Hypercalcemia
Malignant exophthalmos
Neurologic diseases
Pulmonary diseases
- Aspiration pneumonia
- Bronchial asthma
- Infant respiratory distress syndrome (antenatal)
- Sarcoidosis

Renal diseases
- Certain nephrotic syndromes

Transplantation—prevention of rejection
Infections—occasionally helpful to suppress excessive inflammation
Skin conditions
- Atopic dermatitis
- Dermatoses (see above)
- Lichen simplex chronicus (localized neurodermatitis)
- Mycosis fungoides
- Pemphigus
- Seborrheic dermatitis
- Xerosis

[a]This table is not meant to be comprehensive, nor are steroids always indicated in the conditions listed.

This last is the most hedged disparaging opinion, and it applies to the realm in which American use is most common (40% of total administrations outside hospitals are for rheumatic conditions, especially juvenile). In hospitals, the most common indications treated with ACTH are CNS disorders (29%): myasthenia gravis, myoclonic epileptic seizures of infants, and multiple sclerosis. Allergic disorders (22%), especially urticaria, and digestive disorders (15%) also account for a substantial proportion of administration of ACTH to hospitalized patients in the USA.

In contrast to the therapeutic use which involves daily injections, the diagnostic use does not consume much drug per patient. Standard procedure to check out pituitary and adrenal function calls for measurement of normal serum cortisol or 24-hour urinary excretion of 17-hydroxy corticosteroids followed by a test in which the secretion of cortisol is provoked by injection of ACTH or cosyntropin, as the ACTH 1-24 is known. Such a test can be performed in a physician's office. Normally the basal level of secretion will be doubled by a 0.25 mg dose. A confirmatory test would usually be performed if the adrenal gland appeared unresponsive. About 75% of ACTH sales are in the form of gels, not appropriate for diagnostic use.*

Sales of ACTH to drug stores and hospitals in 1978 were $3 million,# to which some sales directly to physicians no doubt need to be added. Armour Pharmaceuticals accounted for 79% of the sales reported. The average price of the

---

* Regular periodic testing for adrenocortical suppression has recently been instituted and probably accounts for part of the 35% increase in ACTH observed in 1978.

For comparison, U.S. corticosteroid sales exceed $240 million.

many dosage sizes and packages offered was approximately 1.5¢ per IU at the manufacturer's level.** Assuming that the total market, including sales to physicians, was $4.5 million - it was almost certainly less - 300 million IU of ACTH were required in the USA in 1978. Since 10 mcg of synthetic ACTH 1-24 is equivalent to one IU of activity, that demand corresponds to 3 kg of the 24 AA polypeptide. Most of the sales are in the form of vials containing 2-4 mg polypeptide, so that a great deal of the product cost is associated with packaging, labelling, sterilizing and distribution (selling price $3-6 per vial). If one assumes one-third of the manufacturer's selling price is the cost of the drug - a very conservative estimate - an estimated cost of $0.5 million per kilogram is established. That (50¢ per milligram) is less than the asymptotic cost suggested for synthetic polypeptides ($1 per milligram). It is, however, substantially higher than the unit market value (12¢ per milligram) of animal insulin, and we would not be far wrong to presume synthetic ACTH 1-24 has a unit value about ten times that of animal-derived insulin.

Use of ACTH is somewhat greater abroad than in the USA: in part because physicians in other cultures make far less use of systemic steroids than their American counterparts and because frequent injections are more common in medical practice in many other countries, whereas the US is a pill-oriented society. Thus, the sales of ACTH in Britain - with its much smaller population - equal American sales. Furthermore, the claims permitted in other countries are far more favorable than those sanctioned for ACTH by

---

** Independent of this calculation, Diosynth, Inc. quoted $12.50 per 1000 units (1.25¢ per unit) as the current market price for bulk, pharmaceutical grade ACTH.

the FDA. A reasonable guess at the world market for ACTH is that it is roughly four times the US sales, or the equivalent of 12 kg of ACTH 1-24, with a bulk market value in the neighborhood of $6 million.*

It appears that the dosage form (injection) is a far more likely deterrent to wider use of ACTH, than is the cost of the drug itself. Reports of nasal insufflation of ACTH, along with a small but increasing use of the nasal route for administration of other polypeptides, suggest that one or more transmuccosal dosage forms may provide for expanded adoption of such agents which have heretofore required injection. Implantable controlled release dosage forms could become available within the next five years.

It must be recognized that, just because it is nature's way of stimulating our adrenal glands to produce steroids, ACTH is not without its drawbacks. Martindale's The Extra Pharmacopeia, an authoritative source on drugs more commonly used outside the USA, points out that hypertension, acne and hirsutism occur more frequently than with corticosteroids, although gastrointestinal effects are less evident; skin pigmentation sometimes increases (melanocyte stimulating activity). Hypertrophy of the cortical cells under the influence of exogenous ACTH is the counterpart of the atrophy exogenous steroids can cause, and symptoms of hypopituitarism (due to suppression of

---

* Ulcerative colitis, Crohn's disease, juvenile rheumatoid arthritis and remission induction in muscular dystrophy are all established indications for ACTH in the UK, and nervous system conditions like Bell's palsy, Guillain-Barre syndrome, myasthenia gravis and multiple sclerosis all appear to be treated with ACTH in Britain, except when adrenal insufficiency makes use of exogenous steroid necessary. A great deal of stress is laid upon the suppression of growth that can occur as a result of pituitary suppression by exogenous steroids and the rapid reversibility of ACTH therapy.

natural ACTH) may occur upon cessation of treatment. However, it is easier to withdraw ACTH after prolonged administration than orally administered corticosteroids. The bottom line to quote Martindale is "corticosteroids by mouth are usually preferred because of the ease of administration."

The British Committee on Safety of Medicines, (Current Problems No. 1, Sept. 1975) is quoted in Martindale as concluding, "Experience has shown that the hope, that tetracosactrin (ACTH 1-24) would cause fewer reactions than natural corticotrophin, might not be realized. There had been two deaths in patients whose treatment had continued after a mild reaction had occurred with an earlier dose, and serious, even fatal, reactions had occurred at various times during courses of therapy with no mild premonitory symptoms." As a consequence, doctors were advised to keep an eye on patients for an hour following each injection and cautioned against permitting home injection.

Ciba-Geigy was the first to develop synthetic ACTH products (1969), and they market Synacthen (ACTH 1-24) widely abroad, but have never done so in the USA. Organon's synthetic ACTH 1-24, Cortrosyn, is identified as "For Diagnostic Use Only" in the USA. Armour has had an approved NDA for its synthetic product Syn-ACTHAR since 1978, but has not yet introduced it commercially. Cost considerations are obviously paramount, and even the many fold increase in the price of porcine pituitary glands has not yet made the switch economical. Representatives of Armour suggested that the experience gained in the synthesis of calcitonin - a much higher unit value product (by a factor of 40 according to our estimates) - may enable them to justify the introduction of Syn-ACTHAR before long.

It is appropriate to move at this point to calcitonin and parathyroid hormone. Both exert profound influences on calcium balance and its many consequences. The simpler of the two, calcitonin, is a polypeptide of 32 AA in length with a single disulphide bridge. The complete sequence of AA is essential for activity, although substitutions are permitted. It is synthesized and secreted by a group of special cells located in the thyroid in mammals.

2.36 Calcitonin

2.361 Physiology and Pharmacology

The principal pharmacological action of calcitonin is inhibition of bone resorption. Like parathyroid hormone (PTH), it promotes urinary clearance of phosphate. However, it antagonizes the action of PTH on urinary calcium (calcitonin promotes Ca clearance) and on bone. Calcitonin secretion is directly proportional to blood calcium level, and once secreted it disappears rapidly from the blood.* It is unknown whether calcitonin actually has any important physiological functions in adult humans; patients who have undergone total thyroidectomy, and who receive replacement therapy only with thyroid hormone (presumably, therefore, lacking calcitonin) do not seem to have any detectable abnormality in calcium or skeletal metabolism. Nor do hypersecretors of calcitonin (with medullary carcinoma of the thyroid) develop perturbations in their calcium metabolism. It is thought that the physiologic action of calcitonin may be expressed during embryonic development or it may be a vestigial hormone in man.

* Half-life 2-20 minutes, depending on the species involved.

Nevertheless, the pharmacologic value of calcitonin is well established and growing. Exogenous calcitonin is used to block pathologic bone resorption in conditions such as Paget's Disease, a chronic, slowly progressive disorder in which an initial phase of bone decalcification and softening is followed by uncontrolled recalcification leading to thickening, abnormal architecture and enlargement of the affected areas. Any bone may be involved, but common sites are the pelvis, skull, vertebrae, femur and tibia. Fractures of the consequently bowed femurs and tibia are common. Impaired hearing is a frequent complication as a result of temporal bone involvement. In a small fraction of patients, deep, dull, aching bone pain may develop and persist, but the onset is insidious, with many localized lesions being diagnosed during an x-ray examination for other reasons. Hence, the data on incidence are not clear. For example, in the USA in 1978, there were only 145,000 office visits in connection with osteitis deformans, as Paget's Disease is known formally, and only one-third of those resulted in a drug being prescribed or administered. Calcitonin was the drug chosen approximately half the time, with aspirin and a variety of drugs accounting for the remainder. That seems very low in light of the observation that, as the Merck Manual of Diagnosis and Therapy says "characteristic changes in bone are seen in about 3% of persons over 50." About 3% would be consistent with estimates that Paget's Disease affects up to 3% of the population over 40 years of age in Western Europe, and that half-a-million suffer from the condition in Britain.* Treatment of the condition with calcitonin has been practiced in Britain for many

---

* However, the disease is practically unheard of in Scandanavia, Asia or the Arab countries. Its etiology is unknown.

years. Initially, porcine calcitonin was employed, then salmon calcitonin and synthetic human calcitonin were introduced.

As stated earlier, the full 32 AA sequence is necessary for activity, with no shortening tolerated, but the substitution of AA other than those in the molecule synthesized in the human body can actually be beneficial - at least in terms of potency under assay conditions. Thus salmon calcitonin is more than 20 times as potent as the human form. However, circulating antibodies to salmon calcitonin are detected in about half of the patients studied (after 2-18 months of treatment), and, although many of these are still responsive to the treatment, neutralizing antibodies may account for the relapses that are observed. Certainly, the patients found with high antibody titers are those who have suffered a biochemical relapse of Paget's disease and are unresponsive to the acute hypercalcemic effects of calcitonin.

2.362 Chemical Synthesis

Although salmon calcitonin derived from the endocrine organs of the fish was clinically tested, the product identified as Calcimar, produced and marketed by Armour Pharmaceutical Company is made synthetically by the solid-state polypeptide synthesis route originally developed by Dr. Bruce Merrified and his associates. It is the largest polypeptide produced by chemical synthesis for commercial pharmaceutical use. Ciba-Geigy, which was the first organization to synthesize human insulin, and which has been marketing synthetic ACTH 1-24 for about 10 years abroad, introduced synthetic human calcitonin in 1978. On the face of it, they might appear to be at a 20-to-1 disadvantage, with respect to raw material cost, relative to the more potent salmon analog produced by Armour. However, the economies of scale at the current levels of production should be substantial, because the current process is quite labor intensive. The companies involved operate internationally and we do not know their worldwide sales, hence any estimate of their volume of production would be shear speculation, but it clearly does not exceed a few kilograms.#

Additional demand will result when there are approvals of osteoporosis and hypercalcemia as indications. At least one of the competitors, Armour, is pessimistic about the value of applying the recombinant-DNA approach to the synthesis of molecules the size of calcitonin (32 AA). Their estimates of the probable economics of the alternative synthetic technologies lead them to favor chemical synthesis for potent polypeptides of less than 50 AA. However, they are quite interested in the recombinant approach for larger molecules and more complex products, e.g., multiple chain polypeptides.

2.37 Parathyroid Hormone

Parathyroid hormone (PTH) consists of 84 AA residues, without covalent cross-links. It is synthesized, through higher molecular weight precursors, by the parathyroid gland. Fragments of parathyroid hormone display comparable activity to the intact hormone, e.g., the sequence of 34 AA beginning at the N-terminal end of bovine PTH is 80% as potent *in vitro* and 130% as potent *in vivo* on a molecular

---

\# If one ignores all the other variables, which is foolhardy to do, and considers the fact that the cost of synthetic salmon calcitonin must be less than $85 per milligram and the factor of 20 difference in potency, then the cost of synthetic human calcitonin must be less than $4 per milligram to compete at the same price range.

basis. Deletion of one or two AA at the N-terminal end is much more devastating.

2.371 Physiology and Pharmacology

A synthetically prepared polypeptide with the sequence identical to AA 1-34 of human PTH has been shown to be as active on a molar basis as intact hPTH in the rat adenyl cyclase assay *in vitro* and as active as bovine PTH 1-34 for chick hypercalcemia *in vivo*. However, USP parathyroid extract - primarily intact hormone - is the only form of the hormone approved for medical practice.

Although a number of fragments can be found in the circulation* there is no direct evidence against the concept that the principal, if not sole, physiologically active molecular species of PTH is the intact 84 AA peptide. With an eye to clarifying this question and making pure hPTH available for experimentation, we understand that a group at Harvard and Massachusetts General Hospital has cloned a gene coding for parathyroid hormone, but the details have not yet been disclosed.

Decreasing plasma calcium concentration (in the normocalcemic range 8-10.5 mg/100 ml blood) elicits a progressive increase in PTH secretion in the calf. There is also a persistent basal secretion rate, independent of blood calcium concentration. PTH accelerates bone breakdown and increases renal calcium retention and intestinal calcium absorption. The latter effects predominate under physiological conditions (i.e., other than when excessive PTH is administered or produced by a

* Four discrete cleavages occur between the initial biosynthesis of pre-proparathyroid hormone and the eventual disappearance of biologically active PTH from the circulation.

tumor). In the face of a low calcium diet, or perhaps every day prior to breakfast, parathyroid hormone may promote bone resorption to prevent the serum calcium from falling but most of the time it is the renal threshold, rather than the bone calcium reservoir, which determines the critical level of serum calcium. A host of physiological functions, including such apparently unrelated processes as blood clotting and heart rhythm, depend upon that serum calcium level.

When PTH is slowly infused to produce slightly supra-physiological blood levels, the effect achieved is the opposite of the formerly accepted picture of parathyroid hormone as an agent of bone destruction and demineralization. In fact, PTH and active fragments are being investigated, alone and in combination with calcitonin, for the treatment of osteoporosis. Osteoporosis is a very common bone atrophy in which there is loss of both mineral and protein matrix components. It occurs most commonly in postmenopausal women (estrogen deficit), and is also a consequence of long-term exposure to steroids from disease (e.g., Cushing's Disease) or medication (exogenous corticosteroids). Immobilization, liver disease, gonadal dysgenesis and osteogenesis imperfecta also contribute to the patient population with osteoporosis. Bone pain and easy fracture are frequent manifestations of the condition.

Before the side effects of estrogens became a dominant concern, osteoporosis accounted for hundreds of millions of dollars of sex hormone products. Presently this major market need is unsatisfied and it could provide the greatest opportunity in the polypeptide hormone field for the development of one or more much needed products.

2.38 Growth Factors

A number of compounds identified as growth factors appear to be simple polypeptides in their active forms, and could prove worthwhile objects of molecular cloning efforts. In particular the proinsulin-like, 118 AA, nerve growth factor (NGF) which influences the development, maintenance, and repair of neurons, including the directed growth of nerve fibers, could turn out to have significant clinical utility - itself or a derivative thereof - since nerve restoration is a major handicap to surgery today.

Other factors, such as epidermal growth factor (EGF) and fibronectin are important for some mammalian cell cultures. Expensive, complex media (such as fetal calf serum) are presently required for growing most mammalian tissue cells *in vitro*. Genetic cloning could enable the synthesis of vital constituents for media at relatively low cost, and thereby, markedly reduce the costs of cell culture for both research and production purposes.

2.4 Glycopeptides

Many polypeptides and proteins are modified by the attachment of carbohydrates, especially oligosaccharides. Glycosylation is believed to designate the destination of proteins within a given cell and to influence the fate of proteins circulating in the bloodstream, but much remains to be learned about most aspects of glycoproteins from their biosynthesis to the function and specificity of the carbohydrate substituents. Many of these side chains turn out to be very similar to one another, and markedly different from other common sugars.

Apparently the carboydrate side chains attached to proteins

with different functions and from widely different sources are synthesized via similar pathways. It is currently thought that oligosaccharides linked N-glycosidically to asparagine residues of glycoproteins are produced by transfer to the protein of carbohydrates preassembled on a carrier lipid, dolicholpyrophosphate. [Schwartz & Datema , 1980] Terminal N-acetylneuraminic acid (sialic acid)[#] residues are apparently added after the precursor oligosaccharide has been attached to the peptide, but the details and differences in prokaryotic and eukaryotic systems remain to be worked out. Removal of one or more sialic acid residues (via neuraminidase) radically shortens the half-life of otherwise long-circulating glycoproteins in the blood stream - actually it seems to be the exposure of the penultimate sugar (galactose) that triggers the disposal mechanism. The important point is that these sugar moieties are involved in a cellular recognition system. Many workers are convinced that the carbohydrates serve as address markers analogous to postal zip codes. [Marx, 1980] Exploitation of the new biosynthetic technologies will almost certainly require the mastering of techniques for achieving appropriate glycosylation of polypeptides, and that seems most likely to involve initially the preparation of glycopeptide hormones. Several of these share a common subunit, the biosynthesis of which is a natural object for academic study.

2.41 Gonadotropins

The gonadotropins are glycoproteins each composed of two subunits (multimeric). The 80 AA α-component is common to luteinizing hormone (LH),

---

N-acetylneuraminic acid is the most common of the five sialic acids, all of which contain 9 or more carbons. Other sugars frequently present in glycopeptides are glucosamine, mannose, galactose, galactosamine, and fucose.

follicle-stimulating hormone (FSH), human chorionic gonadatropin (HCG), and the non-gonadotropin, thyroid-stimulating hormone (TSH).* The β-components are what distinguish the different members of this family. They vary in composition and size (111-145 AA). Carbohydrate content of the human gonadotropins ranges from 14 to 31%, with the points of attachment being asparagine and serine residues. Secretion of each of the anterior pituitary hormones (all except HCG) is under the control of the hypothalamus, through the mediation of releasing factors.

LH and FSH act synergistically to induce ovarian growth and estrogen production and secretion. As the name implies, FSH stimulates follicular development in the female, but in the male it is responsible for spermatogenesis, an effect augmented by androgen and LH. LH affects the ovary by inducing a rapid preovulation enlargement of one or more predisposed follicles, ovulation, transformation of granulosa cells to luteal cells, and thus formation of the corpus luteum, maintenance of the corpus luteum and secretion of progesterone. LH also stimulates the morphological development and steroid secretion of ovarian interstitial cells. In the male, LH is responsible for the morphological development and stimulation of androgen secretion by the Leydig cell. HCG shares many of the biological actions of LH, but with greater potency, probably due to its longer half-life.

---

* The common α-subunit has two carbohydrate-binding asparagine residues (positions 49 & 75). The nucleotide sequence encoding the α-subunit (from HCG) has been cloned into a bacterial plasmid by Fiddes and Goodman at UCSF.

Gonadotropin therapy has been used to induce follicular growth and ovulation in infertile women who have ovaries capable of responding to LH and FSH. A preparation called menotropins, made from the urine of postmenopausal women and containing nanogram quantities of FSH and LH, is administered (by intramuscular injection) for 9-12 days, and usually produces follicular growth and maturation for women in whom the cause of anovulation is secondary and not due to primary ovarian failure. When clinical assessment indicates that sufficient follicular maturation has occurred, HCG or pituitary LH is administered to effect ovulation. The doses of HCG or LH employed to induce ovulation are relatively large - 1 mg or 5 mg, respectively. Multiple births are common (20% chance) with such procedures. HCG is also used to promote testicular descent - in doses of 500-1000 IU ($<$.01 mg) 2-3 times/week for periods up to 6 weeks. HCG is made from the urine of pregnant women and available from many suppliers. Gonadotropins are also used, with limited success, in treatment of male infertility problems.

Given the extra orders of technical challenges involved in the synthesis of multimeric, glycosylated peptides, with many disulfide cross-links (5 in α-, 6 in β-), and the ready availability of material derived from urine, plus the limited uses for the existing products, the gonadotropins are not likely to be attractive candidates for preparation by applied molecular genetics on commercial grounds.

2.42 Thyrotropin

Thyroid-Stimulating Hormone is actually just one of a family of glycoproteins capable of stimulating thyroid function. The thyroid gland secretes the metabolic hormones thyroxine (tetraiodothyronine, T_4) and triiodothyronine (T_3), which regulate oxygen consumption of most cells. Thyrotropin controls the activity of the pump mechanism which concentrates iodide ion in the principle cells of the thyroid gland for the synthesis of T_3 and T_4. The two hormones T_3 and T_4 are released into the blood by diffusion. Normal blood-levels of thyroid hormones enhance protein anabolism and are thus essential for growth. They also influence carbohydrate and lipid metabolism in many ways. Thyroid-stimulating hormone (TSH) from human pituitary has an α-unit identical with that of LH, FSH and CG from the same species. There may be more than one structure for the 112 AA TSH β-component since extracts yield at least 3-4 active polymorphs. Furthermore, there are some immunoglobulins that express thyroid-stimulating activity - giving rise to a hyperthyroid condition known as Grave's Disease. Under normal conditions, the plasma concentration of the thyroid hormones (T_3 and T_4) - free, not bound to plasma proteins - controls the rate of TSH secretion by negative feedback , exerted at the levels of both the hypothalamus, where the releasing hormone TRH is elaborated, and the adenohypophysis, where TSH is produced. The influence of TSH on thyroid activity involves stimulation of at least five processes.

While thyroid disorders are extremely common, some authorities claim more common than diabetes mellitus, treatment with TSH is rarely indicated [Lee & Laycock, 1978]. The majority of patients with hypothyroidism are female and have primary thyroid disease, for which the treatment of choice is oral administration of L-thyroxine. TSH is usually the third adenohypophysial hormone to be affected in hypopituitarism. Patients so affected are given oral thyroxine along with medications to compensate for the other pituitary deficiencies. Bovine TSH is employed for diagnosis and in management of certain types of thyroid carcinoma. Although a number of side-effects are associated with such treatments, the level of use is modest, and it is not an attractive commercial opportunity. However, the uncertainties about the specificity of the β-chain structure and activity*, along with the fact that the α-chain could also be used to prepare the gonadotropins, are just the sort of things that underlie a challenging and worthwhile academic program. In fact, the UCSF group has cloned the α-subunit peptide.

2.43 Research on Glycoprotein Hormones

Government support of basic research on the genetics and expression of the biosynthesis of these hormones can be expected to provide broadly useful information about glycoproteins that will be applied to the preparation of therapeutically and commercially more important agents. All of the more valuable glycoproteins, of which we are aware, are substantially more complex in structure, less well characterized, and (with the exception perhaps, of immunoglobulins produced by myelomas) it will probably be far more difficult to isolate the appropriate mRNA than the corresponding mRNA involved in the synthesis of the glycopeptide hormones.# Much remains to be learned with

---

* The β-subunit of bovine TSH (112 AA) contains 9% carbohydrate, apparently attached as a single oligosaccharide to asparagine No. 23.

\# Interferons are glycoproteins of molecular weight close to the glycopeptide hormones on which intense interest is focused these days. Far less is known of the structure of interferons, because they are so scarce, than is known about the hormones. Fundamental work on the hormones could pay important dividends in interferon production, and might even prevent the establishment of a harmful monopoly.

respect to the possibility of using microorganisms to produce glycoproteins with the appropriate carbohydrate architecture for activity in the human body. Some experts believe that a simple eukaryote, such as a yeast, may prove superior to bacteria for this function, as well as the elaboration, folding and cross-linking of the polypeptide backbone. It could prove desireable or necessary to resort to culture of mammalian somatic cells (or derivatives thereof) for this purpose.

An interesting recent development is the finding of a microorganism, *Progenitor cryptocides*, that produces a molecule very similar to chorionic gonadotropin. [Maruo, 1979] Not only does it appear to have similar polypeptide subunits and sequences, but it is glycosylated to some extent (glucose or mannose moieties). One explanation offered for its low biological potency compared to HCG is its low (if any) sialic acid content; it being well known that desialylated HCG possesses extremely low biological activity *in vivo* due to its rapid removal from the blood.

Exploration of the ability of natural and artificially-recombined microorganisms to make glycopeptides promises to be a very interesting area in the next few years.

2.44 Erythropoietin

Erythropoietin is one of the glycopeptide hormones that does appear to possess significant clinical and commercial potential. It is a relatively poorly characterized glycoprotein, of about 39,000 molecular weight, that is present at very low concentration in normal blood and can be obtained from the urine of anemic individuals.

Erythropoietin (EP) appears to be the substance responsible, in large part, for the regulation of red blood cell differentiation. It causes an increase in the number of immature red blood cell precursors (normoblasts) in the bone marrow, and in hemoglobin synthesis in existing normoblasts. Fat in the marrow is replaced by erythroid cells, and inactive "yellow" marrow becomes active "red." Under maximum stimulation by EP, the marrow is capable of increasing its production of red blood cells 6-8 fold. The kidney is certainly the primary organ involved in Ep formation in adults, but in fetal and neonate animals this renal function is inoperative and the liver is thought to be the source of a precursor, hepatic pro-Ep. The biogenesis of Ep seems to be mediated via synthesis of a pro-Ep molecule (known as REF, renal erythropoietic factor), which is converted to active Ep upon interaction with a serum factor. In adult humans, or rodents, 80-90% of the total Ep activity derives from the kidney, while the extrarenal portion is largely or totally of hepatic origin. REF is produced by the kidney in response to an oxygen deficit, induced by either environmental hypoxia, or decreased red blood cell concentration, or obstructed renal blood flow.

Human Ep has been isolated from the urine of anemic patients and purified to apparent homogeneity. [Miyake, 1977] The apparent molecular weight of the asialo human hormone, which is inactive, is 34,000. The molecular weight of the active native Ep is not precisely known because of questions about the carbohydrate content, which is roughly 13% hexose, 9% hexoseamine and 7.5% sialic acid. The activity of the purified material is about

70,000 units/mg.* It has been estimated that a dosage of 200 units/kg/day (i.e., ~0.2 mg/day) would be required to evoke a favorable response in anemic patients being maintained by hemodialysis. [Gordon, 1973]

The availability of Ep for experimentation could lead to some interesting therapeutic applications, in conjunction with hemorrhage and burns, as well as anemias and other hematologic conditions. Since blood transfusions of renally insufficient patients, currently a common practice complicated by the need for histocompatibility matching, may lead to sensitization to a subsequent kidney transplant, it would seem highly desirable to maintain erythropoiesis in patients on dialysis rather than administer transfusions. Additional possible indications relate to anemias that are not characterized by heightened levels of Ep, such as those accompanying protein deficiency, chronic inflammatory states, hepatic disease and neoplasia. Also of possible benefit in cancer therapy, would be the use of Ep in conjunction with bone marrow grafts where the hormone should result in more rapid recovery of erythropoiesis. It may also find use as a radioprotective agent, and in patients receiving blood transfusions to over-ride the inhibition of erythropoiesis induced by the transfused red blood cells.

Of the glycoprotein hormones currently recognized, Ep appears to have clinical potential that could be far reaching and will take years to ascertain. Despite the formidable challenge that its production represents, it seems likely to receive early attention as genetic

---

* The concentration in the urine from which Ep was extracted was 1-6 units/ml. About 20 mg of glycopeptide was obtained from 2550 liters of urine.

engineering advances from simple polypeptides to more complex molecules.

3. ENZYMES

Considering the facts that enzymes are critical to virtually every biological process and that an astounding amount is known about the specific enzymes for many of these processes, it is somewhat surprising to realize how small a direct part enzymes play in the practice of medicine today. Therapeutic enzymes accounted for U.S. sales of about $70 million (manufacturer's level) in 1978, but one-half of those sales concerned the blood-plasma-derived coagulation factors used to treat hemophilia. Table 8 outlines the major types of enzymes employed in therapy today. The fantastic potency and diversity of enzymes defies categorization in terms of quantity employed. The individuality of contemporary regimens makes it difficult even to estimate the number of patients treated. Given that there are roughly 14,000 hemophiliacs in the USA, the total number of persons receiving any type of enzyme therapy in 1980 probably does not exceed 50,000.

Undoubtedly, the problems (i.e., immune and febrile responses) associated with the use of non-human enzymes, and the scarcity of human enzymes, have impeded both R&D and clinical exploitation with respect to enzymes for therapeutic purposes. The genetic engineering technologies, of recombinant DNA and somatic cell fusion/culture, open the only conceivable routes to relatively inexpensive production of compatible (human) enzymes at the present time. However, the technical obstacles and time to achievement must not be

Table 8. Ethical (Prescription) Enzymatic Products Currently Employed in the United States

Therapeutic Category	Enzymes Involved	Source	Indications	Approximate Mfrs. Sales
Blood Clotting Factors:				
Antihemophilic Factor (AHF)	Factor VIII	Donor plasma	Hemophilia A	$40 million
Plasma Thromboplastin Component	" IX		Hemophilia B	4
Gastrointestinal Digestive:				
Combinations - including phenobarb, simethecone &/or bile-salts	e.g. Pepsin, Pancreatin	Animal organs	"Nervous" or other indigestion	11
Single source	Pancreatin: lipase, trypsin, amylase	Porcine pancreas	Inadequate fat digestion; Cystic fibrosis	5
Wound Debriding Agents:				
Ointment or powder w. or w.o. antibiotic	Fibrinolysin & Deoxyribonuclease	Bovine plasma & pancreas	Removal of purulent exudates & eschar	4
Ointment w. or w.o. antibiotic	Trypsin	Mammalian pancreas	"	2
	Subtilains	*Bacillus subtilus*	"	1
	Collagenase	*Clostridium histolyticum*	"	<1
Topical, intramuscular & oral	Streptokinase & Streptodornase	Group C streptococci	Clot lysis, reduction of edema & inflammation	1
Oral Proteolytic Preparations:				
	Chymotrypsin	Bovine pancreas	"Possibly effective" for episiotomy	2
	Bromelains	Pineapple		3
	Papain	Papaya	"	
Thrombolytic:				
	Streptokinase	Hemolytic strep.	Lysis of intravascular blood clots	<2
	Urokinase	Urine or tissue cult.		
	Fibrinolysin (plasmin)	Donor plasma	"	
Absorbtion Promoter	Hyaluronidase	Bovine testicles	Rarely for IM or SC inj.	<1
Ophthalmic Surgery	Chymotrypsin	Bovine pancreas	Cataract removal	<1
Cancer Chemotherapy	Asparaginase	*Escherichia coli*	Leukemia	<1
Allergic Drug Reaction	Penicillinase	*Bacillus cereus*	Destroy penicillin	<1

underestimated. In our judgment, it seems unlikely that enzymatic products for therapy can be produced in substantial quantities by the new techniques and marketed in the 1980s, with perhaps one or two exceptions.

Enzymes are employed widely as reagents in diagnostic reactions and to a lesser extent in the preparation of medicinally useful products. These processes take advantage of the feasibility of finding and culturing a microorganism that will produce an enzyme with the desired catalytic characteristics. The quantities of enzyme required for such purposes are only rarely large, and the necessity of reproducing enzymes found only in humans does not pertain. Hence, it seems unlikely that the costs of genetically engineering enzyme producing cells for these purposes will be justifiable until widespread application of the approach for enzymes of greater intrinsic value drives down the cost of development.

Intervention to counteract hereditary enzyme deficiencies, by provision of the missing enzymes or the genes which dictate their synthesis, represent major opportunities for enzymes in therapy. Many enzymopathies, such as the glycogen and cerebroside storage diseases, are being studied.(Hughes, 1977; Brady, 1981)

The examples we shall discuss below will serve to illustrate: (1) typical therapeutic enzyme applications, in terms of alternative sources of potentially useful enzymes and the trade-offs such alternatives imply; (2) the sort of technical issues that must be dealt with, when enzymes are the desired product, in contrast to the relatively simple hormones discussed previously; and (3) the limitations of speculating on the potential clinical utility of enzymes (or other substances) which possess activities that are theoretically desirable or have even been accepted as clinically useful in foreign countries. In addition to the enzymes that shall be discussed in this context, we shall later present brief case studies of two products (L-asparaginase and dextrans) which will illuminate the manner in which microorganisms have been developed and are currently being used by the pharmaceutical industry to produce macromolecules that could not otherwise be synthesized for parenteral therapy. One can reasonably assume that within the next decade, one or more enzymes may be identified which will prove to be of far greater potential therapeutic value than any which can be pointed out at this time. The intense efforts in laboratories around the world to isolate, purify, and determine the origin of enzymes involved in biologic processes with broad clinical implications are the basis for this expectation.

Most biochemical reactions take place within membranes that are impermeable to such large molecules. Hence, therapeutic use of enzymes will be limited to extra-cellular reactions unless means are found for delivering them where their activity will be appropriate.

The field of enzymes is probably the best example of a chicken-and-egg problem that will hamper the exploitation of molecular cloning: there is insufficient incentive to develop the capability of producing a particular molecule (enzyme) until a clinical need of a scale adequate to justify the investment is identified, while the clinical investigation of hypothetical therapeutic uses cannot be conducted without adequate supplies of the pure compound.

The agents which will first break out of this cycle will most probably be those which can with substantial effort be produced in quantity from a natural tissue, because the latter will provide both the material for evaluation and the mRNA from which molecular cloning can proceed.[#] Alternatively, someone - perhaps the Federal Government - could initiate the screening of products from human gene clones analogous to the NCI screening of antitumor agents and industrial antimicrobial screening.

3.1 Blood Coagulation Factors. Genetic deficiences of critical enzymes are obvious areas to consider applying genetic engineering capabilities for enzyme synthesis. Replacement of enzymes that should be circulating in the blood is a first order problem, in that it does not require solution of the additional problem of getting the enzyme into target tissues and cells. The only enzymes administered today to replace deficiences are certain factors from the "hemostatic cascade" that are given to hemophilia patients. In truth, these are proenzymes which precursors convert to active enzymes when injury triggers the "cascade." The most common of these agents are Antihemophilic Factor (AHF), which contains Factor VIII, and Plasma Thromboplastin Component (PTC) complex, which contains Factor IX (along with Factors II, VII and X). These agents are currently extracted from human blood plasma. The diseases characterized by deficiency of Factor VIII and Factor IX, which account for over 90% of all bleeding disorders, are called Hemophilia A and Hemophilia B, respectively.

Interferon is, of course, a non-enzyme example of this phenomenon; it took more than 10 years and a reasonably good tissue source (white blood cells) to break the cycle.

3.11 Supply and Demand

The supplies of AHF and PTC were recently (1977) projected to exceed demand through 1980 and well beyond, if the harvesting and processing plasma continues the trends established in recent years. However, "the volume of plasma fractionated is governed by the demand for serum albumin, since albumin accounts for an overwhelming proportion of revenues generated by the sum of end products resulting from plasma fractionation."* Only if the use of albumin were to be drastically curtailed, or if a source of albumin other than plasma were to be developed, does it appear that a shortage of clotting factors could develop in the foreseeable future (Table 9).

Table 9. Supply-Demand Forecasts for 1980, Based on Plasma Production Forecasts#

	Millions of Units	
	AHF	PTC
Projected demand for clotting factors	354	65
Projected potential supply	590	534

Nevertheless, it must be acknowledged that the limitations of techniques for detecting infectious hepatitis B virus cast a persistent cloud of uncertainty over all plasma-derived substances, and motivate continual consideration of alternative sources of production. Both the individual units of plasma collected and each lot of

* A study conducted for HEW, under the auspices of NHLBI by Booz-Allen, estimated that the U.S. incidence in 1980 would be 11,164 hemophiliacs with Type A and 2,791 with type B. Patients with von Willebrand's Disease have a prevalence rate comparable to that of Hemophilia B. [DHEW Publication, No. (NIH) 77-1274].

\# Between 1971 and 1975 the unmet need for AHF was reduced from 125 million units (45% of theoretically desireable therapy, "need") to 11 million units (~4% of "need") and both supply and third-party payment means have increased further since that time. [Ibid] product are tested for hepatitis B surface antigen with the available radioimmunoassay methods, but that is not sufficient to preclude occasional outbreaks, and it does add to the cost of production. However, the triggering event for a shift in the supply and cost picture for blood coagulation factors seems almost certain to occur in the serum albumin sector, which will be discussed later.

3.12 Plasma Fractionation

The plasma fractionation industry is presently characterized by excess capacity, intense competition, high input plasma costs and tighter profit margins. To meet their profit margin and return on investment expectations, fractionators have placed strong emphasis on the marketing of the non-albumin fractions, especially AHF.* Competition between the suppliers of AHF, and the availability from blood banks of less expensive (albeit less pure and less convenient) cryoprecipitate which can be substituted in the treatment of Hemophilia A, have served to keep the price of AHF down - particularly in contrast to its cost abroad, because of the vigorous U.S. export business in human serum albumin. The excessive supply of PTC and plasma exert a similar effect.

Most of this business is based on competitive bidding for large lots in contracts that involve several blood products jointly derived from common raw material pools. In economic terms, the costs of such by-product streams are allocated on the price the market is willing to pay - third party is willing to reimburse in this case - for

* The evaporation of markets for immune serum globulin (ISG) in the early '70s made the contribution of AHF preeminent. ISG revenue as a percentage of total fractionation revenue fell from an estimated 23% in 1970 to about 0.2% in 1976.

each of the products. Furthermore, there is an element of barter involved in many transaction in this field, because fractionators obtain (recovered) plasma from blood banks and hospitals in exchange for finished blood components.

Recovered plasma is a preferred (cheaper) raw material for albumin production, but it is a relatively poor source of AHF, which is unstable in unfrozen blood or plasma. However, the supply of recovered plasma is limited - providing only 25% of the demand for albumin - and (source) plasma from industrial plasmapheresis centers supplies the remaining demand for albumin and virtually all the AHF produced.

3.13 Albumin, The Co-product

So long as the production of albumin from source plasma remains near present levels, the cost of input plasma allocated to AHF will be minimal. The marginal processing costs for AHF are in the range of 2¢/unit. From a unit price of 14¢ in 1971, the manufacturer's selling price of AHF fell to 11¢ in 1975 and was 8-9¢ in 1978. At times, producers have sold AHF at prices as low as 3-4¢/unit to gain market share or to price competitively during periods of falling prices and excess capacity, which has occurred periodically since 1975.

For years, it has been argued that serum albumin is used excessively and, since it is a major purchase item among hospital costs, scrutiny arising out of cost controlling efforts involving drug-use audits could result in contraction of the volume of serum albumin employed. On the other hand, as we shall describe shortly, serum albumin is a rational long-range objective for genetic engineering efforts and some groups appear to be directing efforts toward it. Either a decrease in the use of albumin or a source other than plasma would upset the present supply of AHF. Hence, it is reasonable to consider the nature of the challenge involved in attempting to produce it by molecular cloning. Furthermore, the persistent hazard of hepatitis is a continuing motivation to seek new routes to these agents.

3.14 Hepatitis Risk

The risk of hepatitis associated with AHF is considered high, that with PTC extremely high, and that with serum albumin (which is heat treated to destroy the virus) practically nil. One recent study found chronic persistent hepatitis, chronic active hepatitis, and fatty infiltration with portal fibrosis in a high percentage of asymptomatic hemophilic patients treated with Factor VIII or IX concentrates.*

3.15 Antihemophilic Factor

The product called AHF possesses at least two activities, which are separable under certain conditions: (1) correction of the clotting defect in the sex-linked bleeding disorder known as Hemophilia A, and (2) correction of the abnormalities of platelet function known as von Willebrand's Disease (which is inherited as an autosomal dominant trait independent of sex). Surprisingly, little is readily available in terms of molecular characterization, beyond the fact that glycoproteins of the order of 200,000 molecular weight, which form aggregates with molecular weight up to 2 million, are involved. AHF is highly sensitive to oxidation and its stability is inversely related to

* Medical Letter 11/16/79 is the source of the assessments of risk of heptatitis associated with various blood components.

purity. Stabilizers are needed to protect the activity of AHF when its concentration is enriched by purification.

Its concentration in plasma is in the range of 10^{-7}g/ml and, since the unit of activity is approximately one per ml of normal plasma, that puts the potency in the range of 10 million units per gram. The market value of $1 million per gram makes it the molecule with the greatest intrinsic value we have discussed thus far. Furthermore, the conceivable bulk drug annual market volume, perhaps $7 million in the USA* and $15 million worldwide, is substantial enough to constitute a strong inducement to entrepreneurial enterprise. To be sure, the technical challenges are formidable, but solution of them (mRNA isolation, high molecular weight, glycosylation, purification, etc.) will have broadly useful fallout, and its cost to the entrepreneur could be repaid with the sale of less than 100 grams of product. Its cost in terms of disruption of a balanced blood fractionation industry could not so easily be recouped.

Nevertheless, the use of molecular cloning to make Factor VIII or other clotting factors is not just over the horizon. There is no evidence that any group is actively working on such a scheme - however, one would expect that such an activity would be a closely guarded secret since, in high technology businesses, the very judgment that the possible attainment of an objective may be feasible within the emerging state-of-the-art is regarded as precious proprietary insight. As we shall discuss shortly, the upset seems more likely to come from the side of the by-product (serum albumin) with much lower intrinsic value

* Based on 2¢/unit and the 1980 demand for about 350 million units - a very conservative estimate.

- a million times less - whose synthesis involves somewhat less imposing obstacles.

3.2 Thrombolytic Agents. Intravascular coagulation is involved in many major, and often fatal conditions, that are considered among the greatest challenges to modern medicine (myocardial infarction, stroke, pulmonary and other emboli). Enzymes for the removal of unwanted blood clots have long been of interest but, for a number of reasons, they have yet to impact significantly on the practice of medicine in this country. Recently, one agent, urokinase, has been hailed with growing enthusiasm, but the costs of producing it have limited its use. A brief review of the history of enzymatic thrombolytic agents will provide the necessary context for appreciation of the possible molecular cloning of urokinase, which seems likely to be one of a few therapeutic enzymes that may be made commercially by this means in the 1980s. A bacterial enzyme, streptokinase, was discovered in 1933, purified, and shown in 1945 to be an activator of human plasminogen - the precursor of plasmin, which the body employs to dissolve blood clots by digesting fibrin, the structural matrix laid down by the "clotting cascade" referred to in the preceding section. Despite the impurity of the bacterial extract, the clinical efficacy of streptokinase, was demonstrated by 1949. While the technology for purifying streptokinase was evolving, human plasminogen was purified (1953), and a number of studies were made which led to the following conclusions:

- "The human body regulates endogenous fibrinolysis through control of the activator level and not that of free plasmin

- Thrombolysis was most readily achieved by activation of plasminogen within and in immediate proximity of a thrombus
- The maintenance of levels of free plasmin in the circulating plasma sufficient to lyse thrombi could only be achieved at the expense of extensive digestion of fibrinogen and other clotting components, notably Factors V and VIII, which are also native substrates for the action of plasmin." [Sherry, 1977]

Studies with plasmin as a therapeutic agent continue, but the predominate thrust since the late '50s has been based on the thought that "the key to effective thrombolytic therapy was in the maintenance of a high circulating activator level to effect rapid thrombolysis, and the control or prevention of significant hyperplasminemia so as to minimize the risk of serious hemorrhage." [Ibid]

3.21 Streptokinase

In the late '50s Lederle Laboratories made available for clinical testing streptokinase which was over 95% pure, for intravenous administration. However, they withdrew the product around 1960 because of adverse (pyrogenic) reactions, which varied from batch to batch, despite clinical enthusiasm for its therapeutic potential.* The European companies, Kabi and Behringwerke (Hoechst), continued the development of streptokinase for intravascular use, and both presently market such a product in the USA. "Under certain circumstances, e.g., extensive deep vein thrombosis, life threatening pulmonary embolism, or peripheral arterial thrombo-embolism, streptokinase has proven to be a useful adjunct in the initial management of many patients." [Ibid] Adverse effects are a prime limitation on the use of streptokinase, including pyrogenic and anaphylactic

* Lederle continues to market streptokinase, in combination with streptodornase, for topical and intramuscular use.

responses, as well as bleeding complications, that demand the attention of an experienced physician and considerable laboratory testing.

Streptokinase, which is produced by Group C β-hemolytic streptococci, is also subject to neutralization by antibodies that most people carry from previous exposure to streptcocci. This is generally overcome by using a large loading dose, followed by a high sustaining infusion rate (250,000 units in 30 min., plus 100,000 units/hr).* There is no totally reliable test with which to predict allergic responses to streptokinase. In general, the potent antigenicity of streptokinase is a handicap, and a (human) plasminogen activator that did not evoke immune response would be much preferable.

3.22 Urokinase

The presence in human urine of an enzyme capable of activating plasminogen was discovered in the early '50s, and the substance, urokinase, was purified, crystallized and taken into the clinic in the mid-'60s. From the outset of clinical studies, it was apparent that "an intense thrombolytic state could be achieved with a much milder coagulation defect than occurred with streptokinase; no pyrogenic or allergic reactions were noted; and no antibodies resulted from its administration...there did not appear to be as great variation in patient responsiveness," as there had been with streptokinase. [Sherry, 1977] In 1967-68 (Phase I) and 1970-73 (Phase III) the National Heart and Lung Institute organized and implemented multi-institutional clinical trials which

* Clinical research suggests that, if heparin is used concomittantly, the streptokinase dosage could be reduced by a factor of ten.

compared urokinase with streptokinase and with heparin in the treatment of pulmonary embolism, the conclusion of which was that streptokinase and urokinase were equivalent and superior to heparin in the short term, although long term benefits or diminished mortality were not established. "The clinical investigation of urokinase has been hampered by the high cost of production of this activator and its relative lack of availability as compared to streptokinase." [Sherry, 1977]

3.221 Sources of Urokinase

Abbott Laboratories obtained an NDA on urokinase and introduced the product (Abbokinase) in January 1978, but at that time the sales of urokinase in Japan were already near the $90 million/year level. Recently, Sterling Drug has begun marketing a urokinase product (Breokinase) manufactured for it by Green Cross of Japan. "According to Japanese reports, urokinase is the first Japanese-made drug formulation to receive production and sales approval from the FDA. Green Cross estimates, that within 3 years of the start of Sterling's marketing activities, the value of urokinase exports will reach Yen 500 million ($2.12 million) per month, and considers that its profits from exporting a finished product will probably be better than those from bulk drug sales or the licensing of technology." [Scrip 440, 6, 11/21/79] The Green Cross product is made from human urine collected throughout Korea and Japan, and they will take advantage of technology licensed from Sterling in the process. Abbott's product, on the other hand, is derived from kidney cell culture.

3.222 Cell Culture Production

Abbokinase is the only commercial therapeutic product

derived from mammalian cell culture. The details of the process actually used by Abbott are considered confidential.* However, publications from Abbott and elsewhere suggest what is involved and the production parameters. Embryonic human kidney cells are subcultured, probably selected for productivity, and grown in a medium containing serum until they are established - utilizing the Mass Tissue Cell Propagators developed by Abbott in the late '60s. [Barlow, 1977] When the cells reach confluency, the growth medium is probably replaced with a protein-free production medium (e.g., 0.5% lactalbumin hydrolysate) in which the cells can be maintained for at least 5-6 weeks. [Ibid] Cultures must be maintained at 37°C for periods of the order of a month. Urokinase is reported to be harvested from the medium by use of ion exchange and gel filtration, [Fratantoni, 1978], but it may be practical to employ affinity chromatography methods similar to those that have been published. [Pye, 1977]

The concentration of urokinase in the production medium is probably several mg/l, compared to less than 0.1 mg/l in urine.#

3.223 <u>Molecular Weight</u>

Crystalline urokinase (UK) obtained from urine appeared at first (1965) to be a single polypeptide of 54,000

* Cell culture is employed for vaccine production and is also being used by Collaborative Research in USA and Yeda in Israel to produce urokinase. Experimental production of interferons by cell culture on a large scale is an area of intense interest at this time.

From the yields reported by Abbott (270-1350 units/ml) for cultures only 5-10% of the cells of which are producers, ~10 mg/l would seem to be a reasonably attainable level to be achieved in producer-enriched cultures (assuming 226,000 units/mg). (For comparison, asparaginase production by <u>E.coli</u> runs about 75 mg/l, and the maximum yield of β-galactosidase from <u>E.coli</u> is about 4g/l. Of course, the <u>E.coli</u> production cycle is about one-tenth as long and the growth medium is much less costly.)

molecular weight, but at least two forms are found in current therapeutic preparations. In addition to the higher molecular weight (HMW) form, that predominates in UK, there is a low molecular weight (~33,000) form (LMW), which can be derived from HMW by proteolysis and which predominates in tissue culture urokinase. Immunologically, the two forms are similar. On a molar basis, the activities appear to be comparable under certain *in vitro* conditions, [Johnson, 1977]* and limited studies disclosed similar activity in the clinic. [Fratantoni, 1978] However, it was noted that HMW is more active when assayed with standard enzyme kinetic methods [Ibid], and quite a bit of controversy has arisen over the relative activity of various preparations with respect to the several substrates these enzymes can transform. [Gaffney, 1978; Samama, 1978] The Bureau of Biologics has reported that the greater activity of HMW is observed only when measurements are made using plasminogen concentrations at plasma levels. [Murano, 1978] It has been suggested that HMW has greater fibrinolytic activity, but that it is inactivated more rapidly than LMW in plasma, which may account for the equivalent hemostatic and clinical results.# The inactivation may be due, in part, to antithrombin III, with which both HMW and LMW form irreversible complexes, and which (antithrombin III) is required for the anticoagulant activity of heparin.

The controversy has sparked efforts around the world to establish better measures of the "fibrinolytic enhancing"

---

* This publication reports the highest specific activity that we have been able to find, 226,000 CTA units/mg of protein, for LMW after separation by affinity chromatography of UK derived from urine.

\# Half-lives of both are estimated in the 10 minute range.

activity of such preparations, and to explain alleged differences between materials from various sources. It should be noted that urokinase activates plasminogen by a different mechanism than streptokinase and consequently is supposed to cause less extensive fibrinogenolysis, and thus should cause less interference with normal coagulation.

Tissue activator, isolated from animal tissues, appears to have a higher fibrinolytic to fibrinogenolytic ratio than either urokinase or streptokinase - which is a desireable shift since bleeding problems are significant with both agents. Presumably, this is due to the very high affinity for fibrin displayed by the so called tissue activator. [Sherry, 1977] Thus, perhaps the objective for molecular cloning efforts should actually be human tissue activator, However, that is rather a long-range program, whereas the clinical background is established for urokinase.

3.224 Cost Issues

The economics implicit in 30 day growth of cells on fetal calf serum (or its equivalent), or collection and fractionation of urine, as reflected in the market price ($150/mg at manufacturer's level), should be sufficient incentive to push application of the state-of-the-art to urokinase. A novel production process should also provide patent or trade-secret protection that this relatively-old drug presently lacks. Clinical trials take many years to perform and are extremely costly. The incentive to conduct the necessary studies to prove (if possible) the utility for the many conceivable applications other than pulmonary embolism would be ever so much greater for a pharmaceutical firm with a proprietary, and less costly, manufacturing process. Finally, there must be a great deal of elasticity of demand with respect to price for a product such as urokinase, especially when its relative benefit (e.g., compared to heparin) is uncertain, and many candidates for therapy are patients of advanced age. Without attempting a quantitative analysis, it seems safe to anticipate the cost of treatment could be cut by a factor of 2 to 10, or more, through the use of molecular clones - if they can be made - and that should lead to much wider testing and, probably, use.* Recognition of possible advantages of a "tissue activator" thrombolytic agent provides a sense of possibilities for improvement that define a paradigm of drug development with roots already 50 years deep.

3.225 rDNA Synthesis of Urokinase

In April 1980, Abbott disclosed that *E.coli* had been induced to produce urokinase through plasmid-borne cDNA made by reverse transcription of mRNA derived from human kidney cells. They acknowledged that substantial R&D work would be required for commercialization of the process, but said that it is potentially more efficient than the processes now employed.

Abbott's medical director acknowledges that the use of urokinase is substantially lower in the USA than in Europe and Japan. Although cost may be a factor, he attributed the lag to American physicians' opinion that the use of urokinase should be a rare, "heroic"

---

* Only inside information on costs of the present process would make such an analysis possible, and even then the other side (molecular clone production) would be entirely speculative. However, the gap between $150/mg and the 10¢/mg it costs the Wadley Institute to produce asparaginase defines the incentive.

measure; whereas, Western European physicians accept it as a more common modality. Studies to establish the utility of urokinase for deep vein thrombosis, for example, are now being conducted almost exclusively in Europe. [Janicki, 1980]

4. OTHER PROTEINS

Hormones and enzymes constitute only a small portion of that category of substances - proteins - that makes up approximately three-fourths of the dry weight of animal cells. Of that open-ended set we have designated "Other Proteins," only a few classes and examples can be discussed in sufficient detail to articulate significant implications for the practice of medicine.

In addition to those proteins and polypeptides already mentioned, there are the deceptively-labelled simple proteins such as the collagens (most abundant proteins in the body), elastins and keratins (the water-insoluble, fibrous providers of extracellular structure), albumins, globulins, and a wide variety of so-called conjugated proteins which bear carbohydrate, lipid and other prosthetic groups - many of which act as a transport vehicles.

We shall not dwell upon the structural proteins: for, on the one hand, their size and complexity exceed the synthetic/analytic capabilities that will be available in the next few years, and on the other hand, their utility in medicine has yet to be established, or - as is the case with collagen, for which uses are emerging - abundant material derived from animals appears adequate to supply the perceived needs.

Nor shall we able to devote much space to the transporter proteins containing prosthetic groups, such as hemoglobin and β-lipoprotein. For them, the promise appears to lie primarily in rather radical therapeutic concepts - such as treating sickle-cell anemia or other hemoglobinopathies, by bone-marrow transplants containing the genetic information lacking from a patient's own cells.

Reconstitution of mouse bone-marrow with marrow cells into which a drug-resistance gene has been introduced _in vitro_ so that the proliferation of those cells can be favored by drug treatment _in vivo_ has been achieved. [Cline, 1980] The implications of such procedures are so complex, that it would be a disservice to create the illusion that they had been examined in this assessment. Theoretically, it may be possible within a decade to implant within a person viable cells with a genetic constitution that rectifies certain of that person's genetic deficiences. The cost of forerunner procedures today, (i.e., marrow extraction, irradiation and re-establishment of bone marrow in cancer therapy) is great. The racial, social and political implications are enormous, for each affected group constitutes a special minority interest, and consideration should be given to limiting proliferation of the defect in the gene pool. Certainly the topic deserves thorough assessment as its technical feasibility develops further, but we believe it would be premature to try to define the public-policy issues and options in the context of this study.

We shall have challenge enough to deal with such relatively simple possibilities as the synthesis in molecular clone factories of the major protein constituent of human blood plasma - literally the largest volume product that will be considered in this study. Furthermore, we must intellectually dissect the splendidly complicated immune system of the human body in order to chacterize some of the innovations made possible by new insights into this varied set of genetically-determined mechanisms that have been evolved to fend off invasions through the manifold capacities encompassed by the term protein.

4.1 Plasma Proteins

Plasma, the fluid portion of the blood, contains about 10% solids, most of which are proteins. During World War II, E.J. Cohn and his associates at Harvard worked out a simple low-temperature fractionation of the blood plasma that, with little change, is used to this day. From this stepwise precipitation with ethanol, fraction I yields fibrinogen and AHF; fractions II-IV are primarily globulins, and fraction V contains most of the serum albumin. Because of the time and cost involved in the Cohn process, only fibrinogen and γ-globulin are removed in the production of another product, plasma protein fraction (PPF), that is used to some extent (~15%) as an alternative to the more refined normal serum albumin (NAS). Table 10 outlines most of the proteins dissolved in human plasma with notes about their structure and functions.

4.11 Serum Albumin is a carbohydrate-free polypeptide of molecular weight about 69,000. The smallest of the main plasma proteins, albumin constitutes about half of their total mass. The body contains about 5g albumin per kg body weight (60% extravascular), and the circulating blood level is maintained near 4% by the liver synthesizing some 200mg/day/kg body weight. The main function of albumin is osmotic regulation. It provides 75-80% of the total

Table 10. Major Protein Components of Human Plasma

COMPONENT	ESTIMATED AMOUNT, g/100 ml	SEDIMENTATION CONSTANT, SVEDBERG UNITS	MOLECULAR WEIGHT	ISOELECTRIC POINT, pH	SPECIAL PROPERTIES AND FUNCTION
Prealbumin	0.3	4.1	61,000	....	1.3% Carbohydrate; transport
Albumin	2.8–4.5	4.6	69,000	4.7	Osmotic regulation; transport
Globulins, total	3.0–3.5				
α_1-Globulins	0.3–0.6				
α_1-Globulin (orosomucoid)	0.075	3.1–3.5	41,000	1.8–2.7	40% Carbohydrate
α_1-Globulin (glycoprotein)	0.030	3.5	54,000	....	14% Carbohydrate
α_1-Lipoproteins					
Density = 1.093	0.05–0.13	5.5	435,000	....	Lipid transport; 67% lipid
Density = 1.149	0.3–0.4	5.0	195,000	5.2	Lipid transport; 43% lipid
Haptoglobin, type 1-1	0.1	4.2	85,000	4.1	Binds hemoglobin; 23% carbohydrate
Inter-α-globulin†	....	....	45,000	....	Thyroid hormone binding and transport
α_2-Globulins	0.4–0.9				
α_2-Globulin (glycoprotein)	....	2.6	....	3.8	16% Carbohydrate
α_2-Globulin (macroglobulin)	0.2	19.6	820,000	5.4	10% Carbohydrate
Ceruloplasmin	0.03	7.1	150,000	4.4	7% Carbohydrate; copper transport
Prothrombin‡	....	4.8	62,700	4.2	11% Carbohydrate; blood coagulation
β-Globulins	0.6–1.1				
β_1-Lipoproteins:					
Density = 0.98–1.002	0.13–0.20	....	$5–20 \times 10^6$	....	90% Lipid; lipid transport
Density = 1.03	0.20–0.25	....	3.2×10^6	....	79% Lipid; lipid transport
Lipoeuglobulin III,					
Density = 1.036	....	8.2	3×10^6	5.3	75% Lipid; lipid transport
Transferrin	0.40	5.0	85,000	5.9	5.5% Carbohydrate; iron transport
Plasminogen	....	4.7	90,000	5.6	Precursor of plasmin, a fibrinolysin
Antihemophilic globulin	....	....	2×10^6	6.4	
Fibrinogen	0.30	7.6	341,000	5.8	3% Carbohydrate; blood clotting
Cryoglobulins	....	15.0			
γ-Globulins	0.7–1.5	7.0	150,000	6.3–7.3	Antibodies

† Commonly termed thyroxine-binding globulin.

‡ From bovine plasma.

SOURCE: The information in this table is taken in part from summaries by E. J. Cohn and associates, in *J. Am. Chem. Soc.*, **72**, 465, 1950, and from F. W. Putnam, ed., "The Plasma Proteins," vols. I and II, Academic Press, Inc., New York, 1960.

differential osmotic effect of the plasma proteins, which drives fluid into the tissues on the arterial side and extracts fluid on the venous side. Albumin also plays an important role in the transport of sparingly-soluble metabolic products.

4.111 Physiology and Pharmacology

Total blood volume is regulated in accord with plasma protein concentration. Adding to the albumin pool increases circulating blood volume and improves perfusion of the tissues. Arterial pressure must be maintained at a level sufficient to permit adequate perfusion of the extensive capillary networks in the systemic vascular bed. The body has several mechanisms by means of which it maintains homeostasis and deals with most influences which would otherwise produce marked alterations in arterial pressure. However, under a variety of circumstances, widespread serious reduction of tissue perfusion, shock, can develop, which leads to generalized impairment of cellular function if prolonged. There are several etiologic factors of shock, primary among which is hypovolemia, i.e., subnormal volume of circulatory fluid. hypovolemic shock is one of the major appropriate indications for the use of albumin. Because of the widespread and mutually complicating impacts of the tissue anoxia and compensatory reactions that can rapidly become irreversible, hypovolemic shock must be treated by prompt restoration of circulating volume with infused fluid. Depending upon the degree of hypovolemia and the condition of the patient before the onset of shock, the appropriate fluids range from saline to whole blood, but albumin appears to be the most common choice. Cost seems to be the only contraindication to the use of albumin for shock due to hypovolemia. [Tullis, 1977a] Because it is stable to pasteurization (heating to 60°C), which eliminates virulent hepatitis B transmission, albumin is regarded as safer to use than blood plasma or whole blood.

Solutions of so-called crystalloids - that is small molecules such as saline, glucose, Ringer's lactate - may be adequate for treatment of shock in young patients who are healthy at the time of injury or surgery, e.g., the Armed Forces find that acute trauma and shock can be managed well with saline, although previously ill or elderly surgical patients appear to tolerate poorly even transient postoperative hypoalbuminemia. [Tullis, 1977b] Essentially, water distributes in 60% of body weight, only 1/12 of which is plasma volume, and infusing dextrose (glucose) solution is like adding water. Isotonic saline solutions, including balanced mixtures of chloride with acetate or lactate, distribute throughout the extracellular space, only 1/2 of which is plasma volume. [Moore, 1966] Solutions of higher molecular weight colloids, e.g., albumin, are retained in circulation because such solutes leak relatively slowly through capillary walls. In fact, concentrated (25%) albumin is employed to ameliorate edema (an accumulation of excessive fluid in cells, tissues, or body cavities) in the face of hypovolemia, because of its osmotic activity. Although albumin is the major natural determinant of colloidal osmotic pressure (COP), this "oncotic" pressure is not a linear function of albumin concentration; so that, when the albumin concentration is cut in half, the COP falls to 1/3 normal [Reiff, 1970]

In addition to shock, arising from loss of intravascular fluid through hemorrhage or crushing trauma, hypoalbuminemia is encountered in patients with liver or kidney disease, malnutrition, septic shock (from bacterial toxins that alter vascular permeability), and after infusion of non-colloid solutions. [Thompson, 1975] Such patients with reduced COP, especially the elderly and debilitated, are prone to develop pulmonary and peripheral edema. There is controversy with respect to when such patients require albumin, but there seems to be convincing evidence that post-operative patients showing early respiratory distress are much better protected against pulmonary edema and enjoy faster respiratory function improvement after infusion of albumin rather than saline. [Bushnell, 1975] Shock is but one of the causes of the acute respiratory distress syndrome, for which in addition to ventilatory assistance a balance of diuretic therapy and concentrated (25%) albumin is often appropriate. Other major indications include: burns (where pathological loss of water from the circulation is a critical issue often demanding huge amounts of both albumin and crystalloid solutions, and albumin synthesis is depressed for a long period as well); certain acute, but not chronic, liver and kidney conditions; peritonitis, and after many abdominal surgical procedures (e.g., for intestinal tumors, aortic aneurysm).*

4.112 Cost of Plasma Expanders

The value of albumin in many situations is unquestionable, but it is also argued that the cost of albumin is often incurred when much less expensive fluids could be

* The variation of indications for use of albumin between nations is substantial, as is evidenced by the comments of speakers at one recent international forum. Vox Sanguinus 31: 225-238, 1976 The most comprehensive view of U.S. use and abuse is the Proceedings of the Workshop on Albumin, Feb. 12-13, 1975, ed. by Sgouris, J. T. & Rene', A., Nat. Heart and Lung Institute Pub. No. 76-925.

employed. Thus, one recently published study, conducted in a 360-bed VA hospital, found that 41% of the volume of albumin administered during a three month period was "inappropriate", on the basis of criteria adapted from guidelines proposed by a NHLBI Workshop (Table 11).

Table 11. Indications for Use of Protein Solutions [Alexander, 1979]

Criteria	Conditions
Shock*	1. Due to blood loss 2. Blood pressure <80 mm Hg systolic or central venous pressure <6 mm Hg 3. Administered within first 2 hours after blood loss 4. At least 1 L of crystalloid solution administered between time of diagnosis and administration of albumin solution
Burns*	1. Involve more than 10% body surface area 2. Second- or third-degree burns present 3. Administered more than 24 hours after burn
Retroperitoneal surgery	Aortic aneurysm, vascular graft in abdomen, thoracic duct rupture or transection, or any other procedure involving retroperitoneal dissection
Adult respiratory distress syndrome	Receiving mechanical ventilation
Ascites	More than 1.5 L of ascitic fluid removed or lost
Peritonitis	

* All conditions must be met.

Ninety-one percent of the albumin was given to surgical patients, and only 29% of the time were surgeon's decisions to use albumin deemed appropriate by their peers.# The conclusion was that approximately $40,000 was spent on inappropriately used albumin in that hospital during 1977 (of $101,600 total).

Furthermore, the authors suggest that a similar proportion of the $0.5 million spent on albumin in 1977 at the nearby

Intensive use of albumin for a few patients (e.g., 107 units for one with acute respiratory distress, 87 and 68 units for two with ascites) that were considered appropriate, and for others (e.g., 46, 39, 23, and 21 units for four surgical cases) that were judged inappropriate, suggests how easily slight modification of the criteria or different interpreters might shift the percentages. Also there were no shock or burn cases involved in this study. Thus one should not apply the findings very broadly.

University of Iowa Hospital (1000 beds) might also have been unnecessary, since the physicians involved work in both hositals. [Alexander, 1979] Albumin accounted for 1/10 of the total drug budget of the VA hospital. This was a single, relatively brief study, but its significance is to illustrate the importance that albumin may assume as efforts are made to trim hospital costs wherever possible.

Its polymeric nature is what makes albumin superior as an oncotic agent. A number of other polymers have been evaluated, and some (e.g., dextrans and hydroxy-ethyl starch) are marketed as less expensive substitutes for albumin. Table 12, drawn from the literature [Thompson, 1975], suggests the relative magnitude of the costs involved in the alternatives. Today the price of dextran is about 1/4 and hydroxyethyl starch (hespan) about 1/3 that of albumin.

Table 12. Alternative Polymeric Plasma Expanders

PRODUCT	UNIT SIZE	UNIT COST	COST/GRAM COLLOID
Pasteurized plasma, 5% protein in 0.9% NaCl	500ml	$57.50	$2.61
Human albumin, 25%, (salt poor)	100	57.80	2.31
Human albumin, 5% in 0.9% NaCl	500	49.40	1.98
Single donor plasma	(220)	17.50	(1.60)
Hydroxyethyl starch, 6% in 0.9% NaCl	500	6.60	0.22
Dextran-40, 10% in 0.9% NaCl	500	9.00	0.18
Dextran-70, 6% in 0.9% NaCl	500	4.25	0.14

Because some polymers, e.g., polyvinylpyrrolidone, had been found to accumulate in the body, the synthetic plasma expanders now approved for sale in the United States are rapidly eliminated (only 30% of hespan or dextran remains intravascular after 24 hours). [Ibid] Hence, the effect of the synthetics is much more short term than that of albumin, which has a 20 day half-life. None of these materials facilitates oxygen transport, the function performed by red blood cells in whole blood, but there is experimentation with fluorocarbons and hemoglobin solutions directed toward that objective. However, none of the alternatives seems likely to divert the majority of demand from serum albumin - especially if new technology were to reduce the cost of albumin.

4.113 Value of Albumin Made via rDNA

Serum albumin is a reasonable candidate for molecular cloning, although the fact that its molecular weight is relatively high (which complicates purification), while its unit value is relatively low, makes the economic feasibility extremely doubtful. It may seem contradictory to state that albumin's unit value is low after the preceding comments that cost is the main deterrent to albumin use. The explanation lies in the fact that albumin is the only substance mentioned thus far for which a therapeutic dose is measured in grams. The "potency," so to speak, is thousands to millions of times less than other pharmaceutical products that might be genetically engineered.

The market value of normal serum albumin is approximately $3/g*, but the volume is such that domestic sales exceed $150 million. Including exports, annual production is in the range of 100,000 kg. Two kinds of activities give rise to the starting material that is fractionated to satisfy the demand for albumin:

*List prices (for a unit consisting of 50 ml of 25% NSA) today are in the $44.20-75.00 range, but the $29.50/unit quoted for a typical VA hospital [Alexander, 1979] and other, confidential, sources suggest that the average selling price for the contracts negotiated between processors and hospitals at the present time is somewhere around $40/unit, which corresponds to $3.20/g.

- Recovered plasma is produced from whole blood by the American National Red Cross (472,248 liters in 1976, 43% of the total recovered plasma), community blood centers, hospital blood banks, and not-for-profit blood bank systems. Barter for fractionated products is a feature of the trade in recovered plasma.

- Commercial source plasma is produced by plasmapheresis, a process by which the plasma is separated from about 1200 ml of blood and the red cells are immediately reinfused into the (paid) donor.

Fractionators make use of all the recovered plasma they can get - in keen competition with producers of typing reagents, controls, and standards for clinical laboratories and with exporters - because recovered plasma historically costs at least $10/liter less than commercial source plasma. In 1976 recovered plasma and a small amount of placental plasma provided only 30% of the demand for plasma to produce albumin (NSA) and plasma protein fraction (PPF) for domestic comsumption, 2.3 million liters of plasma. (DHEW Pub, No. (NIH) 77-1274, p.39)

The other 70% of the plasma required was derived from plasmapheresis centers, the number of which expanded phenomenally in the period 1968-1974. Commercial plasmapheresis is a practice almost exclusively restricted to the United States. Many abuses and "fast buck" operations of the early years have now given way to a regulated and respectable component of a vertically integrated industry. [Solomon, 1979] Half a dozen major, diversified companies now dominate both collection and fractionation.

The widespread adoption of blood component therapy, along with special efforts to treat hemophilia and use Rh immune globulin to prevent Rh hemolytic disease of the newborn, and the swelling demand (11.7%/yr. growth) for albumin, fueled an expansion that culminated in overcapacity in 1975. (DHEW 77-1274) After a period of retrenchment and establishment of export channels to bring supply and demand into balance, growth has resumed. In 1976, the Director of the Division of Blood and Blood Products, Bureau of Biologics, FDA, estimated that albumin production was 75,960 kg/annum [Barker, 1976], but that figure is substantially higher than estimates by Booze, Allen, Hamilton (BA&H), (53,462 kg/annum in 1975, [DHEW 77-1274]. The current Acting Director of that FDA division believes about 1.5 million liters of raw plasma (equivalent to about 35,000 kg of albumin) plus roughly 30-40% of the bulk fractions, are exported annually, so that the precise figure for a given segment is of little consequence in the present argument.

The growth of albumin use in the USA, as projected by BA&H in 1977, is outlined in Table 13. Also presented is the average price of a liter of starting material (averaging recovered and commercial source plasma) anticipated in that study. To get a ballpark estimate of the cost of albumin, one can ignore the value of the other fractions (or, to put it another way, assume that they cover the cost of processing - estimated at $10/liter by BA&H), and calculate what we shall call a "Raw Albumin Value." Those figures suggest the order of magnitude of incentive for innovation - certainly over $100 million for domestic use and several hundred million on a worldwide scale.*

---

* Aside from profit motives for development of synthetic albumin, the Department of Defense has long been a sponsor of R&D on blood substitutes. It is reasonable to anticipate that it would endorse if not support the application of molecular cloning for this purpose.

Table 13. An Estimate of the Volume & Value of Albumin Production for U.S. Consumption

	1975	1976	1977	1978	1979	1980
Kg. Albumin[1)]	53,462	55,062	56,725	58,425	60,188	61,975
Plasma Cost [2)] $/liter	35.87	41.36	42.89	44.49	46.02	47.64
"Raw Albumin Value": $/g[3)]	1.53	1.76	1.83	1.89	1.96	2.03
$ millions[4)]	82	97	104	110	118	126

1) Projected production levels, expressed in kg, from Exhibit X of NIH 77-1274

2) Average cost per liter of plasma as projected p.62, Ibid.

3) Average plasma cost divided by the normal yield of 23.5g. of albumin per liter of plasma assumed p.37, Ibid. Note this is only a relative value because of the assumptions referred to in the text. It yields a cost 1/2 to 2/3 the market price in dosage form.

4) The product of the quantities defined by 1) and 3).

As though to endorse the speculation that such a market may be addressable with molecular cloning technology, W. N. Hubbard, President of the Upjohn Company, said recently (1/24/80, London) "it would be reasonable to assume we are working on human albumin" as he explained that Upjohn is concentrating mostly on the production of large globular proteins. However he also said that, apart from the special cases of human insulin and growth hormone which may appear in 2-5 years, pharmaceutical products resulting from recombinant-DNA technology are unlikely to be marketed within the next decade. Dr. Hubbard stressed that Upjohn is working on various yeast strains as potential producers; in particular Upjohn is concentrating on near-anaerobic yeasts which will grow in low pressure fermenters from which ethanol is removed by evaporation, to reduce feedback inhibition of growth. He pointed out that for some cases like insulin, where it is relatively easy to prove chemical identity of the product with the natural substance, registration should be straight forward. However, for more complex products, which are less completely characterized, proof of identity will be difficult and a whole gamut of pharmacological, toxicological, and clinical testing will be necessary, imposing a further considerable delay on commercialization.

4.114 The Technical Challenge Represented by rDNA-Albumin

The complete 581 amino acid sequence of human serum albumin was published in 1975. The 17 disulfide bonds probably cause it to fold into 9 loops as is depicted in Figure 5 for bovine serum albumin. [Peters, 1977] The molecular weight is half again as large as the chicken ovalbumin molecule that Upjohn scientists and others have produced by genetic engineering techniques.

The first report referred to concerning ovalbumin involved the _in vitro_ construction of a multicopy plasmid in which the _lac_ operon control region was fused to the DNA sequence encoding for the 43,000-dalton protein normally synthesized by the chicken oviduct. _E.coli_ containing the fused plasmid synthesized about 1.5% of their protein as full-length, immunoreative ovalbumin and actively secreted it through the cell membrane into the periplasmic space. While impressive for a first try, the level of protein produced was less than 10% of the theoretical level expected, given the fact that there are multiple (25-30) copies of the gene in each cell and one copy of the

β-galactosidase is expressed as 2% of the total E.coli protein.* [Fraser, 1980]

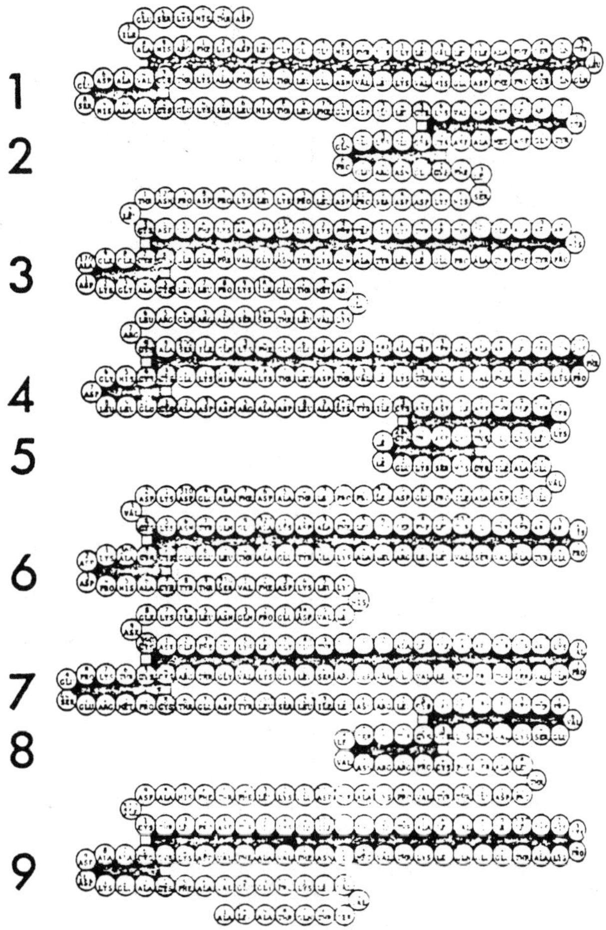

Amino acids are indicated by their official three-letter abbreviations, with the amino-terminus at the top of the page. Disulfide bonds between Cys-Cys residues are designated by double bars. The molecular weight calculated from this sequence, assigning Phe-Gln-Asn to residues 400–402 by analogy with human albumin, is 66 210. The nitrogen concentration, if certain assumptions are made regarding amide assignments (*2*), is 164.59 g/kg. Reprinted from *Fed. Proc* **34**, 591 (abstr. 2105), 1975, by permission

Figure 5. Amino Acid Sequence of Bovine Serum Albumin

---

* One explanation for the discrepancy, which appears to be gaining favor, is that the transfer RNA molecules required to translate the particular codons utilized for certain amino acids in the chicken may not be abundant in E.coli, because its DNA predominantly uses synonomous codons for those amino acids, which would limit the rate of translation of the foreign gene.

The DNA employed in this case was complementary DNA (cDNA), made by reverse transcription from messenger RNA (mRNA) for all of the structual sequences of ovalbumin. However, the original DNA of the chicken from which the mRNA is derived actually contains seven intervening sequences that interrupt the eight segments that determine the amino acid sequence of ovalbumin. The occurrence of such intervening sequences turns out to be a common feature of genes for complex eukaryotic polypeptides.

One of the most surprising discoveries of modern biology was the finding of this profound difference at the genetic translation level between eukaryotes, like chickens or men, and prokaryotes, such as _E. coli_. Eukaryotes possess a mechanism, still not understood, for processing the immediate RNA transcript of the DNA and editing out of the intervening sequences.

The second report referred to above, which involves several workers at Baylor and an Upjohn scientist, illustrates an alternative approach to the synthesis of ovalbumin that makes use of the mRNA processing capacity of eukaryotic cells. [Lai, 1980] In this case the entire chicken ovalbumin gene, accompanied by genomic DNA sequences flanking both termini of the gene, and three copies of a thymidine kinase (TK) gene, was introduced into thymidine kinase-deficient mouse cells in tissue culture.*

(The TK genes simply serve as a marker to permit the selection of transformed cells.) The entire ovalbumin

* The vector employed was a linearized pBR322 plasmid, commonly used in circularized form in prokaryote hosts. (This evidence that a bacterial plasmid can become integrated into a mammalian genome is a source of concern to some assessing potential risks of this technology.)

gene was found to be integrated into high molecular weight (cellular) DNA in the transformants, and a protein identified as chicken ovalbumin by immunoreactivity was detected within the cells of clones of the transformed cells. Between 5000 and 30,000 molecules of ovalbumin were produced per mouse cell. This is a marked contrast to the finding that no ovalbumin was synthesized by E.coli if natural intervening sequences were included in the plasmid-borne structural gene. [Mercereau-Puijalon, 1979]

A number of examples of translation of a gene from one eukaryote being appropriately spliced in cells of another species of eukaryote have now been demonstrated. This suggests that cell culture might be employed for production of a heterologous protein. However, the yields, rate and costs of cell culture hardly compare with what is theoretically possible with microorganisms, and certainly the economics necessary to make biosynthesis of human albumin practical are altogether different from those pertaining to cell culture.

4.115 The Other Challenges Represented by rDNA-Albumin

The important point is that consideration is being given to the manufacturing of albumin, which is presently the principal reason for fractionating blood plasma. Since a number of products are derived from that fractionation, especially the clotting factors, a shift in the need for plasma-derived albumin would have a significant impact on either the availability or cost of those by-products.

Of course, it is conceivable that molecular cloning could be used to produce Factor VIII and other clotting factors, as we have discussed in an earlier section. However, it will be recalled that the present supply and demand situation in the USA provides limited incentive to engineer a Factor VIII-producing scheme, and the technical problems are formidable (i.e. uncharacterized glycoprotein, m.w. 200,000). If the synthesis of albumin, for therapeutic blood expansion, is judged economically viable (by Upjohn, Department of Defense, or others here or abroad) and the product moves toward approval and marketing, then it may be desirable for the government to orchestrate developments with respect to the clotting factors so that the entire by-product spectrum remains available at reasonable prices.

Since the solution of the problems posed by the synthesis and purification of a genetically-engineered Factor VIII may take a decade, and the techniques developed in solving them will be useful for many developments throughout the field, government investment in the basic biochemistry of Factor VIII and the biosynthesis of glycoproteins could provide the necessary underpinnings of entrepreneurial efforts in the late '80s that would produce not only the answer to the potential imbalance caused by albumin developments, but a variety of other agents important to therapy.

The feasibility of making and purifying a protein as complex as albumin for costs in the range of \$2/g is extremely speculative at this point.* Furthermore, normal serum albumin is already regarded as excessively

---

* "Making and purifying" is a brief way of suggesting an enormously sophisticated process through which the proper sequence would be proven and maintained for much higher volume production than is contemplated for any other macromolecule(derived from this technology)that is intended for injection into humans.

expensive, so that lower pricing would be anticipated for expanded usage. On the other hand, the federal government, especially the Department of Defense, might disregard immediate economic prospects, and conclude that a source of human serum albumin that did not depend upon the payment of blood donors would be in the national interest. There are many nations which import albumin and, thus, exports derived from a molecular cloning process might contribute $100 million or more to the U.S. balance of payments. However, such developments must not take place without due consideration of steps to assure the continued availability of other essential blood fractions.

4.2 Immunoproteins

Under this heading we include all of the proteins, both endogenous and exogenous, that relate to the multifaceted immune system. Polypeptides, the primary products of every molecular cloning scheme, are at the heart of each aspect of immunology. Hence, it is reasonable to anticipate diverse, manifold impacts on all aspects and burgeoning new developments only made possible by recent breakthroughs. No area of current therapy nor, especially, category of unalleviated health needs, will be more affected by the technologies of applied genetics than will the field encompassed by the term immunopharmacology.

A familiar salient of immunopharmacology is the realm of vaccination. We shall outline briefly how molecular cloning will facilitate the study of the immune state that can be evoked by various vaccine preparations and will enable the production of antigens useful for classical vaccination or for novel prophylactic or therapeutic measures. An early dramatic benefit should be production of harmless substances capable of eliciting specific antibody-mediated defenses against infectious diseases which heretofore have frustrated efforts to obtain safe vaccines.

Having related the new technology to this established type of immunopharmacology, we shall touch upon a number of areas of experimentation which hold great promise on account of molecular cloning being available at a time when dissection of the immune system has revealed some of the myriad ways that system contributes to both the prevention and cause of disease. In this respect we shall describe work on interferon* and other agents, called lymphokines and cytokines, that activate disease-fighting cells. We shall introduce the recent hypothesis for genetic predisposition to many diseases and, in broad terms, relate suggested means for counteracting such susceptibilities and addressing some of the problems of graft and transplant organ rejection.

At that point it will be opportune to describe the production of antibodies by clones created and selected from the fusion of certain cells. These monoclonal antibodies have had such an impact on research that we could scarcely avoid mentioning them earlier, but the variety of implications for diagnosis and therapy that can be anticipated from monoclonal antibodies will be the subject of the final segment of this brief survey on immunoproteins.

---

*The Office of Technology Assessment commissioned an independent study of the implications associated with interferons during the period this study was being made. Hence, the present review of the subject does not attempt to cover the vast amount of effort of the past three years comprehensively, and no attempt was made to update the study in 1981.

We heartily endorse the view that is expressed as follows by Dr. Robert Good: "Among the looming triumphs of the biologic revolution is the rapidly developing understanding of the mechanics of bodily defense. In the short span of 35 years, knowledge of immunological machinery has progressed from the crudest descriptions to major understanding in cellular and molecular terms... With this surging new information about the immune system comes the possibility of developing a pharmacology which can modulate and control immunologic functions - Immunopharmacology." [Haddon, 1978] There is little doubt that applied genetics will play a critical role in the realization of that possibility, for it provides the only apparent means to synthesize many of the agents that can be anticipated as components of immunopharmacology.

4.21 Antigens for Vaccines

Antigens are the specific counterparts of antibodies, eliciting antibody formation and being the object of the tenacious affinity of the complementary antibody - usually to the benefit of the antibody-forming organism. Surface components of pathogenic organisms, toxins, or other proteins secreted by pathogenic microorganisms act as antigens, drawing attention to themselves from certain white blood cells (β-lymphocytes) capable of synthesizing antibodies uniquely structured to bind to a particular site (an antigenic determinant) on a given antigen.

The enormous variety of lymphocytes, each with the capability of synthesizing only one specific antibody, assures that the body will be able to generate antibody appropriate for a given antigen, but the process takes time - especially when the antigen is novel. Fortunately, the system has a memory so that defenses can be much more rapidly mobilized for a previously recognized invader. We consciously take advantage of this mechanism, called acquired immunity, by vaccination.

4.211 Vaccination

The objective of vaccination is to provide effective immunity by establishing in advance adequate levels of antibody and a primed population of cells which can rapidly expand on renewed contact with the antigen. The initial contact obviously should not be injurious. Hence a number of maneuvers have been developed over the past two centuries to modify the pathogenic effect without losing important antigenic determinants. Vaccines generally contain one or more of the virulence factors (e.g., surface components, toxins) of a pathogenic organism in modulated or attentuated form, in which it is too weak to be functional in a pathogenic sense but is capable of inducing a selective immune response. After the selective response - that identifies and favors the proliferation of those cells capable of producing the appropriate antibodies - when a pathogenic organism containing antigenically similar or identical virulence factors appears in the body, the vaccinated individual will produce antibodies that enhance white cell engulfment (phagocytosis) and destruction of the invader, prevent its attachment to cell surfaces and/or inactivate the toxins it produces. Unfortunately, in the contest for survival some organisms have evolved mechanisms for evading such defense systems.

4.212 Antitoxin Vaccines

Toxins, such as those responsible for diphtheria and tetanus, are proteins produced by bacteria that interfere with mammalian physiology at extremely low concentrations. Detoxification, treatment with formaldehyde for example, is used to destroy their virulence without destroying the immunogenic determinants. Immunization with such a toxoid, (usually absorbed on aluminim hydroxide, an adjuvant that enhances response), provokes the formation of protective antibodies. Molecular cloning could enable the production of toxins, or better non-virulent fragments of toxins, by microorganisms that are incapable of serving as pathogenic vectors and eliminate the need to cultivate the normal pathogen. There does not seem to be very strong interest in applying molecular cloning for the production of toxoids, although work has been authorized on pseudomonas toxin [RAC, Mar. 1980] This may be due to concern about creating organisms of uncertain virulence*, but it is more probably attributed to general satisfaction with the present methods.

Still, some toxin-producing microorganisms, e.g., *Vibrio cholera*, do not yield a toxoid that produces satisfactory immunity. Perhaps, molecular cloning could prove useful to generate a non-virulent modified cholera toxin, with better antigenic characteristics, or a different vaccine more suitable for eliciting antibody (secretory IgA) production by the gut wall where the toxin attacks.

For the moment, at least, this is unlikely to be pursued since international public health authorities reportedly feel that vaccination for cholera is undesirable - preferring that the disease manifest itself by symptoms and be treated rather than chance the carriage by an

---

* Work with the genes of toxin-producing organisms is very strictly regulated under the NIH guidelines.

immunized traveler of sub-clinical infection into unprotected populations.

4.213 Antibacterial Vaccines

There is a good deal of work today on the development of oral vaccines against enteric bacteria: *Salmonella typhi*, *Shigella* (dysentery), *E.coli* (enteritis), and *V. cholera*. Much of this depends upon the observation that the pathogenic forms of these organisms possess surface components, such as minute filaments, called pili, that enable them to attach to human cell surfaces and to resist phagocytosis. Antibodies that bind to pili block attachment and promote phagocytosis. [Buchanan, 1979] The relevance of molecular cloning techniques to producing these bacterial products, whether they involve simple proteins or more complex molecules is obvious. One wishes an organism that is safe to work with and produces an abundance of the virulence factor in the variety of forms it is known to take in nature. (Whether molecular cloning may prove useful likewise in the production of non-protein antigens such as the capsular polysaccharides that are the basis of vaccines against meningococcus and pneumococcus is a matter for speculation).

"Vaccines for pertussis, bubonic plague, typhoid and typhus probably contain one or more of the outer-membrane components of pili, lipopolysaccharide, and major outer membrane proteins. The precise contribution of each of these components to vaccine efficacy has not been determined, since no vaccines containing only one of these antigens have been evaluated. However, if the experience with cholera vaccination is representative, it is likely that several virulence factors are necessary for infection to develop. Therefore, it is possible that maximal protection will be afforded by vaccines containing a mixture of virulence factors, rather than single components." [Quotation from Buchanan, 1979]

Vaccines against the threadlike bacterial structures, called pili, which are essential to colonization of body spaces by several pathogens, are being developed and tested. Diseases as dissimilar as bacterial diarrhea, gonorrhea, strep throat, kidney infections, and tooth decay might be prevented by vaccines designed to block adhesion of the causative organisms. To date the results, with pili extracted from bacteria as antigens, have been mixed - with some successful and some failures in several clinical studies. [Marx, 1980] Of course, it is hoped that molecular cloning will permit the production of those antigenic determinants which are most universally distributed among the pathogenic forms of a given species.

The immunity conferred by live vaccines invariably exceeds that which can be developed by injecting non-living antigenic material, even when the latter is given with adjuvant that enhances the antibody-producing response. "This must be partly because the replication of the living microbes confronts the host with a larger more sustained dose of antigen and also because the immune response takes place largely at the site of the natural infection." [Quotation from Roitt, 1978, p.211] Conceivably molecular genetics could be used to address most of these limitations. Molecular clones could be very productive sources of high potency antigen that would be the basis for larger (without side-effects due to contaminants found in vaccines consisting of killed microorganisms) and/or more sustained doses. The latter could be achieved through the utilization of sophisticated drug delivery technology - such as the bioerodible polymer matrices being developed by ALZA Corporation, which are currently being employed to release contraceptive agents for months at a time. Finally, the issue of site of inducement of antibody formation might also be addressed by delivery systems, either physical-chemical (i.e., dosage forms) or biological (i.e., an innocuous microorganism that would colonize the desired bodily space and express a gene that produces a non-toxic antigen. Such "engineered life-forms," made possible by molecular genetics, represent a logical progression from, for example, the live, attentuated vaccines (e.g., BCG vaccine for tuberculosis made from attenuated *Mycobacterium bovis*, which has been used for decades in many countries).

These novel vaccine-organisms for which we propose the term "vaccinisms," would be safe by design for they would not possess the complete genetic information required for virulence. Vaccinisms made from appropriate bacteria could produce antigens proper to viruses, protozoa or other parasites, as well as pathogenic bacteria.

4.214 Antiviral Vaccines

4.2141 Polio

The contrast observed for polio vaccines provides an example and an introduction to the discussion of viral diseases, for which molecular genetics will probably be more widely important than for bacterial infections. "In contrast with the ineffectiveness of parenteral injection of killed polio vaccine, intranasal administration evoked a good local antibody (nasopharnygeal IgA) response; but whereas this declined over a period of two months or so, oral immunization, with live attentuated virus, established a persistently high IgA antibody level." [Ibid, p.212] Furthermore, the original (Salk) polio vaccine, which was killed by formalin treatment, involved the discomfort of an intramuscular injection, and when some live virus inadvertently was permitted to contaminate the vaccine, there was at least one case of polio caused by the contaminated material. The live attentuated (Sabin) vaccine, which is the standard today is given orally, without need of medical attendants; it induces an inapparent infection of the alimentary tract that almost always establishes the desired resistance, especially with local secretory IgA antibody that serves to block virus implantation and multiplication - protecting both the individual and the community by preventing dissemination. (In the absence of vaccination the ratio of inapparent infections to clinical cases is greater than 100:1, and during active infection the virus is present in the throat and in feces.) However, for rare individuals (approximately 0.06 recipients and 0.2 contacts per million doses) this live vaccine results in paralytic poliomyelitis. [Merck Manual, p.50]

Polio provides the most dramatic model for the benefits of vaccination. In the USA, the number of cases has fallen from an average of 21,000/year in the 5 year period before introduction of the vaccine in 1955 to 9 in 1976. A number of authorities have begun to argue,and many laymen to believe,that the risk of acquiring polio from the environment is less than that of acquiring it in connection with the vaccination program. Many recommend return to a killed vaccine, such as the one which has been used widely in Holland that is said to be far superior to the original

Salk vaccine and requires only a single injection. On the other hand, it should be possible to produce well characterized protein by means of molecular cloning that would be adequate for a maintenance program and be absolutely free from the possibility of contamination with viable virus that is a spector over any killed-virus program. Polio is ideal for such an approach because it changes rarely - three types are included in contemporary vaccines (trivalent).

We cannot begin to discuss vaccines for all viral diseases in this context. Commercially, the most important are the influenzas (20.8 million doses/year annual average from 1973-75, less numerous than 25.0 million doses/year of polio vaccine, but undoubtedly more profitable).

4.2142 <u>Influenza</u>

Flu vaccines are killed or partially purified viruses which give rise to 65-85% protection for 3-6 months. [Buchanan, 1979] Influenza is caused by myxoviruses, which consist of a core of RNA surrounded by a coat made up primarily of two glycoproteins: one having hemagglutinating activity (HA) and one having enzymatic activity as a neuraminidase (NA). These two types of proteins constitute the strain-specific antigens. (The first stage of infection of a cell by a myxovirus involves attachment by means of the HA to a glycoprotein receptor on the surface of the cell; the NA is involved in the budding process by which the offspring are released from the cell after reproduction of the virus.) Three major types (A,B,C) are distinguished on the basis of the antigenic characteristics of a soluble nucleoprotein (NP), associated with the RNA in the core. Different serotypes of Type A influenza viruses are numbered H_0N_1, H_1N_1, H_1N_2, and H_3N_2, according to the major antigenic types represented by the HA and NA components. Host antibodies that bind to the HA can prevent cellular infection, but mutant viruses which produce an HA different from that for which antibodies have been produced may proliferate for the period required to develop the appropriate antibody.

Influenza has not come under control like polio and measles, largely because of the high frequency with which antigenic changes occur in the surface proteins. HA and NA undergo gradual change (antigenic drift) under increasing antibody prevalence in the population. Significant drift in the HA antigen is often associated with epidemics of varying impact, which occur without any regular time intervals. Occasionally, the HA and NA antigens of influenza A virus may abruptly undergo a complete change (antigenic shift) and bring about an influenza pandemic, e.g., in 1957 when both HA and NA changed (Asian flu) and in 1968, when only HA changed (Hong Kong flu). The functional HA subunit is a triangular rod-shaped glycoprotein with a molecular weight about 250,000, comprised of three HA monomers. The HA monomers are synthesized as single polypeptide chains. During maturation and virus assembly, HA is processed into two chains (HA1 and HA2) which remain linked together by disulphide bridges.

HA1, the 262 AA N-terminal portion, has a molecular weight of 47,000 and contains most of the carbohydrate. The molecular weight of HA2 is approximately 30,000. Amino acid (AA) sequence maps of 9 strains of A/Hong Kong/68

(H_3N_2), isolated between 1968 and 1977, indicate that most of the primary structure remained unchanged; in fact, in most cases, only one or two AA changes could be observed between the HA1 chains of neighboring drift strains. About 18 out of the 262 residues of HA1 and two of the residues of HA2 differed between the first and last members of the 10 year series. The specific relationship between the AA sequence changes and the antigenic determinants of HA is not understood, but most of the antigenic variation is thought to involve only HA1, and it has been shown that single AA substitutions in HA1, requiring a change of only one base in an RNA codon, dramatically alter affinity of antibodies. [Laver, 1980]

4.21421 Recombination In Vitro

G. D. Searle has disclosed molecular cloning efforts that successfully produced hemagglutinin corresponding to that in the coat protein of the fowl plague virus (FPV), which closely resembles human influenza virus. The HA of FPV consists of 563 AA, of which 319 are in HA1 and 221 in HA2. The AA composition shows similarities to human flu HA, with average homology around 38% and as high as 55% for the H_3N_2 strain A/Memphis/102/72. The researchers at the Searle Laboratory in High Wycombe, England - where all of Searle's *in vitro* recombinant-DNA work is being carried out - cloned cDNA made by reverse transcription of the virion RNA, and determined the complete (1,742 base) nucleotide sequence of gene 4 which encodes the HA proteins. This is the first gene of an influenza virus to be completely sequenced, and it predicts for the first time the complete AA sequence of a viral HA. The authors conclude, "It will now be possible to determine the primary structure of HA genes from influenza strains that differ only in their reactivity towards monoclonal antibodies. (See p.304 for discussion of monoclonal antibodies.) A single nucleotide change would show which AA had changed and pinpoint the antigenic site. Ultimately, these methods may enable us to predict which AA changes will produce influenza viruses capable of causing epidemics." [Porter, 1980] As the rationale has been paraphrased in the press "Many varieties of the antigenic protein genes could be held "on tap" in bacteria. Their synthesis for vaccine production then would be programmed according to the antigenic profile of the flu making the rounds." [C&EN, 5/28/79] Furthermore, the genetic code for several flu antigens could be introduced into a production organism - E.coli, other bacterium or yeast, for example - so that a multi-valent vaccine might be produced in one fermentation. This has been characterized in press reports as making possible a "universal" vaccine. While that seems excessively optimistic, certainly it should be possible to provide protection against more strains than are economic to cover with the present antigens that are produced by virus culture in eggs, killed, and combined into trivalent vaccines that cover three anticipated strains.

4.21422 Recombination In Vivo

A useful perspective is attained when one perceives that classical viral genetics have been important in the established production of flu antigen. *In vivo* recombinant is used to shuffle genes from the virus that possess the antigen of interest into a virus which grows well under the production conditions. That has facilitated production of more antigen and made possible partial purification so that larger doses could be given, especially to children - who are both less likely to develop protective antibody levels at tolerable doses and more susceptible to adverse reactions to the complex vaccine.

Live, attentuated strains have also been developed by the use of in vivo recombination to exploit temperature-sensitive (TS) mutants which will colonize only the cooler regions of the upper respiratory tract. The TS-mutant donor viruses have been mated with wild-type viruses to produce suitable attentuated virus recombinants that have inherited TS limitations from the donor viruses. Two master donor strains of the influenza A virus and their recombinants have been developed and are involved in clinical trials sponsored by NIAID. Vaccine viruses with two TS lesions (TS1A2 recombinants) have produced the most encouraging results. If this approach is successful - and it is much nearer term than the in vitro recombination work at Searle - two different strains of influenza virus will be used as donors of attentuating genes for vaccine viruses: one attenuated for persons with some NA immunity and the other designed to elicit production of both HA and NA antibodies.

The live vaccines have the added virtue of being suitable for intranasal administration. That way the production of secretory IgA antibody is stimulated at the portal of infection to a degree that cannot be achieved by injection of antigen at a parenteral site. Since many feel this is the reason that parenteral vaccines do not always protect against the flu strains that they ought to, it may be that the genetic information developed by molecular cloning will have to be incorporated in some sort of "vaccinism" to get the antigens where they can do the most good.

4.2143 Other Common Viral Vaccines

As causes of morbidity, measles and mumps rank next to influenza in significance, Table 14 enumerates the expansion of vaccination programs, especially with respect to measles and rubella, in recent years.

Table 14 Percent of Population 1-4 Years of Age Adequately Immunized Against Polio, Diphtheria-Tetanus-Pertussis (DTP), Measles, Rubella, and Mumps Vaccines, United States 1964-1975

YEAR	Measles	DTP	Polio	Rubella	Mumps
1964	24.0	76.0	78.6	**	*
1965	33.2	73.9	73.9	**	*
1966	45.5	74.5	70.2	**	*
1967	56.4	77.9	70.9	**	*
1968	58.8	76.5	68.3	**	*
1969	61.4	77.4	67.7	**	*
1970	57.2	76.1	65.9	37.2	*
1971	61.0	78.7	67.3	51.2	*
1972	62.2	75.6	62.9	56.9	*
1973	61.2	72.6	60.4	55.6	34.7
1974	64.5	73.9	63.1	59.8	39.4
1975	65.5	75.2	64.8	61.9	44.4

Doses considered adequate are:

Measles	1 dose
DTP	3 doses
Polio	3 doses
Rubella	1 dose
Mumps	1 dose

*Mumps data collected for first time in 1973, licensed in December 1967.
**Rubella vaccine was licensed in June 1969.

SOURCE: United States Immunization Surveys, 1964-1975.

Table 15 demonstrates the effect of those immunization programs in decreasing the incidence of those diseases. Unborn children are the real beneficiary of the rubella program, of course. All of these vaccines involve attentuated viruses which confer long lasting protection and are models of what constitutes a satisfactory vaccine. The fact that there are still quite a few cases is an indication of that other crucial factor involved

Table 15 Reported* Measles, Pertussis, Tetanus, Diphtheria, Polio, Rubella and Mumps Morbidity and Mortality, United States 1960-1976

YEAR	MEASLES Cases	MEASLES Deaths	PERTUSSIS Cases	PERTUSSIS Deaths	TETANUS Cases	TETANUS Deaths	DIPHTHERIA Cases	DIPHTHERIA Deaths	POLIO Cases	POLIO Deaths	RUBELLA Cases	RUBELLA Deaths	MUMPS Cases	MUMPS Deaths
1960	441,703	380	14,809	118	368	231	918	69	3,190	230	nn	12	nn	42
1961	423,919	434	11,468	76	379	242	617	68	1,312	90	nn	14	nn	53
1962	481,530	408	17,749	83	322	215	444	41	910	60	nn	8	nn	43
1963	385,156	364	17,135	115	325	210	314	45	449	41	nn	16	nn	48
1964	485,083	421	13,005	93	289	179	293	42	122	17	nn	53	nn	50
1965	261,904	276	6,799	55	300	181	164	18	72	16	nn	16	nn	31
1966	204,136	261	7,717	49	235	158	209	20	113	9	46,975	12	nn	43
1967	62,705	81	9,718	37	263	144	219	32	41	16	46,888	16	nn	37
1968	22,231	24	4,810	36	178	86	260	30	53	24	49,371	24	152,209	25
1969	25,826	41	3,285	13	185	89	241	25	20	13	57,686	29	90,918	22
1970	47,351	89	4,249	12	148	79	435	30	33	7	56,552	31	104,953	16
1971	75,290	90	3,036	18	116	64	215	13	21	18	45,086	20	124,939	15
1972	32,275	24	3,287	6	128	58	152	10	31	2	25,507	14	74,215	16
1973	26,690	23	1,759	5	101	40	228	10	8	10	27,804	16	69,612	12
1974	22,094	20	2,402	14	101	44	272	5	7	3	11,917	15	59,128	6
1975	24,374	na	1,738	na	102	na	307	na	8	na	16,652	na	59,647	na
1976**	39,585	na	927	na	68	na	146	na	9	na	12,090	na	37,743	na

*Data reported to the National Morbidity Reporting System and the National Center for Health Statistics. In some diseases, underreporting may be a problem.
**Preliminary data
na=not yet available
nn=not nationally notifiable at that time

in prophylaxis, utilization. Table 16 indicates what a large fraction of vaccinations are accounted for by public programs (45% of 90 million doses). "Public acceptance of mass immunization in the United States is in a period of decline, sufficiently so as to pose a growing threat to public health."
(Conclusion No. 1 of National Immunization Work Group on Health Information and Public Awareness, 3/11/77)

Table 16 ESTIMATED ANNUAL DISTRIBUTION OF MAJOR BIOLOGICS BY AGE AND PROVIDER CATEGORIES - UNITED STATES[1]

	CIVILIAN			MILITARY
VACCINE	NUMBER OF DOSES (millions)	PERCENT FOR CHILDREN[2]	PERCENT IN PUBLIC PROGRAMS[3]	NUMBER OF DOSES (millions)
DTP	17.4	100	45	0.25
Td	7.6	50	45	0.15
POLIO	21.0	>95	45	4.0
MEASLES	7.1	>95	55	0.2
MUMPS	3.5	>95	<15	0.12
RUBELLA	7.6	>95	55	0.09
RABIES	(9,000-10,000)[4]	50	<15	0.09
INFLUENZA	17.6	20-25[5]	<15	3.2

1 Based on averages from 1973-75

2 Less than 18 years old; data from varied sources and with assumptions

3 Estimates from Federally funded immunization projects with assumptions

4 Patients treated

5 Estimates from U. S. Immunization Surveys

SOURCE: BIOLOGICS SURVEILLANCE PROGRAM, C.D.C.

Table 14 illustrates that even the most widely accepted vaccine is only used to protect about three-fourths of the target population for which it is designed. This despite the fact that most states require immunization prior to school entry (Table 17). These and many other aspects of vaccine utilization are treated in the OTA review

Table 17: Immunization Requirements Prior to School Entry—September 1976

State	Type of Legislation			If no state law at present, are there plans pending for proposing such a law?		If Yes, before September 1977?	Immunizations Required for Specific Diseases							
	Mandatory	Permissive	None	Yes	No		Diphtheria	Pertussis	Tetanus	Measles	Mumps	Polio	Rubella	Smallpox
Alabama	X						X	X	X	X		X	X	
Alaska		X					X	X	X	X		X		
Arizona (2)			X(2)		X									
Arkansas	X						X	X	X	X		X	X	
California	X						X	X	X	X		X		
Colorado	X						X	X	X	X		X	X	
Connecticut (1)	X	X								X		X	X	
Delaware	X						X	X	X	X		X	X	
District of Columbia	X						X	X	X	X		X	X	
Florida	X						X	X	X	X		X	X	
Georgia	X						X	X	X	X		X	X	
Hawaii	X						X	X	X	X		X	X	
Idaho			X		X									
Illinois	X						X	X	X	X		X	X	
Indiana	X						X	X	X	X		X	X	
Iowa			X	X										
Kansas	X						X	X	X	X		X	X	
Kentucky	X						X		X	X		X		
Louisiana	X						X	X	X	X		X		
Maine		X					X	X	X	X	X	X	X	X
Maryland	X						X	X	X	X		X	X	
Massachusetts	X						X	X	X	X		X		
Michigan	X						X	X	X	X		X	X	
Minnesota	X									X			X	
Mississippi		X					X	X	X	X		X	X	
Missouri	X						X			X		X	X	
Montana		X					X	X	X	X		X	X	
Nebraska	X						X	X	X	X		X	X	
Nevada	X						X	X	X	X		X	X	
New Hampshire	X						X	X	X	X		X	X	

(continued)

Table 17: (continued)

State	Type of Legislation: Mandatory	Permissive	None	If no state law at present, are there plans pending for proposing such a law?: Yes	No	If Yes, before September 1977?	Immunizations Required for Specific Diseases: Diphtheria	Pertussis	Tetanus	Measles	Mumps	Polio	Rubella	Smallpox
New Jersey	X						X	X	X	X		X	X	
New Mexico	X						X	X	X	X		X	X	
New York	X						X			X		X	X	
North Carolina	X						X	X	X	X		X		
North Dakota	X						X	X	X	X	X	X	X	
Ohio	X						X	X	X	X		X	X	
Oklahoma	X						X	X	X	X		X	X	
Oregon	X						X	X	X	X		X		
Pennsylvania	X						X		X	X		X	X	
Rhode Island	X						X		X	X		X	X	
South Carolina	X						X	X	X	X		X	X	
South Dakota	X						X		X	X		X	X	
Tennessee	X						X	X	X	X		X	X	
Texas	X						X		X	X		X	X	
Utah	X						X	X	X	X		X	X	
Vermont (3)			X(3)		X									
Virginia	X						X	X	X	X		X	X	
Washington (2)			X(2)	X		Yes								
West Virginia	X						X	X	X	X		X	X	
Wisconsin	X						X	X	X	X		X	X	
Wyoming			X		X									
Guam (2)			X(2)		X									
Puerto Rico	X													X
Virgin Islands	X						X	X	X	X		X	X	X
Total	47	(1)	7	2	5	1	44	37	42	46	2	45	39	3

(1) Connecticut: rubella and measles = mandatory; polio = permissive.
(2) Arizona, Washington and Guam have adopted regulations, with the effect of law, establishing school entry immunization requirements.
(3) Vermont has local option regulations at the school district level, establishing school entry immunization requirements.

Source: Information received from immunization projects.

published in September 1979. Later in this report we shall describe the declining interest of U.S. pharmaceutical companies in the vaccine field and discuss a few of the policy issues that are of greatest relevance to the exploitation of molecular genetics in this realm. Three more viral diseases deserve at least brief comment.

4.2144 Rabies

Rabies was the first human disease for which a systematically developed vaccine was successful, but the vaccines made by killing virus grown in animal brain tissue or duck embryonic tissue are far from convenient or free from adverse effects. Recently approved, is a vaccine based on a virus grown in human embryonic cells, which permits reduction of the number of immunizations from 21 to six after exposure to probable infection. We understand that the new vaccine is very difficult to make - because one cannot build up virus concentration in the extra-cellular fluid - and is subject to substantial fluctuation in antigenic activity from lot to lot. As a result, the cost per dose is in the $50 range, compared to $3/dose for the duck embryo vaccine. When one takes into account all the travel and contact with medical professionals, that does not greatly increase the cost of postexposure treatment, but it would make for very costly prophlyaxis of persons who are frequently placed at risk. (Three or four injections over 6 - 7 months to attain protective antibody production.) Apparently the immunogen is a glycoprotein (Cabasso, 1980), so that as well as the mRNA for the peptide sequence, it may be necessary to have a satisfactory glycosylation system in order to use molecular cloning to synthesize antigen. The fact that there is a substantial market ($10 million in the USA) for rabies vaccine for domestic animals, in addition to the limited demand for vaccines suitable for humans, could make this a somewhat more attractice market than Lilly's one-quarter million dollar annual sales to hospitals and pharmacies would make it appear. Of course, today the production and marketing of animal and human vaccines are totally separated, but that need not be the case were the synthesis of pure antigen the key step.

4.2145 Foot and Mouth Disease

Approximately 800 million doses of foot and mouth disease (FMD) vaccine are used annually worldwide - making it the largest volume vaccine produced. It has to be given frequently to livestock in areas where the disease is endemic, and that includes most of the world outside North America. As described by the Director of the USDA Animal Disease Center at Plum Island, the present methods of immunogen production in bovine tongue epithelium and tissue culture involve containment of enormous quantities of hazardous virus. Dr. Callis says that many outbreaks today result from incompletely inactivated vaccine or escape of the virus from factories. He speculated, in remarks to the Recombinant-DNA Advisory Committee (RAC) that molecular cloning of the antigen could produce vaccine at considerably less expense as a stable product, without risk of the virus escaping factories. On the basis of that potential a joint program between USDA and Genentech has been approved by the RAC to clone segmental transcripts of the FMD picornovirus (RNA) genome with the objective of producing the pure antigen. [RAC, December 1979] This was a landmark decision for it represented the first exception to the NIH prohibition against cloning DNA derived from a Class 5 pathogen. In our estimation, this is the vaccine development based upon molecular cloning likely to first come to commercial fruition - not for use in the USA. However, since the project is just beginning, that hardly could be achieved before 1985. The availability of high purity antigen would facilitate schemes to prolong the interval between immunizations through the use of a very long-acting dosage form based on polymeric drug delivery systems. Certainly there is reason to hope that the world protein supply in the 1990s will be improved as a result of the relief from this wasting disease through the availability of less expensive FMD vaccines that make it easier to achieve satisfactory compliance to immunization requirements.

4.2146 Hepatitis

Viral hepatitis represents that category of infectious diseases which remain unconquered, in part, because the causative agent has resisted attempts to cultivate it in the laboratory. Some 300,000 Americans develop hepatitis each year, with hepatitis B estimated to account for 80-100,000 cases and about a thousand deaths. The incidence in other parts of the world runs ten times as high. Between 3% and 15% of healthy blood donors in Western Europe and the USA show serological evidence of past infection, and 0.1% are chronic carriers of type B virus, but in many African and Asian countries the majority of the adult population have been infected and 5-10% of the population are clinically infected. Recently, very strong association has been demonstrated between the carrier state of hepatitis B and liver cancer. In areas of the world where hepatitis B is endemic, primary liver tumors account for 20% of cancer - in contrast to the 1% level of liver tumor incidence in the USA. [Smuzness, 1978]

The first major clinical trial of a vaccine against the hepatitis B virus took place during 1980-81. The immunogen was obtained from blood donated by chronic carriers of the virus, or rather of a virus-like particle called the hepatitis B surface antigen (HB_sAg), which was first identified in 1964. Production of the vaccine requires 65 weeks - the longest production and testing cycle of any vaccine now manufactured. The process developed by Merck & Co. involves many purification steps, including digestion with enzymes and treatment with urea to remove human liver and blood cell contaminants, and both thermal and chemical sterilization procedures. A six-month safety test in chimpanzees was employed to ensure safety of the vaccine. Three 40 microgram doses were administered over a six month period to achieve immunization.

The results of the initial clinical trial, conducted among 1083 homosexual males in New York City, are quite encouraging, but additional trials by other investigators will be required before FDA approval can be obtained. In the double-blind study, less than 3.4% of those who received the vaccine developed hepatitis, compared to 27% of those who received placebo injections. Dialysis patients and workers in dialysis centers, who are exposed to patients' blood, are at high risk of hepatitis, and they are the subjects of additional trials now underway. Children in Taiwan, where 15% of the population is infected, are the subjects in another test of the vaccine. If the results continue favorable, the vaccine could be available commercially in 1982. Indicative of the differences between regulatory agencies around the world, is the fact that a single trial, conducted among dialysis patients at the Pasteur Institute in Paris, can serve as the basis for registration of a similar vaccine developed by Phillipe Maupas of the Institute of Virology in Tours, for use in France. [Ibid.]

The derivation of vaccine from the blood of chronic carriers is far from altogether satisfactory. Although it remains to be seen if a simple protein would be sufficient to provide protection, a number of groups are racing to achieve production of hepatitis B virus proteins in bacteria through molecular cloning. A team of scientists at the University of Edinburgh and the Microbiological Research Establishment at Parton Down in the UK, working in conjunction with Walter Gilbert, a Harvard professor and founder of Biogen Corporation, incorporated fragments of DNA from viral particles into an E. coli plasmid and achieved expression of some viral antigens in early 1979. [Burrell, 1979]

In 1980, a group at the University of California, San Francisco, induced bacteria to produce substantial quantities of HB_sAg fused to a bacterial protein, which is secreted by the microbe. Dr. Rutter, the leader of that team predicted that antigen produced by fermentation could be used for a commercial vaccine by 1985. [Maugh, 1980] Merck is believed to be supporting the work at UCSF. Several other groups are investigating alternative molecular cloning approaches to hepatitis, employing phage and bacterial plasmid vectors.

The market is unsatisfied, and the race is on to provide AN answer - the first vaccine need not be THE answer. Merck, and possibly Cutter or others in the blood-processing business, will probably market a purified HB_sAg vaccine, but if they are successful it will become progressively harder to find donors to provide the raw material. According to Dr. Hilleman of Merck an estimated 16 million U.S. citizens have a relatively high risk of contracting hepatitis B. Others have estimated that dialysis patients and homosexuals have a yearly attack rate of 12% and that dialysis unit staff members have a 5% annual rate of incidence - these are the high risk populations among whom the clinical studies are being conducted. The cost and safety of the vaccines available will determine where between the latter quite limited population and the much larger potential, the actual level of demand will lie.

Hepatitis A virus resisted attempts at cultivation until 1979, when the Merck group reported success in growing it on cultured cells. Until then, the virus could only be propagated in rare South American marmoset monkeys. Cultivation is still a long way from a vaccine, however, and we believe that the first vaccine against hepatitis A to go into clinical testing may be produced by molecular cloning.

4.215 Parasitic Disease Vaccines

Many of the most devastating infectious diseases remaining unconquered involve complex parasites that have resisted efforts to grow them under laboratory conditions. Only in 1976 was the first cultivation reported of the protozoan which causes the most malignant form of malaria. Workers at the Rockefeller University developed conditions for the propagation of the merozoite phase of *Plasmodium falciparum* in human red blood cells and produced a vaccine suitable for trials in monkeys, but they readily admit they are a long way from having sufficiently pure malarial antigens for clinical trials of a vaccine in humans. Monoclonal antibodies, a related development which shall be discussed shortly, may facilitate this process considerably. Ruth and Victor Nussenzweig, of the New York School of Medicine, reported in 1980 the isolation of a surface antigen of the infectious, sporozoite, stage of *P. falciparum*, and its use to produce an antibody that, on the one hand, could protect against infection and, on the other hand could be used to acummulate a relatively large amount of the antigen. This technique should enable the programming of a bacterium to produce the antigen in quantity if it is a protein. Biogen is known to be endeavoring to synthesize malaria antigens by molecular cloning.

Limited scale efforts are underway with respect to such scourges as: schistosomiasis (bilharzia), filariasis (onchocerciasis and elephantiasis), leshmaniasis, hookworm, and amebic infections. Success with veterinary vaccines against the animal disease analgues of some of these parasites, lends credence to the hopes, although the adverse effects now tolerated for animal treatment clearly must be overcome. Diminution of side effects (by elimination of contaminants) is exactly what molecular cloning offers to these efforts, along with the potential for convenient mass production of the antigen when and if the specific immunogen proves satisfactory. Unfortunately, "over millions of years of evolution, parasites have learnt a variety of tricks to foil the immune system of their hosts...; the most successful parasites are those that evoke a partial immune response, (which) fails to get rid of the parasite altogether." [Nossal, 1978] To be successful, the vaccine designers will have to identify an Achilles' heel, some phase more readily amenable to immune destruction than an established parasite burden, and that will probably require insight into the molecular biology of the initial stages of infection. Molecular clones from the mRNA produced by the parasite at that stage seems likely to be the key. "While the problem will vary from disease to disease, it should not be impossible to define antigens crucially important for some of these steps, and to build a vaccine up around well understood principles." [Ibid]

4.22 Antigens for Other Purposes

4.221 Cancer Antigens

Other roles of the immune system in the battle against cancer will be discussed further hereafter, but it is appropriate to mention here studies which indicate that treatment with vaccines prepared from purified human lung cancer antigens appears to substantially prolong survival in Stage I lung cancer patients. The vaccines, prepared by Ariel Hollinshead at George Washington University Medical Center, were made from squamous cell carcinoma antigens or adenocarcinoma antigens.

4.222 Autoimmune Antibody Scavengers

The Salk Institute is engaged in very limited, exploratory

clinical trials, on the basis of promising results in 11 patients, in which a porcine myelin protein prepared by Eli Lilly is injected into multiple sclerosis patients to mop up the antimyelin antibodies which those patients are producing. Treatments involving 15-42 grams of myelin being injected without adverse effects, suggest an emerging potential therapeutic approach to such auto-immune diseases. The protein supresses the symptoms of experimental allergic encephalomyelitis, an animal disease resembling multiple sclerosis. Should this approach continue to be attractive, the utilization of molecular clones to produce the human protein antigen seems inevitable.

4.223 Contraceptive Antigens

The use of proteins associated with various stages of pregnancy as antigens for a sort of contraceptive "vaccine" is being investigated in a number of laboratories. For example, antibodies have been shown to inhibit the penetration of the coating which surrounds the ovum (the zona pellucida) prior to implantation. Immunization with a peptide sequence unique to a protein of the human zona pellucida might involve the production of the antigen through molecular cloning. Other scarce proteins involved in reproduction that are being considered as targets against which it might be desireable to provoke antibody production are hormones, such as chorionic gonadotropin, and antigens associated with sperm. The latter are being investigated for male as well as female "contraceptive vaccines". The extended period of testing required for such a product can scarcely be imagined. Hence, there is little interest in the private sector in undertaking such development, except insofar as it may be funded by public sector agencies.

4.23 Interferons *

This scarce family of glycoproteins has rececently enjoyed a great deal of popular attention, [e.g., Toufexis, 1980]. Viruses and certain other stimuli induce the production and secretion of relatively low molecular weight (~20,000) glycopeptides by vertebrate cells. In turn, when other vertebrate cells are exposed to very small amounts of these interferons they are profoundly modified in anumber of ways. Induction of an antiviral state in susceptible cells is the definitive capability of all interferons. However, at least fifteen other activities are recognized, most of which involve other elements of the immune system.

Unlike the complementarity of the antibody response to an antigen, interferon responses are generally nonspecific, i.e., interferon induced by one kind of virus impedes infection by that and other kinds of viruses. Production of Type I or "classical" interferon is the initial response to viral infection, and it appears to play a causal role in normal recovery from viral diseases. Interaction with interferon from the appropriate species, probably through cell-surface receptors, results in the establishment of an intracellular antiviral state (for which protein synthesis is known to be required, although the antiviral protein has yet to be identified), that apparently interferes with the production or translation of some mRNA necessary to produce the virus.

The antiviral state diminishes rapidly when interferon is removed and responds proportionately to interferon level with respect to both the rate of establishment and degree of antiviral protection. Hence, consideration is given to, and there are promising preliminary studies supporting, the use

* See note page 257.

of interferon in the treatment of viral diseases such as rabies, hepatitis, varicella-zoster and various herpes infections. To date the effect of interferon has been far more impressive as a prophylactic, rather than therapeutic, agent and for prevention or treatment of viral infections in cancer patients - whose immune responses are compromised by both the tumor and the steps taken to counteract it.

A number of cellular alterations, in addition to the antiviral state, can be induced by interferons. In particular these include regulation of interferon synthesis, inhibition of cell growth and modulation of the immune system. One of the most dramatic of the latter effects involves promoting the expression of the capacity of certain lymphoid cells, referred to as "natural killer" (NK) cells, to selectively destroy cancer cells. This and other activities probably account for the favorable preliminary results that have been obtained in experimental cancer therapy involving interferons.

There are at least two kinds of human Type I or "classical" interferons. These are termed leukocyte interferon and fibroblast interferon, for the type of cell from which they are presently obtained. A third, called lymphoblastoid because it is produced from cells derived from a Burkitt's lymphoma, appears to be a mixture of the former two interferons. All produce the antiviral state, and are induced by viruses. Another kind, known as Type II or "immune" interferon, is produced by lymphocytes in response to certain antigens and, especially, agents that stimulate cell division (mitogens). Some evidence indicates that Type II interferon may be a more potent antitumor agent than Type I, but relatively little is known about the Type II variety.

Activity of interferon preparations is expressed in terms of an *in vitro* assay in which 1,000 units is the amount of interferon that will protect 500 of 1,000 cells from a viral challenge. The specific activity of the purest interferon preparations are estimated at 10^8 to 10^9 units per milligram. The number of units being given in current cancer trials is millions per day, which corresponds to about a microgram per day - astonishing potency! Nevertheless, only infinitesimal quantities of any interferon have ever been available. Most of that tested to date has been leukocyte interferon prepared under the direction of a Finnish virologist, Kari Cantell, in conjunction with his country's Red Cross.

Cantell's process, which utilizes the "buffy-coat" leukocytes from the blood bank in Helsinki that serves all of Finland, has been scaled up to produce about 10^{11} units/yr. - with specific activity of 10^6-10^7/mg protein (i.e., 0.01-1% pure interferon). But that is not enough for all the clinical trials that have been proposed, along with the basic R&D including purification for structure determination which accounts for substantial (>90%) losses.

Fibroblast interferon, which is produced by culture of cells from neonatal foreskin, is scarcer still - even though a phenomenon called superinduction has boosted yields 100 times former levels. New techniques, such as that in which fibroblast cells are attached to high surface area beads rather than to the walls of a bottle (developed by MIT and Flow General), may improve availability. Recently G. D. Searle announced that new technology, developed at its R&D facility in England, has increased the yield of fibroblast interferon by a factor of 60 through the use of a stacked plate tissue culture system. On the basis of this process,

Searle expects to supply material for the first large-scale clinical trial of fibroblast interferon. Abbott Laboratories also recently announced plans to resume production of limited quantities of fibroblast interferon, for clinical studies it plans to sponsor.

Unlike leukocytes and superinduced fibroblasts, which can be used only once, lymphoblasts derived from the tumor Burkitt's lymphoma grow freely in suspension and produce the least costly interferon presently obtainable. However, it is a mixture containing both leukocyte and fibroblast interferons in a ratio of five or six to one. Burroughs-Welcome produces lymphoblastoid interferon in 1,000- liter fermenters and has clinical trials underway in England, but the U.S. FDA has generally resisted efforts to make use of products derived from malignant cells. It is used extensively in research, and FDA is considering evidence from Burroughs-Welcome that may lead to a relaxation of the prohibition, under pressure from NCI.

Recombinant-DNA is widely regarded as the key to mass production of interferons and important initial steps have already been accomplished. Under sponsorship of Biogen, S.A., a group directed by Charles Weissman at the University of Zurich cloned cDNA from leukocytes (provided by Cantell) in E.coli plasmids.*

Michael Revel and colleagues at the Weizmann Institute and Pasteur Institute, along with Tadatsugu Taniguchi and his collaborators, reportedly have accomplished the same feat for fibroblast interferon. Cetus Corporation has established a collaboration with Yeda Research and Development to

* Schering-Plough, a stockholder in Biogen, has been granted a license from Biogen for worldwide marketing rights to the fruits of this work.

improve the yields and scale up the process based on the clones developed at the Weizmann Institute. Other labs at or close to the point of having clones for the peptide sequence of one or another interferon include such major firms as Roche and DuPont, and such collaborations as that between Genentech and the Wadley Institute or between Taniguchi and Ptashne at Harvard.

The importance of the carbohydrate side-chains on interferon is not certain. In the presence of glycosylation inhibitors, leukocytes induced with virus produce a so-called "interferoid" that contains no carbohydrate. It possesses nearly normal interferon activity in several assays and behaves like native interferon with respect to hydrophobicity. Unlike native leukocyte interferon, which contains both 15,000 and 21,000 mol.wt. entities, the sole interferoid species has a molecular weight of 15,000. Whether the carbohydrate is critical to function *in vivo* has not been shown. Fibroblast interferon, (20,000 mol.wt.), certainly does differ pharmacologically from the leukocyte variety, i.e., it is cleared much more rapidly from the circulation and is ineffective when injected intramuscularly.

The true significance of these differences remain to be seen. In fact, there remains an enormous amount to be learned with respect to the interferon system from many perspectives, and, furthermore, it appears the interferons should be viewed as simply one of many families of molecules that are involved in physiological regulation or response to disease, which only now have molecular biology and genetics made it feasible to study and, perhaps, to utilize.

Not only is there hope that genetic engineering will make it possible to produce interferons in quantity, but tools made possible by modern genetics are being applied throughout the effort to comprehend the interferon system. For example, hybrid cell lines created by fusion of human cells with mouse cells are illuminating how the genetic information with respect to interferon production is organized in human cells and called into play by stimuli [e.g., Meager, 1979]. Antibodies with unique specificity and high affinity are becoming indispensible for identification and purification. They will be essential in one of the efforts that seems certain to become very active as soon as sufficient quantities of interferon are available - namely, the dissection of the apparent plenipotential nature of an interferon molecule. There may be different sites for different activities, such as induction of the antiviral state, arousal of NK cells, and modulation of the immune response through the HLA complex.

All aspects of molecular biology may be called upon in programs designed to lead to derivatives or analogues of interferons that will emphasize those particular activities appropriate to a pharmacologic purpose.

The interferons are enjoying great attention right now largely because studies in Sweden, and to a limited extent in the USA, stimulated appropriation of $5.4 million by the American Cancer Society (ACS) for expanded clinical trials. That commitment by the nonprofit ACS - the greatest by far in ACS history - was followed by a boost in NIH funding for interferon research from $7.7 to 19.9 million for fiscal 1980. Much of the expense in both the ACS and NIH plans is for interferon procurement. Initially, ACS bought 40 billion units of leucocyte interferon from the Finnish Red Cross for $50/million units. In March 1980, Warner-Lambert was awarded a contract to supply the National Cancer Institute with 50 billion units of leucocyte interferon within the next two years at an average price of $18/million units - based on a variation of the Cantell process developed in conjunction with Sloan- Kettering researchers. NCI is also planning to purchase 50 billion units each of fibroblast and lymphoblastoid interferons.

The bulk of the NIH funding is included in the NCI's new Biological Response Modifier (BRM) program - interferon accounts for $13.9 million of the $34.1 million allocated for BRM work in fiscal 1980. (NCI expenditures on interferon in 1979 were $2.6 million: 19% of the amount budgeted for 1980.) Other major elements of that BRM program concern other immunoproteins known as lymphokines and thymic hormones for which molecular genetics has major implications.

4.24 Lymphokines & Thymic Hormones

Lymphokines and thymic hormones are categories of regulatory molecules which have begun to emerge from the obscure fringes of immunology in the past ten years. Interferon is generally considered as a lymphokine that has been characterized sufficiently to deserve independent status.*

Lymphokines are biologically-active soluble factors (other than immunoglobulins) elaborated by white blood cells. Studied in depth only within the last 15 years, these molecules are being implicated in virtually every stage in the complex series of events making up the immune response.

Lymphokine-dependent reactions, as well as those involving the direct cytotoxic effects of lymphocytes, are collec-

* There are some researchers who argue that interferon can be viewed as a hormone, citing evidence of common pathways of action between hormones and interferon. [Blalock & Stanton, 1980]

tively known as cell-mediated immunity. Both T- and B-lymphocytes can be induced to produce lymphokines, when stimulated with appropriate agents. They are not present, in subtantial amounts, in resting lymphocytes, but are normally synthesized and released as the result of stimulation by agents interacting with membrane receptors. Though often induced by antigens, the vast majority of lymphokines do not have specific antigen-binding sites, do not exhibit specificity for the inducing antigen, and do not require antigen for performing their function.* Even when cell activation is induced by stimuli other than antigens, treatment of the cell donors with immunological adjuvants often enhances lymphokine production.

Lymphokines affect a wide variety of cells, including lymphocytes; macrophages; neutrophil, basophil, and eosinophil polymorphonuclear leukocytes; thymocytes and bone marrow cells; fibroblasts; lymphoid cell lines; osteoclasts; and endothelial cells.

Lymphokines are believed to attract macrophages and other cells to sites where helper T cells are interacting with antigen, to activate these macrophages to a phagocytic stage of differentiation, and to prevent their departure from the site. [Hood, 1978] Some lymphokines apparently also affect cell-free substances such as immunoglobulin, complement components, and clotting factors. Although there is little doubt as to the *in vivo* significance of the lymphokines, very little is known about their precise chemical nature and their properties have not been explained in terms of underlying biochemical events. Even

---

* A number of factors that are involved in the afferent arm of the immune response (cell cooperation and suppressor factors) do show antigen specificity, and some investigators would exclude such antigen-specific factors from the lymphokine category, as a result.

less is known about regulatory mechanisms in lymphokine production and activity. About 100 named factors are now encompassed by the term lympnokine. [Cohen, Pick & Oppenneim, 1979]

All lymphokines appear to be proteins. Most, if not all, are glycoprotein of a molecular weight larger than 10,000. Unlike immunoglobulins, steroid hormones or prostaglandins, lymphokines do not demonstrate a unique molecular structure or shared amino acid sequences. As far as is known, they do not possess similar sugar residues. For the most part, they are not known to be enzymes (although polymorphonuclear leukocyte migration inhibitory factor (MIF) may act as an esterase).

Most are thought to react with specific membrane receptors on their target cells and to exert their intracellular influence through the mediation of cyclic nucleotides. Certain low molecular weight compounds have been shown to interact with some of these membrane receptors and to cause effects indistinguishable from those of the lymphokines.

The BRM subcommittee cites three broad reasons lymphokines may prove important in the cancer treatment program:

- Some lymphokines are directly cytocidal or affect proliferation of tumor cells, e.g., Tumor Necrosis Factor (TNF), a glycoprotein apparently produced by macrophages, which causes acute hemorrhagic necrosis of certain tumors. Definitive assessment in the clinic awaits a more plentiful source.
- Purified lymphokines could direct host tumor responses via specific immune responses, via selectively activating macrophage or other tumoricidal cells, and via altering host inflammatory processes.
- Lymphokines affect the important "non-tumor" systems which are critical to total therapy. Molecules under study have important effects on marrow proliferation, on ossification in and around tumors, on vessel proliferation,

> and on healing of normal tissues. [Subcommittee on BRM Interim Report, Appendix 2, pp. 109, 118, 1979].

Nevertheless the BRM subcommittee concludes "that some of these agents probably have great potential (for cancer treatment), but no agent is ready for clinical development (except interferons, TNF and, possibly, transfer factor *). [Ibid., p. 110] Adequate quantities for laboratory and clinical testing, and use perhaps, will probably not be attainable until the problem of producing glycoproteins by molecular cloning is overcome. When that possibility is fulfilled, a host of studies will be set off that seem likely to lead to a few more compounds with the sort of potential that today is assigned to interferons.

While lymphokines are, by definition, the products of lymphocytes, cytokine has been suggested as the term that would encompass all soluble substances produced by lymphoid or nonlymphoid cells that are characterized by their ability to cause the same effects on target cells as classical lymphokines. Cytokine would then include the several compounds associated with the thymus, which are now being referred to as thymic hormones. Table 18 indicates the cells which give rise to some forty of the best characterized cytokines in the alphabet soup of extracts that comprise cellular immunology today.

The thymus gland processes bone marrow stem cells into "T cells," those lymphocytes primarily responsible for cellular immunity, and it produces a number of proteins which influence T cells elsewhere in the body. Several laboratories have succeeded in isolating and purifying factors from thymus tissue and from blood that are considered thymic hormones.

---

* Transfer factor is a leukocyte extract containing many components.

Table 18 Cellular Sources of Lymphokines and Cytokines

Cells	Examples
Thymus Epithelium	Thymopoietin, Thymosin, TF, THF
Macrophages	GRF (antigen-specific)
	CSF, CMRF, MP, TDF, LAF, BAF, Transferrin EP, CTF, ACT, AIM, SMC, MCF, C, enzymes
T_1	IDS, other
T_2	Antigen-specific helper (IgG&IgE),AEF, SMAF and suppressor (IgG, IgE, contact &GAT) factors, MLR suppressor
	Non-specific MF, NSF, TRF
B	Antibodies (antigen-specific)
T- and B-cells	MCF, LF, MIF, LIF, OAF
Macrophages, T-& B-cells	LT(MIF), Interferon
Kidney cells & Fibroblasts	MIF, MCF, NCF

From Cohen, 1979, p. 595.

Many appear chemically distinct and act selectively on the development of immune functions. Among the preparations, at various stages of characterization, thymosin (fraction 5), thymic humoral factor (THF), thymopoietin II (TP), thymic factor X (TFX) and factor thymic serique (FTS) are the most thoroughly studied.

In connection with NCI's BRM program, there is enthusiasm about the potential usefulness of such compounds to specifically increase host resistance of cancer patients and reverse the immunosuppressive effects of chemotherapy, radiotherapy, surgery, and of the tumor itself. Phase I toxicity studies and a limited number of Phase II trials in patients with far advanced malignancies have been carried out over the past five years with thymosin, TFX and THF. In general these studies showed only that the thymic factors are relatively nontoxic and they induce increased T cell numbers and function.

Calf thymosin (fraction 5) has been found to contain many active peptides with molecular weights ranging from 1,000 to 15,000, several of which have been purified to homogeneity and sequenced. Some of the thymosins promote early maturation of bone marrow stem cells into T cells; others help the stem cells to mature into T cells later, and some apparently direct the transformation of stem cells into specialized T cells, e.g., killer, helper or suppressor. Thymosin fraction 5 has been the subject of more clinical trials than any other thymic hormone, and it is proving (over 6 years, ~100 children) to be of significant benefit for children with primary immunodeficiency diseases.

One randomized Phase II study in which thymosin fraction 5 was given in conjunction with intensive chemotherapy for oat cell (small cell) lung carcinoma, has recently been reported to indicate that thymosin significantly prolongs survivial in patients who achieved eradication of all detectable tumors by chemotherapy. Thymosin had no detectable direct antitumor effect and the patients who benefited most from it were those with relatively low immune reactivity. The BRM subcommittee says this, "is the first positive controlled trial involving a thymic hormone preparation, or for that matter any BRM, in oat cell carcinoma. If this study can be confirmed, it will be a major step forward in the treatment of this disease." [Ibid., Appendix 3, p. 2]

One of the more than 30 polypeptides in thymosin fraction 5, thymosin-α_1, which has activities associated with helper and amplifier functions was characterized and its 28AA sequence was synthesized, first by solution chemistry and then by molecular cloning, within the past two years.

Hoffman La Roche synthesized thymosin-α_1 and provided it to NIH investigators for a small Phase I clinical trial in patients with brain tumors during 1979, and other trials are underway in several centers.

Genentech achieved the molecular cloning of thymosinα_1 in August, 1979 - although they did not disclose it until March of the following year, when they applied to the RAC for permission to scale up beyond 10 liters. That project was completely funded and carried by Genentech, except for the biological assay work on the final product which was done at George Washington University under the direction of Dr. Allan L. Goldstein, who originally discovered the existence of thymosin. The approach employed was the same as that first used in the microbial production of human insulin, i.e., the "gene" was synthesized on the basis of the known AA sequence. Thymosin-α_1 was produced fused to β-galactosidase and liberated by treatment with cyanogen bromide. Approximately one mg/l (100,000 molecules/cell) was synthesized, of which roughly 50% was isolated with a purity greater than 95%. [Wetzel, 1980]

Thymosin-α_1 has been shown to have a 100-fold greater effect than thymosin factor 5 on some immune reactivities associated with T-cell helper functions; it can be obtained in 0.6% yield from fraction 5 of calf thymus.

Thymosin-α_1 is but one of about 30 peptides in this extract, and hypothetical roles in T-cell maturation have been assigned to a number of them (Figure 6). Thymosin-α_1 is the only thymosin that has been sequenced to date. It shows no homology with the AA sequence of the other thymic factors that have been sequenced (the thymopoietins and facteur thymique serique) and little with any of 957 protein sequences that have been published to date. Thymosin-α_1 from humans, pig, sheep, chinchilla and mouse all appear to be identical to the molecule obtained from the calf.

Figure 6: Proposed Role of Thymosin Peptides in T-Cell Maturation

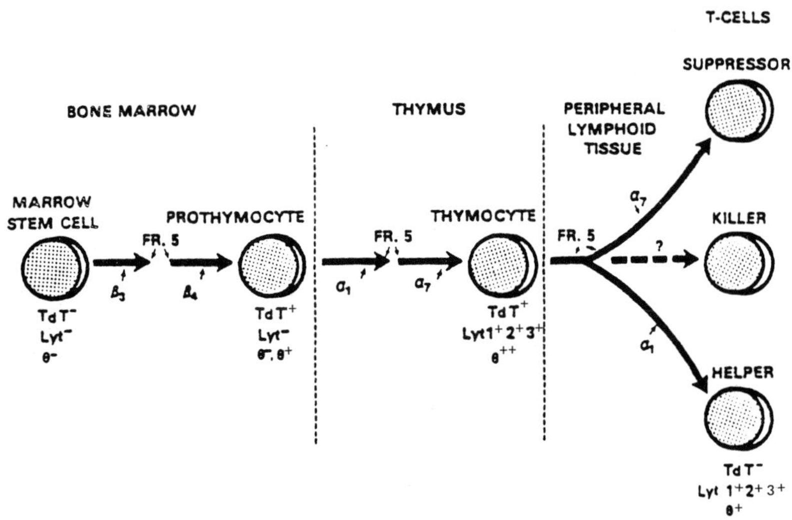

On the basis of the observations that the immune system loses its potency as a person or animal ages, the thymus atrophies, and blood levels of thymosin decline with aging (Figure 7), it has been suggested that exogenous thymic hormone might counteract the hypothetical deficiency. Dr. Allan Goldstein says that a major clinical trial is going to be conducted at George Washington University School of Medicine to determine the effect of thymosin on the immune system of older people.

An encouraging Phase I trial has provided the background for a Phase II randomized trial by Dr. J. C. Daniels to determine the efficacy of thymosin-α_1 therapy in systemic lupus erythematosus (SLE), an autoimmune disease.

Figure 7: Thymus Dependent Senescence of Immunity in Man

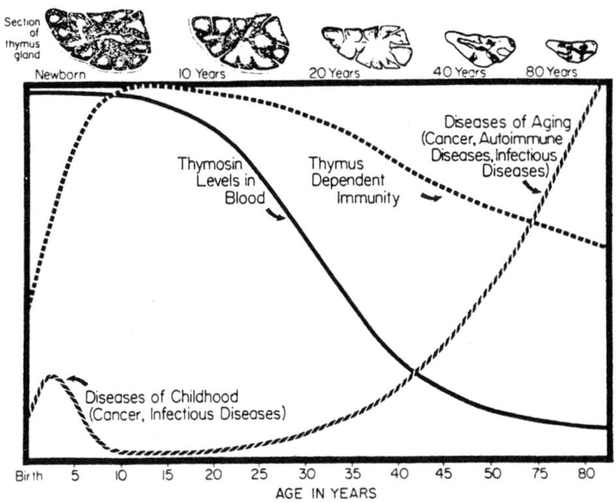

Relatively little is known about the other thymic factors, but the NCI's BRM program and widespread academic and, to a lesser extent, industrial interest should rectify that situation before long. Nathan Trainin, who proposed an endocrine function for the thymus in the early 1960s, and his colleagues at the Weizmann Institute tend to concentrate on thymic hormone factor (THF), the predominant activity of which is to increase the effectiveness of helper and killer T cells after they are made. THF has been given to several dozen children suffering from viral infections as a result of taking immunosuppressant drugs for cancer, with quite favorable results. Now THF is being given to ten youngsters with acute sclerosing panencephalitis and juvenile arthritis (which appear to have an autoimmune basis), but beyond restoring T-cell function, the results are not in yet. [Arehart-Treichel, 1980]

Thymic Factor X (TFX), which is the favorite of Aleksandrowicz and his colleagues in Krakow, is the outcome of a line of research in which thymus taken from patients with myasthenia gravis was transplanted into cancer patients. TFX differs from other entities, other than transfer factor, in that it restores the delayed hypersensitivity mechanism. TFX is a simple polypeptide with a mol.wt. of 4,200 and AA composition quite dissimilar from the other putative thymic hormones.

Facteur thymique serique (FTS) was ultrafiltered from pig serum in the laboratory of Jean-Francois Bach of Hospital Necker in Paris. A nine AA peptide, it has been synthesized and tentatively identified in thymosin fraction 5. Among other biological activities in common with other thymic hormones, FTS has the ominous properties of increasing anti-DNA antibody production* and increasing glomerulonephritis in B/W mice, under conditions that suggest a decrease in suppressor cell numbers or function. On the other hand, moderate doses of FTS improved autoimmune conditions in mice. Many analogs of FTS have been synthesized to try to sort out the activities of this simple peptide.

A primary activity of both FTS and the thymopoietins concerns the differentiation in the thymus of stem cells into specific kinds of T cells.

The authority on thymopoietins is Gideon Goldstein, formerly of NYU. He is working at Ortho Pharmaceuticals in conjunction with the synthetic chemists of Dr. Paul Janssen on two natural 49 AA polypeptides and their derivatives. They have found that a fragment as small as five AA in length (residues 32-36 of thymopoietin) mimics the intact molecule in many ways. Like the thymopoietins, TP-5 induces the

---

* Antinuclear antibodies are thought to be involved in several autoimmune diseases.

differentiation of stem cells (prothymocytes) *in vitro*, and inhibits differentiative induction of B cells *in vitro*. It reverses certain immune deficiencies of aging mice *in vivo*, and in at least one case of a human infant with DiGeorge syndrome (absence of thymus and T cells). [Lancet *1*, 91 (1980)] In human plasma TP-5 has a half-life of only about 30 seconds - it is hydrolyzed - nevertheless, biological effects can be detected 24 hours after a single intravenous injection. Thus, one can see that it may not be necessary to use large natural peptides to achieve the potential of thymic hormones.

Another small compound in this category was found in human serum by Alberto Astaldi of the Netherlands Red Cross. Human serum factor (SF) is a four AA peptide. It specifically increases the intracellular cyclic AMP level of thymocytes; no effect is observed on normal peripheral blood lymphocytes, or other lymphoid or nonlymphoid cells. Furthermore, it turns out that SF acts only on hydrocortisone-sensitive thymocytes and it induces their transformation to hydrocortisone-resistant cells. SF apparently acts on T cells during the latter stage of maturation - inducing, among other changes, the expression of an Fc receptor for IgM but not for IgG. Astaldi has discovered that when thymosin is administered to a patient, SF appears in the blood and the treatment results in the disappearance from the blood of lymphocytes that will respond to SF *in vitro*. This suggests a mediator role for SF.

Hybrid cell lines have been established that secrete "suppressor factor" as a result of fusing cells of a thymus tumor with T lymphocytes artificially induced to make the factor. Thus somatic cells, as well as bacterial factories and chemical synthesis may be exploited to obtain adequate quantities of immune modulators.

The preceding account appropriately depicts the murkiness of the waters in this area of immunology right now. Yet there is little doubt there are big fish in this pond; they have begun to nibble, and molecular genetics may provide the right bait. When they will strike, how big they will be, and how many will be caught is anybody's guess, but this is the water that the interferons came from and they look like record catches.

The immune modulators may be the key to understanding, and possibly treating, the autoimmune components of common but uncured diseases like rheumatoid arthritis (which affects 3.2% of adults aged 18 through 79, (5 million people in the USA), as well as less prevalent conditions such as lupus (SLE), ankylosing spondylitis, scleroderma and Reiter's syndrome - perhaps even multiple sclerosis. And, because they may overcome some of the major limitations on cancer therapy, it appears that basic research, including clinical trials when appropriate, will be much more intensively funded than ever before. The heightened interest in products for cancer therapy and continuing interest in major markets, like arthritis therapy, on the part of industry should assure rapid follow-up on promising leads. However, it must be appreciated that, with the exception of cancer - the therapy of which already wreaks havoc on the immune system - development of products to the point of FDA approval is likely to move very slowly because, until the system is much more clearly understood or long-term adverse effects have been exhaustively sought for, there will be concern that the damage one could do by tampering with the immune system might far exceed the benefit.

A decade or more could be consumed examining the interactions of the antiinflammatory agents that are now used so widely in arthritis therapy, for example, with novel immune modulators. Of course, successes where today there is nothing but hope - like multiple sclerosis - if they were observed would move much more rapidly to utilization. All that can be said with certainty is that recent developments in molecular biology, including molecular genetics and molecular immunology, have provided the intellectual framework and the kind of powerful tools that will be necessary to understand such difficult medical problems.

The importance of immunopharmacology to many diseases will be influenced by two factors that deserve mention, although it would not be appropriate to elaborate on either in this context. The first is another aspect of genetics and immunology, namely, the hypothetical relationships between gene products of the human leukocyte antigen (HLA) region of chromosome 6 and a number of chronic diseases. The second concerns the roles that have been proposed for infectious agents, e.g., viruses, in such diseases.

Three products (A, B, & C) of the HLA region are 44,000 mol.wt. glycoproteins which appear on the surface of all nucleated cells.* The fourth product (D) is comprised of two peptides (34,000 and 29,000 mol. wt.), and is involved in the regulation of the production of helper and suppressor T cells.

---

* These turn out to be the targets of cytotoxic "killer" T cells, and their involvement in the specificity of cytotoxic killer cells for foreign tissues gave rise to the definition of a genetic region called the Major Histocompatibility Complex (MHC), which came to be referred to as HLA in the case of humans because the surface of leukocytes is especially rich in these glycoproteins. Graft acceptance depends in large part on matching these proteins for donor and recipient.

There are a number of alleles in each gene - heritable local AA differences - and each identified allele is assigned a number (for example, B27 is one of 30 B alleles).

Certain combinations of these alleles are being found in association with specific diseases. Thus, more than 95% of patients with ankylosing spondylitis carry HLA-B27. However, only 20% of individuals bearing B27 in the population have any manifestations of ankylosing spondilitis.

It appears that, in order for the disease to occur, the genetic predisposition must be combined with some environmental factor, and/or genetic factors other than the ones studied. That seems to be the role of infections in some cases.

Individuals with genetic predisposition, e.g., HLA-B27, when they are exposed to a particular infection, e.g., <u>Shigella flexneri</u>, have a high probability of developing the disease, e.g., Reiter's syndrome. That is the one system in which the inciting agent, the susceptible genotype, and resulting disease are clearly identified. Table 19 presents representive evidence of the correlations between HLA types and various rheumatic diseases.

"The association of a wide variety of autoimmune diseases, including rheumatoid arthritis, with the HLA-D locus raises the speculation that a hyper- or hypo-normal immune response to an exogenous antigen or to a self-antigen might underlie the association," as Prof. Hugh McDevitt puts it, but, he points out, the evidence for that mechanism is still lacking.

Table 19 Correlations of Disease Incidence with Human Lymphocyte Antigen Types [McDevitt]

Disease	*Racial Group*	*HLA Type*	*Antigen Frequency*		*Relative Risk*	*P Value*	*Number of Studies*
			Patients	*Controls*			
Ankylosing spondylitis	Caucasian	B27	89.8	8.0	87.8	1×10^{-10}	17
	Japanese	B27	66.7	0.0	305.7	1×10^{-10}	1
	Haida Indians	B27	100.0	50	34.4	1.7×10^{-5}	1
	Bella Coola Indians	B27	100.0	20.2	20.2	1.9×10^{-2}	1
	Pima Indians	B27	36.0	18.0	2.6	1.0×10^{-3}	1
Reiters	Caucasian	B27	78.2	8.4	35.9	1.0×10^{-10}	8
Yersinia arthritis	Caucasian	B27	79.4	9.4	24.3	1.0×10^{-10}	2
Salmonella arthritis	Caucasian	B27	66.7	8.6	17.6	3.6×10^{-10}	2
Psoriatic arthritis	Caucasian	B13	19.8	5.5	4.8	9.6×10^{-10}	5
Central joints		B27	40.2	8.7	8.6	1.0×10^{-4}	5
		Bw17	11.6	5.5	2.5	4.6×10^{-3}	5
		Bw38	22.7	2.9	9.1	5.6×10^{-6}	5
Peripheral joints	Caucasian	B13	9.9	5.5	2.3	9.5×10^{-3}	5
		B27	15.5	8.7	2.5	3.5×10^{-4}	5
		Bw17	24.8	5.5	5.8	1.0×10^{-10}	5
		Bw38	12.6	2.9	4.5	3.2×10^{-4}	5
Juvenile rheumatoid arthritis	Caucasian	B27	26.4	8.5	4.7	1.0×10^{-10}	7
Rheumatoid arthritis	Caucasian	Dw4	42.2	15.7	3.0	1×10^{-5}	2
	Caucasian	Cw3	30.0	17.0	2.7	1×10^{-2}	1
Sjogren's	Caucasian	B8	50.6	24.0	3.2	1.0×10^{-7}	1
		Dw3	53.0	17.0	5.2	1×10^{-3}	1

4.25 Antibodies

Antibodies are the best known and most exploited components of the immune system. Until recently all antibodies were obtained by drawing blood from humans or animals. Within the past five years it has become possible to produce them from cells in culture and to obtain specificities previously unattainable. As has been the case with previous advances in antibody technology, basic researchers are imaginatively examining the ways that this new level of specificity can be put to use, and there have been hundreds, if not thousands, of examples published within the first three years that the technique has been available.*

Industrial endorsement lags a bit behind academic embrace, but a number of new ventures have been founded on monoclonal

---

* "Fully half the grant applications coming to certain NIH sections include monoclonal antibody projects." [Fox, 1979]

antibody technology, while the firms already established in antibody businesses have been identifying their proprietary interests.

Antibodies are glycoproteins with molecular weight of 150,000 or greater (Table 20) and complex multimeric composition that would seem to defy direct synthesis - nature seems to select and favor those cells which happen, by random combination of a number of variables, to produce the proper molecular structure to suit a given situation. That such a synthesis would be superfluous now is due to a happy marriage of immunology and that branch of molecular genetics which has developed techniques for creating hybrid cells by fusion.

Table 20. Classes of Human Immunoglobulins

	γG	γA	γM	γD	γE
Heavy chains					
Class	γ	α	μ	δ	ε
Subclasses	γ1, γ2, γ3, γ4	α1, α2			
MW	53,000	64,000	70,000	58,000	75,000
Light chains					
MW 22,500	κ, λ	κ, λ	κ, λ	κ, λ	κ, λ
Formula	$\kappa_2\gamma_2$ or $\lambda_2\gamma_2$	$(\kappa_2\alpha_2)_n$ or $(\lambda_2\alpha_2)_n$ where n = 1, 2, 3, 4	$(\kappa_2\mu_2)_5$ or $(\lambda_2\mu_2)_5$	$\kappa_2\delta_2$ or $\lambda_2\delta_2$	$\kappa_2\epsilon_2$ or $\lambda_2\epsilon_2$
$S_{20,w}$	6.5–7.0	7, 10, 13, 15 . . .	18–20	6.2–6.8	7.9
MW	150,000	180,000–500,000	950,000	160,000	190,000
Carbohydrate, %	2.9	7.5	11.8		10.7
Conc., g/100 ml	0.6–1.7	0.11–0.12	0.05–0.19	0.003–0.04	0.0_41–0.0_314
Half-life, days	23	5.8	5.1	2.8	2.5
%, Plasma pool catabolized/day	6.7	25	18	37	89
Synthetic rate, mg/kg/day	33	24	6.7	0.4	0.016

4.251 Hybridomas and Monoclonal Antibodies

Monoclonal antibodies were the unanticipated fruit of fundamental immunologic research. Drs. Caesar Milstein and Georges Kohler at the Medical Research Council (UK) wanted to determine whether segments from different antibodies could be combined by cells into hybrid molecules. In 1975 they were fusing different cell types that made distinguishable antibodies, and for convenience fused mouse spleen cells with mouse myeloma cells. "The experiments were incredibly successful," Milstein has been quoted as saying. "Only later did we realize the potential of the procedure. It could be done to make any antibody, and it's no longer a mixture, but a well-defined chemical." [Ibid.]

Myeloma cells grow vigorously in culture - a feature that distinguishes cancer cells from ordinary tissues - and they have the unique characteristic of producing large quantities of immunoglobulins, the proteins that comprise antibodies. However, these immunoglobulins display no useful specificity. Individual spleen cells of the immune type, on the other hand, each produce a specific antibody but they do not grow well in culture. When normal, immune spleen cells are fused with myeloma cells, the resulting mixture of genetic capacities forms a cell, called a "hybridoma," which displays the desired characteristics of the parent cells:

- secretion of the antibody specified by the genes of the spleen cell
- vigorous growth, production and longevity that is typical of the myeloma cell.

In practice the procedure involves challenging the donor animal with an antigen, which elicits the production of antibodies by certain cells which are concentrated in the animal's spleen. After the production of antibodies against the target antigen is established, cells are taken from the spleen and fused, in the presence of polyethylene glycol, with myeloma cells.

Some of the cells that survive fusion produce antibodies against the target antigen. The spleen cells that go into the fusion are not all alike. Even those producing antibody against the antigen of interest differ from one another with respect to that segment of the antigen (antigenic determinant) to which they correspond and with respect to the strength of their interaction (affinity) with that antigenic determinant.

Hence, various means are employed to sort out the fused cells, by use of markers and techniques analogous to those employed in genetic selection among antibiotic-producing organisms, or by adaptation of novel cell sorting technology.*

It turned out that the answer to Milstein and Kohler's original question about the merger of parts of different antibodies was affirmative. That complicated the potential of hybridomas for production, until Kohler and his collaborators at Roche's Institute for Immunology in Basel, Switzerland manipulated the genes of a myeloma to produce a strain which does not by itself produce antibody chains, although it does still grow vigorously _in vitro_ and fuses readily with spleen cells. This strain, thus, provides a convenient silent partner for creation of highly specific hybridomas.

---

* One powerful cell-sorting methodology utilizes a number of sophisticated developments in engineering that exemplifies the way that modern bioscience draws upon the capacities of other disciplines to address its problems. The fluorescence-activated cell sorter (FACS) enables the analysis and sorting of several thousand cells per second. The FACS sprays cells in a stream of fine droplets past a laser. Depending upon the fluorescence spectral characteristics of a given droplet (as measured by the laser spectrophotometer component of the FACS), that droplet is given an electric charge and deflected electromagnetically into a specific well on a culture plate. A clone is grown from the contents of each well into which a desirable droplet was placed. Using this technique, workers - such as Leonard Herzenberg at Stanford who developed it - can rapidly isolate clones to produce antibodies for specific determinants of complex and impure antigens.

As cultured cells go, myelomas are copious producers of protein. Milstein estimates that at least one-third of their energy is expended by myeloma cells in tissue culture to make and secrete antibody. [Miller, 1978] Hybridomas grown in mass-culture produce about 50 mg/l, a concentration comparable to weak serum obtained from an animal. However, the volume of serum one can draw from an animal is quite limited* (sheep and horses have often been employed to scale up and mice are normally out of the question), while tissue culture can be conducted on a scale of hundreds of liters.

If that is not enough, there is another alternative. One can inject the fused cell clone into a mouse, where tumors will be induced. In a short time these tumor-bearing mice will develop 5-20 mg/ml of antibody in their blood, 95% of which will be the antibody specified by the clone.

The availability of so much of a specific antibody exceeds the dreams of immunologists of just five years ago. This technique and concurrent insights into the genetic mechanisms that underlie the incredible diversity of antibody specificity are the foundation of a revolution in immunology.

In turn, immunology is providing both the tools and insight which will influence bioscience broadly, especially: embryology and cellular differentiation, tissue/organ development and interaction, cancer, allergy, aging, memory, and many others, as well as infectious disease - the area most people think of in connection with antibodies. We can sketch the

---

* Bear in mind that each individual animal, or even the same animal at different times, will produce somewhat different antibodies in response to the same antigen.

applications in medical practice only very superficially in this context.

4.252 Applications of Monoclonal Antibodies

4.2521 Diagnostic Reagents

The most immediate medical application for monoclonal antibodies, beyond the research laboratory, is in diagnostic testing. Over the past twenty years, substantial segments of the diagnostics and clinical laboratory industries have developed around antibodies derived from antisera. In order to detect and determine the level of a particular substance in a specimen taken from a patient, antibodies for which that substance is an antigen are added to the specimen and, by one of several means, the extent of antigen-antibody reaction is assessed. This technique is used, not only for diagnosis - to aid in the identification of disease - but also to monitor the levels of drugs administered and the degree of success achieved in treatment. The degree of sophistication ranges from the agglomeration tests employed in routine blood typing and pregnancy tests to tests that use radioactivity, fluorescence or enzymatic amplification to enable the detection of extremely small amounts of antigen.

The more sophisticated systems, which provide quantitative information, involve reagent/instrument systems engineered to build the complex functions into the machines and molecules, so that the generation of meaningful information does not take a great deal of sophistication on the part of the operator. This is where the high specificity of monoclonal antibodies is likely to have an impact.

A large antigen, such as a protein, may have dozens of antigenic determinants (and cells, such as microbes, may have hundreds), so that a whole population of antibody-producing cells are turned on when an animal is innoculated with an antigen. Although each cell produces only one type of antibody, the antiserum drawn from the animal contains many specificities. If some of these antibodies happen to cross-react with other antigens which have similar antigenic sites, the antiserum will react positively with more than one antigen. Small molecules often will not elicit formation of antibody unless they are attached to a larger carrier. The antibodies formed in response to such a combination range from those specific to the carrier to those specific for the "non-immunogenic" molecule (hapten) of interest. When it comes to determining the level of a particular protein, hormone, or drug in a patient specimen that may contain all sorts of other proteins and metabolities, it is of the utmost importance to have antibodies with as high specificity and affinity as can be achieved.

Monoclonal antibodies exceed any others in this respect and the employment of hybridomas seems certain to replace animals as the source of antibodies for virtually all diagnostic and monitoring purposes. Not only will the accuracy of tests be improved and, perhaps, development costs decreased, but the reproducibility of product should become much greater.

Right now individual animals, e.g., a sheep, may represent the world's supply of raw materials for a specific product. In contrast to the product mortality that implies, the perverse vitality of cancer cells has been turned to good advantage - rendering the antibody-producing hybridoma immortal and making it possible to maintain identical clones in "cold storage."

The kinds of tests for which such assays are used today include:

- hormone levels - to assess the proper functioning of an endocrine gland or the inappropriate production of a hormone by a tumor
- certain proteins the presence of which has been found to correlate with a tumor or with a specific prenatal condition, for example,
- drugs - to detect the presence of illicit drugs in a person's blood, or to monitor the blood or tissue level of a drug in order to assure that the dosage achieves the necessary therapeutic level without exceeding the limits that could cause toxic effect.
- Precise identification of microbial pathogens

Tables 21 and 22 from a recent review on the immunoassay of drugs suggest the breadth of this application and of the biochemical parameters that antibodies have been used to quantify. No one assay constitutes the major market, and rapid obsolescence of products has been characteristic of this business.

Table 21: Immunoassays for Drugs and Toxins

- Analgesics and narcotics
  - Anileridine
  - Antipyrine
  - Codeine
  - Etorphine
  - Fentanyl
  - Meperidine
  - Methadone
  - Morphine
  - Pentazocine
- Antibiotics
  - Amikacin
  - Chloramphenicol
  - Clindamycin
  - Gentamicin
  - Isoniazid
  - Penicillin
  - Sisomycin
  - Tobramycin
- Anticonvulsants
  - Clonazepam
  - Phenytoin
  - Primidone
- Antiinflammatory agents
  - Colchicine
  - Indomethacin
  - Phenylbutazone
- Antineoplastic agents
  - Adriamycin
  - Bleomycin
  - Daunomycin
  - Methotrexate
- Bronchodilators
  - Theophylline
- Cardiovascular drugs
  - Cardiac glycosides
    - Acetylstrophanthidin
    - Cedilanid
    - Deslanoside
    - Digitoxin
    - Digoxin
    - Gitoxin
    - Methyl digoxin
    - Ouabain
    - Proscillaridin
  - Dihydroergotamine
  - Propranolol
  - Quinidine
- CNS stimulants
  - Amphetamine
  - Benzoyl ecgonine (cocaine metabolite)
  - Methamphetamine
  - Pimozide
- Diuretics
  - Bumetanide
- Hallucinogenic drugs
  - Lysergic acid diethylamide (LSD)
  - Mescaline
  - Tetrahydrocannabinol
- Hypoglycemic agents
  - Butylbiguanide
  - Glibenclamid
- Insecticides
  - Aldrin
  - DDT
  - Dieldrin
  - Malathion
- Narcotic antagonists
  - Cyclazocine
  - Naloxone
- Sedatives and tranquilizers
  - Barbiturates
    - Barbital
    - Pentobarbital
    - Phenobarbital
  - Chlordiazepoxide
  - Chlorpromazine
  - Desmethylimipramine
  - Diazepam and *N*-desmethyldiazepam
  - Glutethimide
  - Methaqualone
- Skeletal muscle relaxants
  - *d*-Tubocurarine
- Synthetic peptides
  - DDAVP
  - Saralasin
- Synthetic steroids
  - Anabolic steroids
    - Trienbolone acetate
  - Androgens
    - Fluoxymesterone
  - Estrogens
    - Diethylstilbestrol
    - Ethinylestradiol
    - Mestranol
  - Glucocorticoids
    - Dexamethasone
    - Methylprednisolone
    - Prednisolone
    - Prednisone
  - Metyrapone
  - Progestins
    - Medroxyprogesterone acetate
    - Norethindrone
    - Norethisterone
    - Norgestrel
- Toxins
  - Aflatoxin B_1
  - Genistein
  - Nicotine and metabolites
  - Ochratoxin A
  - Paralytic shellfish poison
  - Tartrazine

Table 22: Other Immunoassays of Pharmacological Interest

Bile acid conjugates	Peptide hormones
Cholylglycine	Angiotensin
Cholyltaurine	Anterior pituitary
Catecholamines	Bradykinin
Epinephrine	Gastric
Norepinephrine	Hypothalamic
Tyramine	Intestinal
Fibrinopeptides	Pancreatic
Fibrinopeptide A	Parathyroid
Fibrinopeptide B	Posterior pituitary
Indolealkylamines	Thyroid (calcitonin)
Melatonin	Plant hormones
Serotonin	Indole-3-acetic acid
Insect hormones	Gibberellic acid
Ecdysone	Polyamines
Nucleosides and nucleotides	Spermine
Cyclic AMP	Prostaglandins
Cyclic GMP	Steroid hormones
N^2-Dimethylguanosine	Thyroid hormones
7-Methylguanosine	Thyroxine
Pseudouridine	Triiodothyronine
Thymidine	Vitamins
	Vitamin B_{12}
	Vitamin D

The use of high-correlation antibody assays in oncology has only just begun. The implications of monoclonal antibodies can be approached by reflecting on the following words of the editor of the Journal of the American Medical Association: "Once it is is established that a given clone produces antibodies that recognize a cancer cell and nothing else, the clinical potential is enormous, most biologists believe. For example, tagging the antibody with a radioactive [Ballou, 1979] or fluorescent label might allow it to be used diagnostically to seek out cells containing a tumor antigen at an early stage in the disease's progress. It could also provide an early indication that a tumor has metastasized or recurred."

Considering the fact that some 785,000 new cancer cases are estimated for 1980 - with current diagnostic methods; that there will be 405,000 deaths due to cancer; that only heart disease causes more deaths, and that early detection is commonly agreed to be the major key to improving survival, the implications are indeed enormous!

"In addition," again in the words of the JAMA editor, "specific antibodies could be used to monitor the response of a cancer patient to various forms of therapy. That is, if the antibody finds little tumor antigen to bind to in a patient's blood, the tumor is probably regressing. An increase in binding over a previous occasion would suggest that "an alternative form of therapy be adopted. [JAMA 242, 2161-3 (1979)]*

The linch-pin in these predictions is the identification of high-correlation assays. So far there are only a few assays in use for this purpose: carcinoembryonic antigen (CEA), alpha fetoprotein (AFP), and prostatic acid phosphatase (PAP) - all based on conventional antibodies - but many are under study.

What may be a landmark patent has been issued to Koprowski and Croce of the Wistar Institute (for work done under HEW funding) on the production of monoclonal antibodies against tumor cells. For a number of examples, they demonstrated that an animal can be immunized with tumor cells, and hybridomas derived from that animal will produce antibodies which demonstrate a specificity for the tumor. Antibody-producing clones can be selected that produce antibodies with varying degrees of selectivity for tumor cells. Some react only with cells of a specific tumor while others may

---

* Although it does not relate to antibodies, another development of molecular genetics deserves mention here because of its relation to cancer. Work on the genetics of tumor cells has disclosed the fact that certain strains develop resistance to anticancer agents by multiplication of genes that facilitate their survival in the presence of the drug. [Schimke, 1979] The discoverer of this phenomenon has proposed that screens, analogous to those used to determine the susceptibility of bacteria to antibiotics, be used to screen the cells from a tumor to assess ahead of time whether they have this capacity to escape (relapse), so that the appropriate drug regimen can be implemented early.

react with more than one type of tumor.* The final sentence of the patent text leads into the next topic, "If the (tumor) antigen is present, the patient can be given an injection of an antibody as an aid to react with the antigen." [U.S. 4,172,124]

Erlich's concept of a magic bullet - that will kill only the cause of the disease - has a new incarnation in the monoclonal antibody. The use of animal antiserums in connection with infectious disease has a long history, but the occurrence of serum sickness - in which antibodies are produced by the host against the donor's antibodies - and the availability of human immune serums have made their use rare. Horse antiserum is still used in the management of diptheria, gas gangrene, botulism and venomous snake bites, however. Mixed gammaglobulins (immune serum globulin, ISG) came into use in the post WWII period when blood fractions became available and then faded from use almost entirely in the 1970s, except for the specific ISG from donors with antibodies against tetanus, rabies, and for the Rh factor. Almost all such use of exogenous antibody, called passive immunization, is prophylactic in nature. However, the exciting prospect revived by monoclonal antibodies is that of using antibodies as therapeutic magic bullets.

A monoclonal antibody to an antigen commonly associated with human acute lymphoblastic leukemia (ALL) cells, that has been identified by a number of investigators, has been reported by workers at the Sidney Farber Cancer Institute. [Ritz, 1980]

---

* Dr. Irving Weissmann of Stanford reports that radioactively-tagged monoclonal antibodies can detect as few as one million leukemia cells in mice, which is a thousand times more sensitive than any test now in use.

We understand that this antibody, which is derived from mouse x mouse cell fusions has already been entered into preliminary clinical studies. Of course, it is premature to assess the utility of this particular approach, but the pace at which it has been moved into the clinic indicates the optimism of the researchers. Most important, the monoclonal antibody developed at Sidney Farber does not react with any normal cells tested (including those of regenerating bone marrow and fetal hematopoietic tissues, with which antibodies prepared by previous workers had reacted). The antibody, identified as J-5, does indeed appear to be leukemia specific and reactive against the leukemia cells from 21 out of 34 patients with the condition designated non-T cell ALL. One hopes that other varieties of leukemia and other tumors will turn out also to display such identifiable characters. Reports of antibodies against human neuro- blastoma cells, and mouse teratocarcinoma cells certainly support this hope. [**Kennett and Gilbert**, 1979; Ballou, 1979]

Besides the possibility that antitumor antibodies may in themselves bring about the demise of their targets by normal immune processes, it has been proposed that they could serve as guided missiles to carry an antineoplastic drug warhead direct to target cells. Given the high toxicity of many antineoplastic agents, which limits their dosage level, that might prove to be precisely the way to seek and destroy the cryptic cancer cells that are the origin of relapses and metastases.

Other applications suggested by reports of preparation of monoclonal antibodies include:

- o the improvement of kidney (and other organ) acceptance by injection of the recipient with antibodies against certain histocompatibility antigens

- the use of substrate-bound antibodies for affinity chromatographic purification of biological materials such as products made by genetic engineering
- passive immunization against an antigen, e.g., chorionic gonadotropin, involved in reproduction as a reversible immunologic approach to contraception.

5. ANTIBIOTICS

Antimicrobial agents for the treatment of infectious diseases have consistently led the sales of prescription pharmaceuticals in the United States and worldwide for the past three decades. Most of these agents are antibiotics produced, at least in part, by microorganisms rather than chemically synthesized or isolated from higher organisms. Antibiotic production will be the source of most of the precedents for the employment of microbial fermentation to produce substances useful in medicine. The United States has been prominent in the development, production and marketing of antibiotics, with the result that American companies account for about one-half of the roughly $5 billion worth of antimicrobial agents sold worldwide each year, and their share has been growing. New antibiotics continue to be developed and introduced year after year, originating in the USA or abroad (especially from Japan in recent years) and usually being licensed from the inventing company, country by country, to other firms when the inventor lacks marketing strength in a particular region.

5.1 General

Three groups of microbes produce medically important antibiotics:

- actinomycetes of the genera *Streptomyces*, *Micromonospora*, *Nocardia*, *Streptosporangia*, and *Actinomadura*

- o filamentous fungi of the genera Penicillium and Cephalosporium
- o eubacteria of the genera Bacillus and Pseudomonas.

Despite the preeminent role of antibiotic production in industrial microbiology, and the extent to which what we know about molecular genetics is based upon the study of microbes, the influence of applied genetics on the field of antibiotics has been highly empirical to date. The words of J. Langrish have been appropriately applied to this field. "Once a new area has been established, the aim of science is to understand, the aim of technology is to make it work, and industry has been very successful at making things work without too much reliance on understanding." [Quoted in Sermonti, 1979]

Ample benefit has been derived from knowledge of molecular details in metabolism, although the complete biosynthetic pathway has not been elucidated for a single antibiotic [Elander, 1979c]. For over thirty years advantage has also been taken of the fact that most microorganisms exhibit a great capacity for natural variation, but only recently has it been possible to consider genetic processes in molecular terms.

5.2 Classical Genetics

The following is as brief a summary as possible of the state-of-the-art of microbial genetics as it has been applied in the search for and production of antibiotics, including a few examples and an excerpt from a case study prepared for OTA. Reference is made to a number of germinal developments that suggest the future consequences of molecular genetics. Some authorities in the field think: "It is no exaggeration to say that a revolution has occurred in our subject since the Second International

Symposium on the Genetics of Industrial Microorganisms (in 1974)." [Hopwood, 1979] "Rather than relying on evolutionary diversity, research scientists will utilize genetic techniques - such as protoplast fusion, *in vitro* and *in vivo* recombinant DNA, transposable elements and plasmids - to construct or discover novel antibiotics." [Nash, 1980]

5.21 Mutagenesis and Selection

Selection of mutant strains has constituted the predominant means to identify high-yielding antibiotic producers for 30 years. Initially, organisms which produce novel antibiotics are isolated by soil sampling and other broad screening efforts. These are then cultured in the laboratory, and efforts made to optimize productivity. Antibiotics are secondary metabolites, that is molecules which have no general function in integral life processes - although they probably provide survival benefits to an organism in its natural environment. They are usually produced as mixtures of members of a closely related chemical family; e.g., there are at least three neomycins, five mitomycins, 10 bacitracins, four tyrocidines, eight aflatoxins, 10 polymyxins, over 10 natural penicillins, and more than 20 actinomycins. It is characteristic of secondary metabolites that they usually are produced, not during the phase of rapid growth (trophophase) but, during a subsequent stage known as the idiophase - at the end of trophophase marked changes occur in the enzymatic composition of the cells and enzymes specifically related to formation of secondary metabolites appear.

Nutrients, such as glucose, that are readily consumed in trophophase often suppress antibiotic production. Many secondary metabolites exert feedback regulation on their own synthesis, and primary metabolites derived from branches of pathways leading to antibiotics are known to limit formation of a secondary metabolite by inhibiting production of a common precursor, e.g., lysine depresses β-lactam production by negative feedback that prevents synthesis of α-aminoadipate, a common precursor. It is the fact that production of secondary metabolites is controlled by these genetically determined mechanisms which makes it possible to achieve major effects by mutation.

Indeed, the chief factor responsible for the 100 to 1000-fold increases in production of antibiotics, from the time of their discovery to the present, is mutation. [Elander and Demain, 1980] The low productivity of the original strains that produced penicillin and cephalosporin antibiotics led to investigation and "exploitation of the genetic potential of beneficial mutations to generate superproductive commercial strains of those β-lactam-producing fungi." [Elander, 1979a]

Two types of mutation are distinguished in terms of their practical consequences:

1) Major qualitative mutation involves a prominent change of a biochemical character, e.g.

   - non-pigmented variants of _Penicillium_ that produce colorless penicillin [Backus, 1955]
   - mutants of _Streptomyces griseus_ that do not produce low-activity by-product which closely resembles streptomycin, the desired antibiotic [Alikhanian, 1962]
   - mutation of tetracycline-producing organism to produce a novel and valuable analogue, 6-demethyltetracycline [McCormick, 1957]
   - a mutant strain of _Streptomyces fradiae_ that cannot synthesize the deoxystreptamine subunit of neomycin but will incorporate other

diaminocyclitols to produce novel antibiotics termed hybrimycins [Shier, 1969]

2) Minor mutations affect only the quantity of product synthesized, and the variants are usually similar phenotypically to the parental form, e.g.

- a double-revertant strain of Streptomyces viridifaciens produces six times more tetracycline than the original parental strain - many reverse mutations are pleiotropic (have multiple effects) because they often involve mutation of suppressors which affect genetic characteristics other than those for the product of immediate interest. [Dulanhey, 1967]

- a 10-15 percent increase in yield in normally considered significant, but further mutation of successive minor positive variants may produce substantial yield improvements, e.g., stepwise selection of 10-12% greater penicillin productivity led to a variant which exceeded its original parent by 64% [Alikhanian, 1962].

- without such improvements, the cephalosporins, especially cephalosporin C, its nucleus (7-ACA) and derivatives, would still be laboratory curiosities. [Florey, 1955]

- variant strains often require special propagation and preservation procedures, including modified fermentation conditions with continued selection and purification to maintain the gains achieved. [Elander, 1976]

The most common method used to obtain high-yielding mutants is that of treating a population of uninucleate vegetative cells or spores with a mutagenic agent until a certain "desired" degree of killing is obtained, plating out the survivors and testing each resulting colony or a randomly selected group of colonies for product formation in flasks. Another common procedure is to select mutants that are obvious through altered colony morphology or color for testing in flasks. Other sub-populations that are sometimes chosen for liquid fermentations are nutritionally deficient (auxotrophic) mutants, their revertants, and mutants resistant to: toxic endproducts, toxic precursors, or toxic analogues of precursors. Often preferred are revertants of non-producing mutants and mutants producing higher antibiotic concentrations on agar plates (Demain, 1973). All of these screening methods are rather indirect, since no successful selection method has ever been devised for producers of secondary metabolites (i.e., metabolites not required for growth). Furthermore, they all depend on effective mutation and the ability to separate the mutated nuclei from all the other nuclei in the population. [Demain, 1980]

Classical mutation and selection techniques, important in the development of highly productive mutants in the absence of fundamental knowledge, have now been largely replaced by more rational (directed) selection techniques grounded on more scientific bases. In particular, these involve fermentation prescreening of mutagenized cells on solid media prior to studies in liquid broth. [Elander, 1979a]

5.22 Directed Selection

The major emphasis now is on finding the methods for inducing mutations which are most suitable for particular types of microorganisms - often involving cyclical use of mutagens having different modes of action - in combination with screening assays designed to enhance the probability of selecting superior mutants.

5.221 Gentamicin Case Study

The following description of developments with respect to the important antibiotic, gentamicin, prepared by Arnold Demain, an MIT professor and scientific consultant to the companies involved, will illustrate what is involved in a modern mutagenesis and selection effort to improve production yields.

> During the early years of gentamicin at Schering-Plough, Inc., strain improvement did not produce significant yield improvements. As a result,

gentamicin has been one of the most expensive antibiotic products on the market (about $50.00/ gram in parenteral form). The major reason appears to be the lack of techniques available in the 1960s for mutation of filamentous and poorly sporulating microoganisms such as Micromonospora purpurea. This lack of a uninucleate cellular stage, (i.e., individual vegetative cells or spores), coupled with extremely low growth rates, made strain improvement by mutagenesis very difficut. Even if a mutation to higher antibiotic production had occurred in such a multinucleate filament, it probably was masked by the other parental nuclei in the strand. To make the mutated nucleus the only viable nucleus would probably have required excess mutagenesis which could cripple the vigor of the culture. Apparently strain improvement efforts at Schering had yielded only marginal results into the early 1970's. Schering management opted for outside help to improve its production organism.

A company named Cetus Corporation, had been formed in 1971 in Berkeley, California. Cetus was dedicated to improving existing biological processes and devising new biological processes with an emphasis on genetics. A contract was signed in 1973 between Schering-Plough and Cetus to genetically develop a number of Micromonospora organisms. At this point, Cetus Corporation (1) devised methods of properly handling Micromonospora organisms and 2) used their automated large-scale screening capabilities.

The aspect of proper handling is exemplified by a patent which states "For sporulating microorganisms..., it is usually possible to use a single spore as the colony-forming unit in genetic experiments. Since most spores are presumably uninucleate, it is relatively easy to obtain genetically pure colony forming units." In non-sporulating filamentous organisms "each colony-forming unit may contain many tens or hundreds of nuclei which may or may not be genetically identical. If a filamentous culture is treated with a mutagen to introduce genetic inhomogeneity into the culture, any genetic variations having significantly better productivity in the culture are typically masked." In the patented procedure, the filaments are fragmented and filtered. The colonies grown from fragments that will pass through a filter of specified pore size are found to be predominantly genetically pure. These fragments are then subjected to normal mutagenesis procedures. The patent points out that such treatment can produce a 100-fold increase or more in the viable count of Micromonospora purpurea (the species which makes gentamicin). (Ward and Bruner, 1976) Since all of the data in the patent deal with Micromonospora purpurea, it is reasonable to surmise that this work was done during the Cetus/Schering-Plough relationship and was important in the improvement of gentamicin production.

The second aspect of the Cetus effort deals with automated large-scale screening. This is best understood by quoting from Lewin (1978). "Research labs in the industry spend much of their time searching for new strains of microorganisms that, for instance, will give better yields of antibiotics. By conventional techniques, the best equipped labs can screen up to 500 colonies a week, and since mutation rates giving rise to the hoped-for new strain may be as low as one in 10,000, or even one in ten million, such a search can take a very long time. With Don Glaser's inspiration, Cetus developed an automated system that can zip through 50,000 tests at a time"....

It is of interest that although there had not been significant funding of the screening aspect of microbiology, government funding of a very basic project led in an indirect way to Cetus' ability in this respect. For a number of years, the government funded the development of Dr. Donald Glaser's "Dumbwaiter' a huge instrument (40 feet long and two stories high) which is capable of detecting "needles in haystacks," i.e., detecting a few mutant colonies of particular type out of millions of colonies. The Dumbwaiter, described by Bylinsky (1974), "is designed to process automatic surveillance and screening, in accordance with instructions programmed in a computer." This instrument was used by Glaser and his group to isolate many new thermosensitive DNA replication mutants in *Escherichia coli* despite their extremely infrequent occurrence (Sevastopoulos *et al.*, 1977). The machine is in Glaser's laboratory at the University of California in Berkeley and is available to academic investigators from other institutions but has not been used by industry. Although the machine was never used by Cetus, Glaser is a founder of Cetus and Chairman of its Scientific Advisory Board. Many of Glaser's students and postdoctoral associates have obtained permanent employment at Cetus. Although the Dumbwaiter is many times more costly than anything that Cetus could affort in the early 1970s, the know-how and experience of its builders became available at Cetus and could be put to work on the Schering-Plough project. This episode shows how government funding can stimulate innovations in industry even thought it was rather indirect in this case.

Apparently, Cetus' yield improvements had a significant impact on corporate planning at Schering-Plough as evidenced by the following summary of remarks by Richard Bennett, President of Schering-Plough about the consequences of the Cetus yield improvements (12/7/77). "Fermentation yield improvements were characterized as "dramatic' in recent years reducing the company's need for additional manufacturing capacity.

Besides strain improvement for production yield, mutation has been used to:

- develop genetically blocked mutants that enable the elucidation of the metabolic pathways which result in the biosynthesis of antibiotics.

- discover new molecules that, as such or when converted into derivatives by synthetic operation *in vitro*, are antibiotic candidates not otherwise produced by the wild-type isolates.

- create mutants which, when fed unnatural precursor analogues, may synthesize novel antibiotically-active molecules through mutasynthesis.

5.3 In Vivo Recombination

Hybridization between strains to yield diploid and recombinant strains has also been widely investigated in the pharmaceutical industry to develop vigorous high-yielding antibiotic producers. However, incompatability in mating types of many industrially important higher fungi, the presence of chromosomal aberrations in mutationally improved microorganisms, and a number of other problems limit the application of hybridization. Furthermore, natural recombination is most advantageous when strains of extremely diverse origins are mated, and the treatment of commercial strains as proprietary secrets usually prevents bringing together the sort of divergent "competitor" strains most likely to produce vigorous hybrids. Finally, genes determining increased antibiotic yields are usually recessive and independently induced mutations may be allelic, i.e. involve the same gene. [Elander, 1979b]

5.31 Actinomycetes - Sexuality

In principle, sexual transfer and recombination is the most efficient spontaneous means for introducing new attributes and increases variability. Table 23 lists members of the important families of antibiotic-producing prokaryotes for which gene exchange, by conjugation (with the single exception indicated in the table), has been reported.

Table 23: Reported Genetic Exchange in Actinomycetes (Elander, 1979c)

Organism (antibiotics produced)
Streptomyces
S. coelicolor (actinorhodin and methylenomycin)
S. achromogenes var. *rubradiris*[a] (rubradirin)
S. acrimycini[a] (unidentified)
S. aureofaciens (chlortetracycline)
S. bikiniesis[a] (zorbamycin and zorbonomycin)
S. erythreus (erythromycin)
S. fradiae (neomycin)
S. glaucescens[a]
S. grisoflavus
S. griseus (streptomycin)
S. olivaceus[a]
S. rimosus[a] (oxytetracycline)
S. scabies
S. venezuelae[a] (chloramphenicol)
Nocardia
N. erythropolis[a]
N. mediterranei[a] (rifamycins)
Micromonospora
M. chalcea
M. echinospora (gentamicins)
M. purpurea (gentamicins)
Mycobacterium smegmatis
Thermoactinomyces vulgaris[b]

[a] A circular map with close resemblance to *S. coelicolor*.
[b] *Thermoactinomyces vulgaris* has a transformation system; in other cases a "conjugation" mechanism of recombination is implicated. (After Hopwood, 1976.)

The genetic mechanism of actinomycetes is essentially similar to that of E.coli. Their major difference is that in the vegetative mycelium of actinomycetes non-identical (heterogenomic) states are sustained, (as was referred to in connection with gentamicin), whereas the relatively regular cell division process of E.coli establishes homogeneity more rapidly. S.coelicolor, also known as S.viocelaruber, is by far the most thoroughly studied actinomycete, and rudimentary circular gene linkage maps have now been published for it, for S.rimosus, and for Nocardia mediterranei. No doubt interested firms are constructing genetic maps of the organisms responsible for their products.

Structural genes for some antibiotics or subunits reside in extrachromosomal plasmids, which probably accounts for the production of the same or related antibiotics by many species. For example, it has been suggested that a plasmid-borne set of genes for biosynthesis of deoxystreptamine could account for its appearance as a constituent of aminoglycoside antibiotics produced by diverse actinomycetes. Control of the synthesis of certain antibiotics is plasmid-borne, and a hypothesis has been proposed that some plasmids may non-specifically enhance antibiotic excretion, which "could provide a method for rapidly increasing the yield of a new antibiotic and, in turn, facilitate recovery of the active molecule." [Nash, 1980]

Plasmids are prevalent in Streptomyces, and at least four plasmids have sex factor activity. They resemble the F factor of E.coli in many respects, and can exist either as autonomously replicating plasmids or integrated into the host chromosome. Two of these factors, SCP1 and SCP2, have been shown to be highly transmissable to members of the same species and can be transferred between different species of Streptomyces. [Hopwood, 1976 These factors also promote chromosome transfer to an extent that depends upon the plasmid used and upon its state in both parent strains. [Hopwood, 1973] The frequency of recombination between strains that have recently diverged from a common progenitor is of the order of one in a million. [Hopwood, 1977]

For some time now, people have been examining genetic transfer, including syncytic recombination between different species, in the genus Streptomyces (Table 24).

Table 24 Species of *Streptomyces* in which genetic recombination has been demonstrated

Species	Reported by	Antibiotic
Intraspecific		
S. violaceoruber	Sermonti and Spada-Sermonti (1956), Hopwood (1973), Braendle and Szybalski (1959)	Actinorhodin
S. fradiae	Braendle and Szybalski (1959)	Neomycin
S. rimosus	Alikhanian and Mindlin (1957)	Oxytetracycline
S. aureofaciens	Alikhanian and Borisova (1961)	Chlorotetracycline
S. griseoflavus	Saito (1958)	Novobiocin
S. griseus	Braendle and Szybalski (1959)	Streptomycin
S. scabies	Gregory and Huang (1964)	
S. erythreus	Huang-Lo (1962)	Erythromycin
S. sp. (8182H)	Fantini and Wallo (1967)	Unidentified antibiotic
S. antibioticus	Vladimirov (1966)	Oleandomycin
Interspecific		
S. rimosus × *S. violaceoruber*	Alacevic (1963)	Tetracycline and actinorhodin
S. rimosus × *S. aureofaciens*	Alacevic (1963)	Tetracyclines
S. aureofaciens × *S. violaceoruber*	Alacevic (1963), Bradley (1964)	Tetracycline and actinorhodin
S. griseus × *S. viridochromogenes*	Bradley (1964)	Streptomycin
S. aureofaciens × *S. rimosus*	Polsinelli and Beretta (1966)	Tetracyclines

The following antibiotics have been reported to be plasmid-determined:

chloramphenicol by S.venezuelae

kasugamycin and aureothricin by S.kasugaensis

leucomycin by S.reticuli

methyleneomycin by S.coelicolor

oxytetracycline by S.rimosus

turimycin by S.hygroscopicus. [Elander, 1977]

Of the studies listed in Table 24, examples of success were those with S.rimosus, which increased yield of oxytetracycline and decreased troublesome foaming in the production medium, and work with S.erythreus that produced recombinant strains yielding excellent titers of erythromycin. [Elander, 1976]

New antibiotics have been reported for recombinations between: S.coelicolor and S.lividans [Danilenko, 1975]; S.fraidiae and S.bikiniensis [Godfrey, 1978], and S.violaceus and S. hygroscopicus[Fleck, 1978], according to A. Demain. [1980a] Hybrids, both intraspecific and interspecific, are generally detected by selecting for viable recombinants in media that will not support the growth of nutritionally-deficient (auxotrophic) parents. [Elander, 1976]*

5.32 Fungi - Parasexuality

Fungi are the source of many antibiotics. They are eukaryotes with definite nuclei, but they are primarily asexual and possess limited genetic diversity when compared to organisms possessing true sexual cycles. However, a phenomenon known as parasexuality has been discovered in a number of industrially important fungi, including Penicillium chrysogenum, which produces penicillin.# From the latter, it has been determined that there are a number of genetic determinants, recessive in

* One of the main problems impeding antibiotic geneticists has been the lack of suitable markers and the finding that the introduction of biochemical markers (lesions) almost invariably results in sharp decreases in fermentation productivity. This has been the case for both actinomycetes and penicillin producers. [Elander, 1976]

Most industrially important fungi grow normally as haploid, asexual organisms with mitotic nuclear division. Parasexuality is based on mitotic recombination during normal cell division. It consists of 1) the formation of heterokaryons, in which two different nuclei share a common cytoplasm; 2) rare nuclear fusion within the heterokaryon to form a diploid nucleus, and 3) mitotic recombination in diploid nuclei by genetic crossing-over or loss of chromosomal segments through nondisjunction. Mitotic nonconformity, which stems from the existence of duplicate segments of small fractions of the genome, is an additional mechanism allowing for genetic recombination and may explain the culture instability of many fungal organisms. [Elander, 1979a] The parasexual cycle has been especially useful for mapping the chromosomes of fungi.

their expression and located on more than one chromosome, upon which penicillin yield depends.

To date, the technique of mutation and selection has been the most reliable procedure for improving penicillin titers. Figure 8 outlines the stages in development of one line of penicillin producers, and Table 25 presents some corresponding productivity data.

Figure 8 Strain Lineage in <u>Penicillium</u> <u>chrysogenum</u>

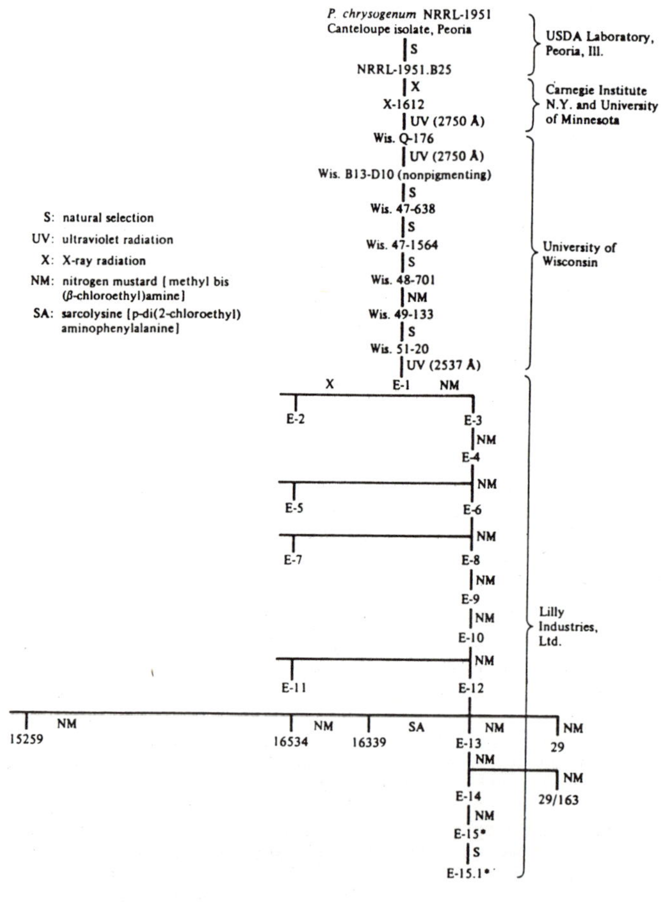

SOURCE: R. P. Elander (1967). Enhanced Penicillin Production by Mutant and Recombinant Strains of *Penicillium chrysogenum*, *Abh. Dtsch. Akad. Wiss. Berlin, Kl. Med.*, 3:403–423.

Table 25 Penicillin Productivity (in shake flask fermentation)

Strain	Improvement in productivity (fold) over			
	Previous strain (Conditions of Moyer and Coghill, 1946)	Fleming strain (Conditions of Moyer and Coghill, 1946)	Previous strain (Modern laboratory conditions)	Fleming strain (Modern laboratory conditions)
Original Fleming				
NRRL-1951	0.25	0.25	0.50	0.50
NRRL-1951.B25	0.80	1.25	2.70	4.05
X-1612	0.63	2.66	0.06	4.90
Wis. Q-176	0.70	5.25	0.52	9.00
Wis. 47-1564	0.88	10.75	−0.04	8.55
Wis. 48-701	−0.14	9.00	−0.14	8.15
Wis. 49-133	0.41	13.08	0.68	12.70
Wis. 51-20	−0.37	7.83	0.73	22.60
E-15	2.73	32.00	1.05	50.00
E-15.1	−0.23	24.16	0.09	55.00

Breeding through the parasexual cycle has rarely produced stable strains, but at least one example of a diploid strain of P. chrysogenum has been described that was more stable than its haploid precursor - a Lilly production strain, E-15. [Elander, 1967] Diploid strains with stable homogeneous population patterns are desirable for production fermentations. It is the view of Dr. Elander (now at Bristol-Myers) with respect to penicillin production that: "The first 30 years of mutation selection and scientific breeding have demonstrated the potentials of mutation breeding and the dangers present in utilizing mutagens which induced chromosomal aberrations. The next 25 years should produce more extensive application of the parasexual cycle to breeding." [Elander, 1979a]

The commercial strains of cephalosporin-producing fungi lack true sexual reproduction, but they too can be manipulated by parasexual recombination. No commercial impact has resulted, but a number of relatively stable diploid strains of Cephalosporium have been reported. [Neusch, 1973]

5.33 Viral Transfer of Genes

Virus-mediated transduction has been reported for S.olivaceus, S.griseus, and other streptomycetes. In the S.griseus system, streptomycin productivity was transferred to a non-producing mutant by actinophage, a virus that infects members of the genus. It is not clear whether the "transactive variants originated through mutation and selection or by phage-mediated recombination." [Elander, 1976] It has been claimed that 35% of the known streptomycetes are lysogenic, i.e., carry viral DNA integrated into their genomes. Periodically there are outbreaks of actinophage infection, especially in connection with streptomycin production. In one such outbreak it was observed that an occasional colony resistant to the phage produces more streptomycin than the sensitive production strain. In such a case, it is usually assumed that the phage are acting as "mutagens" more than vectors. However, one clear case of transduction involved the transfer of the ability to synthesize thiolutin from S.pimprina to S.aureofaciens to create a transformant which showed simultaneous synthesis of the antibiotics normal to both species. [Ramachandran, 1965] To date, no commercially significant use has been attributed to actinophage.

Mycoviruses are common in industrially important fungi, but their roles in antibiotic production are uncertain. The wild-type strain of P.chrysogenum, for example, and its improved-yield descendants harbor viruses containing double stranded ribonucleic acid (dsRNA). Russian workers reported a correlation of bacteriophage viral titer with the yield of penicillin production, and it has been suggested that the genetic basis for the partial synthesis of such secondary metabolites in fungi may involve extrachromosomal determinants [Elander, 1977] However, other work indicates that P.chrysogenum and P.notatum production of penicillin is independent of viral content. [Lemke and Nash, 1974] The presence of viruses containing dsRNA in cephalosporin C-producing fungi has been reported but their significance is unknown at this time. [Elander, 1975]

It has been pointed out [Elander, 1979b] that the desireable objective of increasing the number of copies of limiting genes by phage or episome (plasmid) transfer has yet to be achieved in antibiotic producing organisms because little is known of the potential vectors.* "The prospects for increasing gene copies by episome or phage-transfer probably has current applicability for primary metabolites, i.e., amino acid and organic acid fermentations, but little is known, for example, about the nature of plasmid vs chromosomal synthetase genes in filamentous bacteria and in higher fungi. Once genetic systems are well understood for commercial strains, then these newer sophisticated techniques of molecular genetics may be applicable." [Elander, 1979b] Fulfillment of this objective will mark an important transition from methodologies that depend upon spontaneous rare events to premeditated manipulation analogous to the methodology of chemical synthesis. Just as the organic chemist changes reaction conditions to avoid by-product synthesis, Johnson

* Johnson and Burnett have written, "Overall increases in yields of existing antibiotics by increasing the gene dosage for rate-limiting enzymes should be possible." (Johnson, 1978 p.220], and other antibiotic researchers have expressed the same hope, but no example has yet been published.

and Burnett have suggested, "It should be possible to obtain higher yields of active factors of interest in those cases where fermentation leads to production of several factors, only one or two of which may be pharmacologically useful." [Johnson, 1978] Two techniques introduced in the late 1970s are certain to prove valuable for achieving this objective, as well as opening approaches to new molecules that may be useful antibiotics or for other therapeutic purposes.

5.4 In Vitro Recombination

Manipulation of DNA, to achieve by design what has been slow to develop from spontaneous sexual and parasexual mating among antibiotic producers, is an obvious application of the emerging feasibility of molecular cloning in a variety of organisms. The primary delay is for the development of suitable vectors. Since 2000 of the 2400 known antibiotics are produced by streptomycetes, that is the genus of greatest interest to the pharmaceutical industry. [Elander, 1979b]

It is probably safe to say that every company conducting research on streptomycetes is developing vectors, but the industrial work has yet to be described in the literature.* Most of the published work with streptomycete plasmids has come from the John Innes

---

* Claude Nash of SKF has said privately and in oral discussion at an ICAAC symposium (10/79) that they are developing, in E.coli, a plasmid from Streptosporangium brasiliensis for the purpose of moving genes in streptomycetes to obtain altered aminoglycoside antibiotics and otherwise modify genetic capacities, e.g., improved precursor or primary metabolite production, decreased sensitivity to antibiotics produced, capability of secreting products. Upjohn has published a restriction map of a streptomycete plasmid - an essential first stage in vector development.

Institute in England and a researcher (Bibb) recently moved from there to Stanford.

The previously mentioned plasmid, SCP1 of *S.coelicolor* bears the genetic information for the synthesis of the antibiotic methylenomycin and for resistance to methylenomycin. [Kirby, 1975] (Antibiotic resistance has a special significance for antibiotic producers, since they need to resist self-inhibition by their own products.) The SCP1 plasmid induces methylenomycin production when it is transferred by conjugation to species such as *S.lividans* or *S.parvulus*. [Hopwood, 1977] Unlike all the other plasmid-linked antibiotics studied thus far, the structural genes of methylenomycin and not only of a regulatory molecule, seem to be borne on the plasmid. (Another antibiotic produced by *S.coelicolor* is independent from the plasmid.) SCP1 and SCP2 of *S.coelicolor* and a plasmid of *S.reticuli* involved in leucomycin production and resistance have been observed to promote the transfer of chromosomal genes, by recombination *in vivo*. Attempts to isolate SCP1 for characterization have been frustrated, but several other streptomycete plasmids have been observed, with varying degrees of characterization, as shown in Table 26. [Chater, 1979]

Covalently closed circular (ccc) DNA of a high fertility variant of SCP2 has been shown to be taken up efficiently be streptomyces protoplasts in the presence of polyethylene glycol. Their incorporation is readily detected when the organisms are spread on an agar plate, because the transformed colonies produce a so-called "lethal zygosis" in which a narrow zone of recipient cell growth inhibition

Table 26: Closed Circular DNA Isolated from Streptomycetes

Organism	Plasmid designation	Functions	Mol wt	Restriction enzyme analysis	Aberrations observed
S. coelicolor A3(2)	SCP2 (pSH1)	Transmissible sex factor	$18–20 \times 10^6$	Map	Dimers, trimers, internal duplication
S. coelicolor sp.	pHS2	Unknown	55×10^6	—	—
	pHS3	Unknown	8×10^6	—	—
S. lividans 66	SLP1	Causes "pocks" ("lethal zygosis"?)	7.9×10^6	Map	Dimers, trimers
S. reticuli	—	Transmissible sex factor; sporulation; leucomycin production and resistance; melanin production	48×10^6	Cleavage patterns	Deletions, fragmentation
S. venezuelae 3022a	—	Unknown	18×10^6	—	—
S. kasugaensis	—	Aureothricin production	$\sim6.7 \times 10^6$	—	Deletion?
	—	Kasugamycin production	$\sim30 \times 10^6$	—	—
	—	Unknown	$\sim1.2 \times 10^6$	—	—
S. fradiae sp.	—	Unknown	22×10^6	—	—
		Unknown	15×10^6	—	—
S. fradiae sp.	—	Unknown	55×10^6	Map	Deletion
S. virginiae	—	Unknown	Unknown	—	—

[a] A = M. J. Bibb and J. M. Ward, unpublished data; B = H. Schrempf and W. Goebel, personal communication.

surrounds the transformants. These "pock marks" make it possible to identify a single plasmid-carrying spore in populations of at least 10^9, but in fact, under optimal conditions, transformation of Streptomyces protoplasts can give rise to transformants at the level of 20% of the regenerated population. [Bibb, 1978] Derivatives of this plasmid should be quite suitable as vectors for introduction of foreign genes into those streptomycetes that will maintain them - no streptomycete plasmid has yet been shown to have a wide host range. [Chater, 1979]

The methylenomycin resistance gene of SCP1 has been incorporated into the high fertility variant of SCP2 (and into other streptomycete plasmids) to create practical vectors which enable the identification of transformants on the basis of methylenomycin resistance. The increase in chromosomal recombination frequency for matings involving such plasmids is of the order of 100,000; apparently, all representatives of the recipient strain are "fertilized" with plasmid when a mating experiment is performed. One factor PSCP 111, was demonstrated to replicate and express methylenomycin resistance in at least three streptomycete species, each of which produces at least one antibiotic. Each vector has a number of single cleavage sites in nonessential regions for several different restriction endonucleases - which are important features for facile genetic manipulations. [Bibb, 1980]

For some purposes (e.g., gene amplification and high levels of transcription of introduced DNA), bacteriophage may be preferred vectors, and a number have been identified - some with relatively wide streptomycete host range, heat inducibility, and other desirable characteristics. [Chater, 1979]*

In prokaryotes, functionally related genes tend to be arranged in clusters, e.g., those for oxytetracycline in S.rimosus and for actinorhodin in S.coelicolor are strongly clustered in their respective genomes. [Hopwood, 1977] That would make it relatively easy to transfer all of the genes necessary to produce an antibiotic or a segment of an antibiotic in attempts to create new antibiotics or to improve production by transferring structural genes into a new host. Of course, it will be necessary to learn how plasmids mobilize chromosomal genes in streptomycetes and to overcome or utilize restriction-modification systems present in specific streptomycetes - restriction and modification of phages have been detected. [Chater, 1979] Perhaps it will prove feasible to make use in streptomycetes of transposable elements analogous to the insertion sequences, transposons, and bacteriophage Mu that enable so-called illegitmate recombinations in other prokaryotes.

Little is known about the existence or functionality of plasmids in fungi, or other eukaryotes. To date there have been no reports of plasmid genes controlling secondary metabolite formation in any eukaryotic organism. However, plasmids are found in fungi and yeasts and, probably more important, plasmids can be moved between yeast and bacteria. [Elander, 1980] At least three yeast transformation systems have been developed, as

* Workers at the Institute of Genetics and Selection of Industrial Microoganisms, Moscow, USSR have dominated the investigation of actinophage viruses as vectors for transfer of genetic material between various actinomycetes. [Lomovskaya, 1979]

has one for Neurospora, but no attempt has been described to make antibiotic-producing fungi act as hosts, nor is there much information on the genetic loci involved in fungal biosynthetic pathways leading to antibiotics. "A genetic study of penicillin mutants in Aspergillus nidulans (which was underway [Holt, 1976]) will undoubtedly illuminate the biosynthetic pathway but probably, like nearly all metabolically related sets of genes in eukaryotes, they will be scattered over the chromosomes...the homologous genes will need be identified afresh in P.chrysogenum for manipulation purposes." [Hopwood, 1977]*

5.5 Protoplast Fusion

This technique appears to hold greater immediate promise to antibiotic researchers than does DNA recombination based upon the manipulation of plasmids or phages. Procedures for obtaining protoplasts from yeasts and other filamentous fungi have been known for 20 years, and the use of lytic enzymes to dissolve the mycelium of Penicillium and Cephalosporium was described in the early '70s as a way to elucidate the final stages of antibiotic biosynthesis. [Fawcett, 1973] In the latter part of the decade, the technique was applied to Bacillus and Streptomyces, and the enthusiasm for its practical application is very high throughout the antibiotic industry. [Hopwood, 1979] Protoplast fusion provides a very convenient method for establishing a recombination system in strains, species and genera that lack an efficient natural means for mating.

* A.nidulans (the "academic" penicillin organism) has eight chromosomes, and P.chrysogenum (the industrial workhorse) has seven or eight.

Protoplasts can be prepared from complementary biochemically-deficient (auxotrophic) strains by treatment of whole cells with lytic enzymes (e.g., snail digestive juice). After the cell walls have been removed, and osmotic stabilization provided for the fragile protoplasts, equal quantities of the different strains are combined, centrifuged, and treated with a fusogen, usually polyethylene glycol (PEG). It was the finding that PEG promoted the fusion of protoplasts by a factor of a thousand - which was initially discovered in connection with plant protoplasts - that makes this approach so productive. In the presence of PEG (4000-6000mw, 30% aqueous solution) at optimal calcium ion concentration and akaline pH, cell membranes coalesce and multi-genomic fusion products form readily. These are washed, to remove PEG, and spread on a nutrient medium. A nutrionally minimal medium will not permit the growth of the parents, because of their biochemical deficiences, but cells that possess the necessary genes as result of recombination (prototrophs) flourish.

In *Streptomyces* recombination frequencies of 20% are common. Furthermore, this frequency can be increased even more by additional steps, e.g., UV irradiation of the protoplasts immediately before fusion can raise the frequency from 20% to 40% or, if it is normally only 1%, to 10%.

One is not limited to fusing two kinds of protoplasts. At least four strains of *Streptomyces* have been fused together in a single step to yield recombinants inheriting genes from four parents. [Hopwood, 1979] The technique is applicable to both the prokaryote and eukaryote antibiotic producers, and particularly useful - immediately - to combine the benefits developed in divergent lines by mutation and selection.

Bearing in mind that the first publication on fungal protoplast fusion appeared in 1975 [Anne' and Peberdy, 1976], consider the results published by Glaxo researchers not long after [Hamlyn and Ball, 1978]. The latter made comparative attempts at genetic analysis of Cephalosporium acremonium by the previously available methods of mating and by protoplast fusion. They concluded protoplast fusion to be far superior for that purpose. What is more, the protoplast fusion method gave rise to hundreds of recombinants, including one isolate that consistently produced cephalosporin C in 40% greater yield than the best producer among its parents - without losing that parent strain's rare capacity to utilize inorganic sulfate, rather than expensive methionine, as a source of sulfur - and which had acquired the rapid growth and sporulation characteristics of its less productive parent. This combination of desireable attributes from different strains was readily accomplished with an important industrial organism that had proven recalcitrant in conventional crossing because its hyphal cells are uninucleate, migration of nuclei is restricted and, hence, juxtapositioning of nuclei for recombination is rare: "evidence for haploid recombinants was obtained for all fusion experiments, but in only 25% of conventional mating experiments." [Elander, 1980]

Even more significant are the possibilites for preparation, by means of protoplast fusion, of interspecific and intergeneric hybrid strains that should have unique biosynthetic capacities. Dr. Nash at SKF has told us that they have isolated a novel antibiotic, clearly not produced by either of the parents produced by an organism created through fusion of actinomycete protoplasts; but the work has not yet been published.

Recombination has been demonstrated between two closely related fungi, *Aspergillus nidulans* and *A.rugulosus*, subsequent to protoplast fusion. Many kinds of heterokaryotic colonies are formed, some of which are stable hybrids, identified as interspecific diploids resulting from nuclear fusion. Analysis of recombinant progeny from an extensively marked *A.nidulans* strain, showed that there was no apparent selectivity for any particular *A.nidulans* chromosomes. [Peberdy, 1979] Hybridization is relatively low for interspecific protoplast fusion, and may be limited to regions of homology. "Understanding of the behavior of the hybrids following protoplast fusions is limited and will undoubtedly have to await further developments in this rapidly developing area of fungal genetics." [Elander, 1980]

The only report of a successful intergeneric cross using protoplast fusion technology has been between the yeasts *Candida tropicalis* and *Sacchariomycopsis fibuligera*, which took place at low frequency (10^{-5}) and gave rise to types intermediate between, as well as many similar to, the parents. However, the instability the uninucleate intermediates displayed suggests "they were probably aneuploid with respect to genome composition." [Elander,

1980] Intergeneric hybrids are extremely interesting possibilities; for example, it would be very beneficial to cephalosporin process technology to combine in one organism the acyltransferase from P.chrysogenum and the enzymes of C.acremonium required for synthesis of the cephalosporin nucleus. As it is, C.acremonium does not incorporate side chain precursors onto cephalosporin like P.chrysogenum does for penicillins. [Elander, 1980; Chang and Elander, 1979; Queener and Baltz, 1980]

"The extent to which recombination can be attained between fungal species is an unknown quantity, but the fusion technique at least provides the opportunity to overcome any preconjugation barriers. The applications for interspecific genetics and breeding are potentially very wide. On the one hand, the technique provides the opportunity for a broadening of the gene pool resulting from general mixing and segregation of the genetic material. Alternatively, attempts at more controlled hybridizations are equally possible with a view to modification of existing products, synthesis of new products, or simply changes in the physiology of the producer organism which are ultimately reflected in improved product yields or utilization of new substrates." [Quote from Peberdy, 1979]

Protoplast fusion has major advantages compared to mutation, conjugation or parasexual breeding. The frequency of recombination is so high, for many organisms of interest, that it is practical to screen random samples of the total progeny from a fusion in search of strain improvements. In contrast to the laborious, time-consuming and often crippling necessity of introducing counter-selectable markers into each parent strain before each series of recombination experiments, so that the rare recombinants produced by the other processes can be isolated, protoplast fusions can be performed with the unmodified best candidates. The full genomic potential of parent strains can be combined without any loss due to selection, allowing rapid construction of complex genotypes.

"Protoplast fusion may be regarded as complementary to gene cloning: instead of being applicable to qualitative changes involving few genes, one of its chief uses is likely to be in breeding for a complex quantitative character such as yield of an antibiotic," as David Hopwood has described it. [Hopwood, 1979]

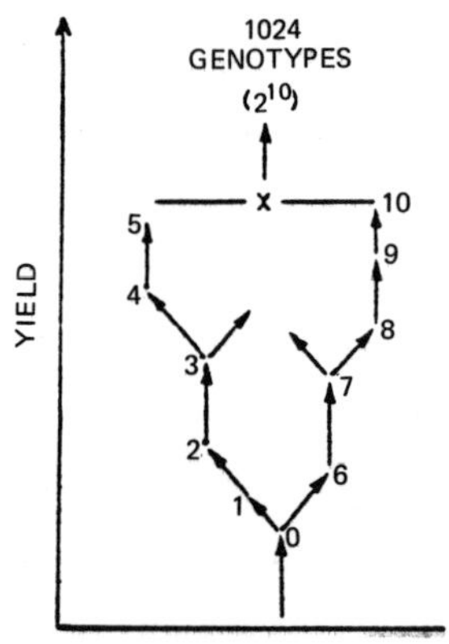

Figure 9: Part of an Idealized Strain-Improvement Scheme

The mutation-selection screening pedigree shown in Figure 9 is based on the assumption that each yield-enhancing step represents mutation in a different gene. "Recombination between the topmost strains in the two branches will generate a very large number of genotypes with respect to yield mutations (1024 in this example, for which five stages of mutation are postulated in each of the divergent lines, i.e., 2^{10}), many of which will not have arisen before, and some of which may show synergistic interactions: that is, particular new combinations will have higher yields than those predictable from the effects of the individual mutations." [Ibid.]

An example of the speculative vistas opened by this technique is provided by a final quotation from Dr. Hopwood. "We could imagine cloning a restriction-modification (DNA cutting and splicing) system into two species that did not possess it, before making an interspecific protoplast fusion, relying on the _in vivo_ operation of the restriction enzyme to achieve desirable new combinations of nonhomologous DNA segments from the two species. In this way, with sufficiently powerful selection, we could hope to produce a complex new genotype which could not be constructed rationally because its components could not be identified in advance... However, we must guard against the danger of disillusionment when it becomes apparent that revolutionary new processes are still going to take some time to reach commercial fruition." [Hopwood, 1979] That juxtaposition of optimism and caution represents the appropriate perspective for the technologies and applications we are discussing.

No doubt a universe of opportunities for drug discovery has been opened and, when one considers the host of products that have been derived literally from the soil of one of its worlds, the imagination boggles at the variety of new entities and means for enhancing their production that probably exist in that universe. Success is not assured, e.g., a weakness today is the inherited instability of the "fused" clones - culture uniformity and reliable productivity are essential characteristics for production strains. Preservation of traits and long-range stability are yet to be resolved. [Elander, 1979b] It seems the most daunting problems lie in the realm of screening - determining what to look for and how to recognize it.

"The new β-lactams are very good examples of the importance of the recognition methods. No new β-lactam was reported from 1956 to 1971, and it seemed that all the β-lactams in existence had been found. The scene changed completely in 1972 after the introduction of β-lactam-super-sensitive strains into the screening and the search for inhibitors of β-lactamases and for sulfur-containing compounds. A new series of β-lactams was discovered and the pool of β-lactams is by no means exhausted." [Elander, 1980] Within the past year, half a dozen new β-lactams have been commercialized and at least a dozen more are in clinical trials around the world. If one added up all the sales forecasts for these new agents, it would be well over a billion dollars.

The simple question is, "What are the unsatisfied needs ?" Most of the new entities must be injected, so that hospitalization or, at least, contact with a medical attendent is implied. No doubt, much attention will be directed toward those properties that make an antibiotic suitable for safe oral administration. Many of the new agents will be promoted on the basis of their activity against organisms that are resistant to established antibiotics; fortunately, such resistance is still quite limited. Because their market is constrained in volume by the generally safe, effective and cheap established antibiotics, and because their production processes are more costly than those of the older agents, the new antibiotics tend to be quite a bit more expensive. The new genetic techniques, especially protoplast fusion, may relieve the latter problem, on a case-by-case basis, but we forsee continuing segmentation and a declining growth rate overall in the developed countries for antibacterial agents - which is the primary market today for antibiotics. Antiviral agents and, especially from a global perspective, antiparasitic agents are areas of enormous unsatisfied need. Today, however, microorganisms have not yielded as much promise for control of viruses and parasites as has chemical synthesis, but we may simply have been digging in the soil of the wrong planet of the universe to which the new technologies provide access.

One should reflect upon the fact that, practically speaking, antimicrobial activity has preoccupied our attention with respect to the molecules we could obtain from microorganisms until now. We shall discuss a few agents (such as the cancer chemotherapeutic agents, adriamycin and asparaginase, and some fungal alkaloids) later, but a review of "Pharmacologically active compounds from microbial origin" [Mathews and Wade, 1977] shows very little application or pursuit of other pharmacologic effects. One can anticipate a conjunction between: 1) the capacities to vary structure and yield of compounds produced by the genera that have been the sources of antibiotics, and 2) the animal, systemic, cellular, and molecular models of diseases made possible by other advances in biomedical technology. That conjunction should open a realm of new drug candidates, more closely related to antibiotics, that may rival or surpass - certainly complement - the realm of agents derived from the knowledge of human biochemistry we have associated with the categories of drugs described in the earlier sections of this assessment.

The Pharmaceutical Industry

6. INDUSTRY PROFILE

The latter half of the twentieth century has witnessed the flourishing of a multifaceted industry that serves the health needs of the population, primarily through providing products to healthcare professionals. Commonly called the pharmaceutical industry, its products now extend far beyond pharmacologically-active agents (drugs) to diagnostic agents and instruments, and to materials and supplies for health care, including agents that are disease-preventative (prophylactic) or health-maintaining (e.g. vitamins) in nature. The molecular genetic technologies that set the perspective for this study have implications in practically all of these areas of pharmaceutical industry activity--although by no means will all, or even most, pharmaceutical company product lines be affected in a substantial way in the foreseeable future by these technologies.

The greatest impact of molecular genetics on the pharmaceutical industry in the next ten years will be in research and development, and that impact is profound. For the first time one can confidently predict that almost any biological phenomenon of medical interest at the cellular level can be comprehended by the diligent application of the scientific tools already in hand! Furthermore--and this is perhaps the bridge between research and development--one can amplify almost any entity involved in the cellular context in which the battle between disease and therapy is waged. At the cellular level, molecular genetics provides to researchers the sort of enhancement in information made possible by "spy-in-the-sky" satellites on a global plane. Just as the terrestrial monitors can provide little information on the network that relates activities in various theaters, so medical researchers are handicapped with respect to the interrelationships between cells, tissues, organs and systems.

New products will come forth as a result of molecular genetics. Some will be produced by molecular genetics. Diseases will be conquered, and the costs that those diseases have imposed upon society will be relieved, which could not be accomplished until this technology became available. Only the most foolhardy prognosticator would try to assign a dollar value to this limb on the tree of healthcare or to plot a growth path. Like a new shoot on a tree, it will tend to grow toward the sunshine, and like the living tree, it will put out some twigs that will wither for a variety of reasons. Like a tree, it may not put out branches exactly where one person or another would wish. The arborist knows that he can influence where the branches develop by pruning. Although trees can be perverse in this respect, if one understands the growth habits of the tree, judicious pruning and supporting may yield a more attractive tree, but woe betide the ill-informed pruner. His tree will not bear fruit! He would be far better off to let nature take its course, or at most to moderately fertilize and water his charge. It would risk wearing out a useful metaphor to consider the equivalents of overshadowing, weeds, and invasive roots, but the reader may find the exercise beneficial. Our purpose, then, in this section of the study, is to describe succinctly the tree that provides the fruits called healthcare products, its growth habits, and its responses to environmental conditions--all with an eye to clarifying where there are public policy issues peculiar to this technological branch we identify as applied molecular genetics.

6.1 Products and Markets

In general several categories of products are usefully distinguished in any characterization of what we shall loosely call pharmaceutical products. The largest segment comprises therapeutic drugs, and in this segment one distinguishes between ethical pharmaceuticals, which are available only by physician's prescription, and proprietary products, which can be freely purchased "over the counter". This report shall concern itself only with the "ethical" sector, for the only foreseeable impact of molecular genetics in the proprietary sector might be in the bulk manufacture of some compounds or intermediates--an impact more chemical than medical in nature. (The sole exception to this exclusion will be a brief comment with regard to vitamins.) Within the "ethical" product sector it is customary to distinguish between prescription legend products, biologicals, and some non-prescription, over-the-counter ethical products. Most of the prescription legend products reflect the influence of synthetic organic chemistry on the industry since the "sulfa drugs" emerged from dye chemistry.

The term biologicals encompasses most of the variety of medicinal products which are produced from living organisms (i.e. blood, plants, microorganisms), except antibiotics.* Antibiotics

*Traditionally, many substances have been "misclassified". For example, microbially-produced ergot alkaloids and plant products such as digitalis, as well as the animal extracts like insulin and heparin, have been included in the category with the non-biologic, chemicals. In practice, the term biologics is applied in the U.S.A. to vaccines and other antigens, antitoxins, therapeutic serums, human blood and blood components--the unifying characteristic of which is that they are complex, incompletely defined natural mixtures, as opposed to the well-defined and reproducible (usually single component) chemical compounds in the other two categories.

are compounds made by microorganisms (and derivatives thereof) which can be administered safely to a person in order to kill another microorganism with which that person has become infected. They are included in the anti-infectives class among the prescription legend agents. A sense of the relative importance of the categories can best be derived from a perusal of the data presented in Table 27 based on the domestic sales of the members of the Pharmaceutical Manufacturers' Association (PMA) for the most recent year upon which such data is available.

Table 27. Domestic Sales ($ millions) by Product Class of PMA Companies in 1978

Product Class	Prescription Drugs	Over the Counter Ethicals
Central Nervous System	$2,052	$105
Anti-infectives	1,339	30
Neoplasms and Endocrine	881	9
Cardiovascular System	861	0
Gastrointestinal & Genitourinary	776	308
Respiratory System	465	246
Dermatologicals	153	110
Other (except below)	439	38
Subtotal	$6,966	$ 846
Vitamins and Nutrients	330	553
Biologicals	255	0
Diagnostic Agents for In Vivo Use	91	
Total	$7,642	$1,399

Actually, that $9 billion in sales is only a fraction of the total healthcare sales of the PMA companies (who do account for the lion's share of the sales in the itemized categories) for it does not include, in addition to the proprietary products, the sales of diagnostic reagents to laboratories and other users, or any sales such as supplies (e.g. bandages, hospital and other equipment) or instruments, which many of these same companies, and others, market for health care. Furthermore, the global sales of U.S.-based firms in these same categories totalled $17 billion and are estimated to have increased to over $19 billion in 1979.

An independent source of overall perspective is the U.S. Department of Commerce, which tabulates $10.44 billion in manufacturer's shipments of pharmaceutical preparations, excepting biologicals, in 1978. The aggregations are usually even less useful than those of the PMA, lumping together products for neoplasms, endocrine system and metabolic disease, for example ($965 million). On the other hand, the Census Bureau breaks out some interesting categories such as $5.573 million shipments of ACTH and $183 million worth of insulin <u>and</u> antidiabetic agents in its detailed data compilation. (In addition to the products for human health already referred to, the Census Bureau compilation indicates manufacturers' shipments of veterinary pharmaceuticals of $451 million, $220 million of which is classified as prescription legend, i.e. dispensed only by or on order of a licensed veterinarian.)

6.11 <u>Biological Products</u> (Biologics), is the term that encompasses bacterial and virus vaccines, antigens, antitoxins, toxoids and analogous products (such as allergenic extracts), serums, plasma and other blood derivatives. In addition to the $255 million domestic sales of biologics for human use reported by the PMA for 1978, there were approximately $75 million worth of biologics sold for animal use. Blood products account for about 60 percent of the total $330 million biologics business, and a single product--human serum albumin--is responsible for 45 percent of that total.* Vaccines are the major non-blood biologics, but no one vaccine's sales approach the sales level of albumin or of leading pharmaceutical products. Most biologics represent specialized, low-sales-volume (< $5 million/year) products, which must bear the same sort of costs of development/registration and liability as more lucrative pharmaceuticals. More so than in any other realm of medical agents are the interests and activities of public sector and non-profit agencies interwoven with those of the pharmaceutical industry in the field of biologics. Each year in the United States there are approximately 90 million new vaccinations and 8.5 million transfusions of blood and components.

Whole blood and red blood cells for therapeutic purposes are not marketed by the pharmaceutical industry--that is the province of the American National Red Cross and public and private blood banks. In general, the pharmaceutical industry has focused its interest on plasma and derivatives. The dominance and scale of the markets for serum albumin and the blood clotting factors used to treat hemophilia have been described in the chapters on plasma proteins and enzymes in the first section of this report: sales of albumin, and the less refined product for the same purpose known as plasma protein fraction, are in the range of $150 million/year and the sales of the two principal plasma fractions for hemophilia amount to approximately $45 million. At least three major competitors, plus several minor ones, participate in these markets and there are alternative sources of supply available in many areas from blood banks and centers, as well as substitute products (e.g. cryoprecipitate) which can

---

* There is, in addition, a more than $100 million trade in whole blood, red cells and platelets in which the pharmaceutical industry does not participate.

be substituted for clotting factors and plasma expanders that are not derived from blood. In large part, the marketability of products from the pharmaceutical industry in this field is based upon the convenience of the dosage forms and their reliability. Sales are almost exclusively to hospitals in the albumin area and primarily so in the hemophilia area.

Immunoglobulins are the next big group of blood products, with sales estimated in the $30 million range domestically. Presently, a single product and marketer dominates the immunoglobulin scene - Rh_o(D) immune globulin, which is used to prevent immunization in Rh-negative mothers following delivery of an Rh-positive infant, in Rh-negative women undergoing abortion, miscarriage, or amniocentesis, or when serious antepartum hemorrhage occurs. The success of the leading marketer, Ortho, is enlightening in that it reflects, not a monopoly position, but a broad involvement in the provision of products for the medical specialties of obstetrics and gynecology. By its users Rh_o(D) immune globulin is viewed as a component of the ob/gyn armamentarium, not as a blood product or an immunologic agent. This is characteristic of the pharmaceutical industry: that products are best comprehended and marketed in terms of the medical practices in which they are used, not the sources from which they are derived or the mechanisms by which they act. The most successful companies focus their R & D and marketing along the lines of the medical specialties. Since discoveries often are revelant to fields remote from the emphasis and expertise of a given company, licensing of products is extremely common.

Reproduced on Page 354 are two tables from the authoritative publication The Medical Letter, which outline the blood products currently in use and their primary indications. (Tables 29 & 30).

Table 29: Indications for Blood Products

INDICATIONS	FIRST CHOICE	ALTERNATIVES
Severe anemia with hypovolemia (acute hemorrhage)	Whole blood	Red blood cells plus a plasma expander
Severe anemia without hypovolemia	Red blood cells	Washed or frozen red cells if high risk of febrile or allergic reaction
Severe anemia with hypervolemia or diminished cardiac reserve	Slow infusion of red cells	Red cells with plasmapheresis; exchange transfusion
Severe thrombocytopenia	Platelet concentrate	
Coagulation Defects		
Hemophilia (Factor VIII deficiency)	Cryoprecipitate or Antihemophile Factor concentrate	Fresh-frozen plasma
Von Willebrand's disease	Cryoprecipitate	Fresh-frozen plasma
Parahemophilia (Factor V deficiency)	Fresh-frozen plasma	Fresh single-donor plasma
Factor II, VII, IX or X deficiency	Factor IX Complex	Single-donor plasma
Congenital hypofibrinogenemia	Cryoprecipitate	Any preparation of plasma
Severe hypogammaglobulinemia; exposure to hepatitis A	Immune serum globulin	
Prevention of Rh-immunization	$Rh_o(D)$ immune globulin	
Prevention of hepatitis B after exposure	Hepatitis B immune globulin	

Table 30: Blood and Blood Components

MATERIAL	USUAL PACKAGE	CONTENT	MAJOR COMPLICATIONS
Whole Blood	unit	450 ml blood	Hepatitis; fever; chills; allergic reactions; circulatory overload; rare hemolytic reactions
Red Blood Cells	unit	220-300 ml RBC - yield from one unit whole blood	Same as for whole blood but with reduced risk of circulatory overload
Washed or Frozen		Same	Same as for RBC but fewer febrile and allergic reactions
Platelet Concentrate	unit	5.5-10.0 x 10^{10} platelets in 30-50 ml plasma	Hepatitis; fever; chills; allergic reactions; development of antiplatelet antibodies; graft-versus-host disease
Granulocyte Concentrate	unit	10-20 x 10^9 granulocytes in 500 ml plasma	Hepatitis; fever; chills; allergic reactions; pulmonary infiltrates; graft-versus-host disease
Fresh-frozen Plasma	unit	200-250 ml plasma - yield from one unit whole blood	Hepatitis; fever; chills; allergic reactions
Cryoprecipitate	unit	80-120 units Factor VIII in 15-25 ml plasma	Hepatitis; fever; chills; allergic reactions; hemolysis due to anti-A or anti-B
Plasma Protein Fraction	50, 100, 200, 500 ml	5% protein solution	Hypotension
Albumin	20, 50, 100 ml or 250, 500 ml	25% albumin solution 5% albumin solution	Overexpansion of blood volume
Antihemophilic Factor	vial or "bottle"	100-1,000 Factor VIII units in 25-30 ml	Hepatitis; fever; chills; allergic reactions; hemolysis due to anti-A or anti-B
Factor IX Complex	vial or "bottle"	450-500 Factor II, VII, IX, X units in 20 ml	Hepatitis; fever; chills; rare disseminated intravascular coagulation
Immune Serum Globulin	2, 10 ml	16 grams/dl IgG	Rare with IM administration
$Rh_o(D)$ Immune Globulin	1 ml	300 or 50 mcg anti-$Rh_o(D)$	Rare
Hepatitis B Immune Globulin	1 ml	IgG with high titer anti-HBsAg	Rare

Vaccines are the other large segment of the biologics field. A recent OTA report lists 51 types of human vaccines (and the establishments licensed to produce them) which are approved in the U.S.A. There are probably an equal number of vaccines for animals, considering combinations in both cases, and a global survey would include 150-200 vaccine products. A beneficial perspective is gained from perusal of Table 31, which is drawn from the OTA report to illustrate the chronology and origin of development for most of the human vaccines used in the U.S.A. American pharmaceutical companies were issued 89 percent of the original licenses for the 42 products for which the establishment that received the original product license still holds that license. The 18 establishments currently producing vaccine products in the U.S.A. hold a total of 143 vaccine product licenses issued by the federal government. Eight American pharmaceutical companies currently hold 70 percent of those licenses; 17 percent are held by foreign-based institutions; two state governments (Michigan and Massachusetts) hold 18 percent, and one is held by the University of Illinois.

Actually, the 143 human vaccine product licenses is only one-fourth the peak that was reached in 1946, (Figure 10) a decline accounted for by increasing reliance on antibiotics, better hygiene and assessments of the efficacy of some vaccines.

Despite the declining number of product licenses in the 1950-60s, the number of manufacturers increased and vaccines against several important viral diseases (e.g. polio, measles and mumps) were developed. The OTA study reports that since 1968 the number of licensed products dropped by 60 percent.

Nevertheless, new or improved vaccines have been introduced

Table 31: Chronological Introduction of Types of Vaccine Products Still Licensed in the U.S.

Year	Type of Vaccine Product	Establishment with Oldest Product License Still in Effect*
1903	Diphtheria antitoxin	Massachusetts Public Health Biologic Lab. (1917)
1907	Tetanus antitoxin	Parke, Davis and Company (1915)
1914	Pertussis vaccine	Lederle Laboratories**
	Typhoid vaccine	Massachusetts Public Health Biologic Lab. (1917)
	Rabies vaccine	Eli Lilly and Company**
1917	Cholera Vaccine	Eli Lilly and Company**
1926	Diphtheria toxoid	Parke, Davis and Company (1927)
1933	Staphylococcus toxoid	Lederle Laboratories**
	Tetanus toxoid	Merck Sharp and Dohme**
1941	Typhus vaccine	Eli Lilly and Company**
1942	Plague vaccine	Cutter Laboratories**
	Rocky Mountain Spotted Fever vaccine	Lederle Laboratories**
1945	Influenza virus vaccine	Lederle Laboratories Merck Sharp and Dohme Parke, Davis and Company**
1946	Diphtheria and tetanus toxoids and pertussis vaccine adsorbed	Parke, Davis and Company**
1947	Diphtheria and tetanus toxoids	Parke, Davis and Company (1949)
1948	Diphtheria and tetanus toxoids and pertussis vaccine	Parke, Davis and Company (1952)
	Diphtheria and tetanus toxoids adsorbed	Parke, Davis and Company**
	Diphtheria toxoid adsorbed	Parke, Davis and Company**
1949	Pertussis vaccine adsorbed	Parke, Davis and Company (1952)
	Gas gangrene polyvalent antitoxin	Lederle Laboratories**
	Tetanus toxoid adsorbed	Parke, Davis and Company (1952)
1950	BCG vaccine	University of Illinois**
1951	Antirabies serum	Lederle Laboratories**
1953	Yellow Fever vaccine	Connaught Laboratories, Inc. (1978)
	Tetanus and diphtheria toxoids adsorbed (for adult use)	Eli Lilly and Company** Wyeth Laboratories
1955	Poliomyelitis vaccine	Merck Sharp and Dohme** Parke, Davis and Company**
1957	Adenovirus vaccine	Parke, Davis and Company**
1959	Adenovirus and influenza virus vaccines combined aluminum phosphate adsorbed	Parke, Davis and Company**
	Diphtheria, tetanus toxoids, pertussis, poliomyelitis vaccines adsorbed	Parke, Davis and Company**
	Diphtheria and tetanus toxoids, pertussis vaccine adsorbed, and poliomyelitis vaccine	Parke, Davis and Company (1963)
1960	Poliomyelitis vaccine adsorbed	Parke, Davis and Company**
1961	Poliovirus vaccine, live, oral, Type 1	Pfizer, Ltd.**
	Poliovirus vaccine, live, oral, Type 2	Pfizer, Ltd.**
1962	Poliovirus vaccine, live, oral, Type 3	Pfizer, Ltd.** Lederle Laboratories**
	Measles virus vaccine live, attenuated	Merck Sharp and Dohme**
	Poliovirus vaccine, live, oral, trivalent	Lederle Laboratories**
1967	Measles-smallpox vaccine, live	Merck Sharp and Dohme**
	Mumps virus vaccine	Merck Sharp and Dohme**
1969	Rubella virus vaccine, live	Merck Sharp and Dohme**
1970	Anthrax vaccine adsorbed	Bureau of Laboratories, Michigan Department of Public Health**.
	Rubella and mumps virus vaccine, live	Merck Sharp and Dohme**
1971	Measles and rubella virus vaccine, live	Merck Sharp and Dohme**
	Measles, mumps, and rubella virus vaccine, live	Merck Sharp and Dohme**
1973	Measles and mumps virus vaccine, live	Merck Sharp and Dohme**
1974	Meningococcal polysaccharide vaccine, Group C	Merck Sharp and Dohme**
1975	Meningococcal polysaccharide vaccine, Group A	Merck Sharp and Dohme**
	Meningococcal polysaccharide vaccine, Groups A and C combined	Merck Sharp and Dohme**
1977	Pneumococcal vaccine, polyvalent	Merck Sharp and Dohme**

*Dates in parentheses indicate dates of product licensure for vaccine products for which original license holders no longer hold licenses.

**Establishment issued original product license.

Source: OTA's interpretation of data provided by the Bureau of Biologics, 1979.

at a rate of 3-7 products every five years since 1940. (A total of 39 added since that date which, along with 10 introduced before 1940 and the two broad categories with "no

Figure 10: Total Number of Vaccine Products Licensed in the U.S.A. by Year (1903-79)

The number of licenses shown on this graph represents the net number of active vaccine *product* licenses for each year. Totals for each year were calculated by tallying a running total of the number of vaccine *product* licenses issued over time and subtracting the number of vaccine *product* licenses revoked over time.

Source: OTA's interpretation of data provided by the Bureau of Biologics, 1979.

U.S. standard of potency",* constitute the 51 currently licensed vaccine products.) The licensed products are intended to provide immunity against 23 different types of infections, itemized in Table 32 [OTA, 1979].

Table 32: Diseases Against Which There Are Currently Licensed Immunizing Agents in the United States (1979)

General population (7 diseases)	Special populations (16 diseases)
Diphtheria	Adenovirus
Pertussis	Anthrax
Polio	BCG
Measles	Cholera
Mumps	Gas gangrene
Rubella	Influenza
Tetanus	Meningococcal diseases
	Plague
	Pneumococcal pneumonia
	Rabies
	Rocky Mountain Spotted Fever
	Smallpox
	Staphylococcal disease
	Typhoid
	Typhus
	Yellow Fever

Source: OTA's interpretation of data provided by the Bureau of Biologics, 1979.

Table 16, in an earlier section (p. 272), presented data with regard to the extent of use and mode of dissemination for the eight most important vaccines in the United States as of the mid-1970s. There, it was noted that 45% of the 90 million immunizations administered are sponsored by public sector agencies. Table 15 (p.271) vividly documents the benefits accrued from some of these programs in the U.S.A. Sales of biologicals overall to government amounted to $44 million (17.3% of total) in 1978, according to the PMA, including blood products as well as vaccines.

On a global scale, the benefits of biologic prophylaxis have been significant, but much remains to be done. The most important

* The designation "no U.S. standard of potency" is applied to certain polyvalent bacterial antigens and polyvalent bacterial vaccines (12 of the latter) that are produced in limited quantities.

vaccines universally are the same ones employed in the U.S.A., especially measles, polio, typhoid and diphtheria. Surprisingly, there has been recent controversy about the efficacy of such common immunizations as the vaccines against whooping cough (pertussis) and tuberculosis (BCG). These and other aspects of worldwide immunization programs that exceed the scope of this study are described in the literature. (WHO, 1977, for example.)

In the U.S.A., the biological products sector is defined, rather arbitrarily, by the purview of the Bureau of Biologics (BOB) within the FDA. The BOB emerged in 1972 from the National Institutes of Health's (NIH's) Division of Biologic Standards, which itself had evolved from the NIH's National Microbiological Institute at the time (mid-50s) when the polio vaccines were coming into prominence--BOB's ancestry goes back to 1902, even farther than FDA's drug involvement (1938). BOB reviews clinical and other studies, grants licenses, and monitors each product under its jurisdiction through lot-by-lot testing.

Veterinary vaccines include rabies and distemper immunizations for dogs, cats and other small animals, accounting for roughly $30 million sales, about $15 million worth of poultry vaccines and at least another $30 million worth of large animal vaccines of many types and combinations. Many more producers are involved in animal vaccine production than in the human vaccine sector, often small regional operations. However, the most important animal vaccine worldwide--hoof-and-mouth disease vaccine--isn't even used in the U.S.A.*, as is the case for many vaccines upon which the availability of animal protein and labor depends

*The U.S.A., Canada, Australia and New Zealand are free of foot-and-mouth disease, but the rest of the world either suffers significant loss due to that disease, is waging a battle to eradicate it by immunization, or is struggling to maintain disease-free status. An estimated 800 million doses of vaccine are used annually worldwide.

in other countries. A list of vaccines against animal viruses alone illustrates the diversity of the products and markets that may be impacted by new developments in vaccine technology# (Table 33).

Table 33. Antiviral Vaccines Available for Use in Animals

Disease Prevented		Virus Group	Forms of Vaccines	Source of Antigen	First Reported
Avian:	Encephalomyetis	Picornavirus	IN	Chick Brain	1959
			LA	CE	1961
	Fowl Plague	Myxovirus	IN	CE	1947
			LA	CE,TC	'49,'60
	Fowlpox	Poxvirus	LA	CE,TC	1939
	Infec. Bronchitis	Cornavirus	LA	CE	1952
	Laryngotracheitis	Herpesvirus	LA	CE	1933
	Newcastle Disease	Paramyxovirus	LA	CE,TC	'46,'50,'53
			IN	CE	1954
Canine:	Distemper	Paramyxovirus	IN	Canine Tissues	1928
			LA	Ferret Tiss., CE,TC	'39,'51,'60
	Infec. Hepatitis	Adenovirus	IN	Canine Tissues	1952
			LA	TC	1956
	Rabies	Rhabdovirus	IN	Nervous Tissues	1881
			LA	CE or TC-LEP Strain	'50,'56','66
Bovine:	Papillomatosis	Papovavirus	IN	Bovine Warts-TC	?
	Vesicular Stomatitis	Rhabdovirus	LA	CE	1963
	Infec. Rhinotracheitis	Herpesvirus	LA	TC	1957
	Parainfluenza	Paramyxovirus	LA	TC	1962
	Viral Diarrhea	RNA-unclassified	LA	TC	1961
	Foot & Mouth Disease*	Picornavirus	IN	TE	'38,'51
			LA	Mice, TC, CE	1960
	Rinderpest*	Paramyxovirus	IN	Bovine Tissues	1928
			LA	CE	1946
	Rift Valley Fever*	Arbovirus	LA	MB	1957
	Wesselsbron Disease*	Arbovirus	LA	MB or TC	1956
Porcine:	Hog Cholera	RNA-unclassified	IS	Hog Blood	1908
			IN-E	Hog Tissues	1933
			IN-CV	" " & Blood	1936
	Transmissible Gastroenteritis*	Unclassified	LA	Rabbit " & ", TC	'46,'53
	Teschen	Picornavirus	IN	TC	1965
			LA	TC	1959
Equine:	Eastern Encephalo.	Arbovirus	IN	CE,TC	1938
	Western Encephalo.	Arbovirus	IN	CE,TC	1938,'68
			LA	CE	1963
	Influenza	Myovirus	IN	CE	1965
	Rhino pneumonitis	Herpesvirus	L	CE	1955
	African Horse Sickness	Reovirus	LA	MB,TC	'58,'65

*Vaccine not available in USA.

Abbreviations: IN = Inactivated; LA = Live, Attenuated; L = Live, Non-attenuated
IS = Immune Serum; IN-E = Inactivated, Eucalyptol;
IN-CV = Inactivated, Crystal Violet;
CE = Chick Embryo; TC = Tissuse Culture; MB = Mouse Brain
TE = Tongue Epithlium, *in vivo* & *in vitro*

Compiled from: "Viral Vaccines Against Animal Diseases - Past, Present & Future," Cabasso, V.J. in *Viruses Affecting Man and Animals*, Sanders, M. ed., W.H. Green, St. Louis.

One major difference between animal and human viral vaccines is that it is permissible to use continuously-multiplying cell substrates for the propagation of animal viruses--a major economic advantage compared to the normal (untransformed) cells to which the tissue culture of human viruses is confined.

In addition, there are many rickettsial/bacterial vaccines and bacterins in use, and antiparasitic biologicals under investigation around the world. While the FDA oversees veterinary pharmaceuticals, the USDA is charged with the regulatory responsibility for animal health biologics.

6.12 Antibiotics, like biologicals, are distinguished on the basis of their origin, i.e. their derivation from microorganisms They make up such a large part of the antimicrobal sector that the term "antibiotic" is often applied indiscriminately to any agent used to combat infection. Actually, antibiotics generate 94 percent of dollar sales in the anti-infective category, which, in turn, accounts for 18 percent of prescription drug sales in the U.S.A. Worldwide, they account for about 11 percent of therapeutic product sales according to one estimate--$6.8 billion of a $62.5 billion total projected for 1980.* (Gargiulo, 1977.)

One finds considerable variation in the relative importance of the different families of antimicrobials when one compares regions of the world, as in Table 34. Penicillins account for roughly one-third of antibiotic usage and over half of antimicrobial sales are attributed to the β-lactams, the chemical family which includes both penicillins and cephalosporins. A variety of other structural types constitutes the remainder of the more than one hundred antibiotics now used throughout the world. It is customary to categorize antibiotics in these chemical terms, but in practice, they are prescribed on the basis of their appropriateness for treating a particular infection in a specific kind of patient.

---

*A number of estimates place the worldwide sales in the $5 billion range. The most recent, by Information Research Ltd, estimates world sales of antibiotics at $8.25 billion, in a total of $75 billion, as of July 1980.

Table 34. Estimated Manufacturers' Sales of Antimicrobials in Regional Markets. (Adapted from Kidder, Peabody & Co. Projections for 1979, as of 1975.) (Abeles, 1975)

	Manufacturers Sales in Millions				
	United States	Western Europe	Japan	Rest of World	Projected World Total
Penicillins	$260	$445	$305	$510	$1520
Cephalosporins	280	255	275	80	890
Tetracyclines	115	170	110	335	730
Sulfonamides	115	150	75	155	495
Aminoglycosides	130	100	100	55	385
Macrolides	95	55	65	70	285
Chloramphenicol	5	5	70	75	155
Other	45	35	60	45	185
Total	1,045	1,215	1,060	1,325	4,645
Region as percent of total	26.2%	22.5%	22.8%	28.5%	100%

Everyone knows the story of Fleming's discovery of penicillin, however, few realize that within a few days of his discovery in 1928, Fleming had determined that the unusual substance produced by his adventitious fungus killed streptococci, micrococci, gonococci, the spirochetes of syphilis and the bacilli of diphtheria, tetanus and anthrax, as well as the staphylococci upon which it first displayed its activity. He also noted its limited activity against *coli* bacteria and the bacillus then thought to be the cause of flu. Though the scaling up and clinical testing had to wait about twelve years, until the team led by Florey did the necessary development work, the activity limits were rapidly established by simple tests *in vitro*. The existence of such tests provided the key that underlay the exploitation of microorganisims for the following 40 years. Even before penicillin had been tested in people, tyrothricin and gliotoxin were recognized as antimicrobials from other organisms, and shortly after the clinical success of penicillin, Waksman and his colleagues identified a product of *Streptomyces griseus*, streptomycin, that kills microorganisms impervious to penicillin, e.g. the cause of tuberculosis. However, tyrothricin could only be used topically, for it was toxic if injected, and streptomycin was also more toxic than penicillin. Thus began the ongoing efforts to find new antibiotics and modify the ones in hand in order to extend the range of activity and decrease side effects. A worldwide survey of soil samples launched in 1945 yielded the first tetracycline three years later. Not only did it suppress or kill the same bacteria that were sensitive to penicillin, but it proved active against other organisms.

Furthermore, since Aureomycin was essentially bacteriostatic rather than bacteriocidal, it turned out to be even more effective *in vivo* than *in vitro*. Its toxicity was very low, and it did not provoke the allergic response that patients occasionally showed to penicillin. Thus, it has gone on for decades--new antibiotics finding niches not filled by existing antimicrobial agents. For example, the tetracycline could be given orally (penicillin-G, as the first penicillin came to be called, and streptomycin had to be injected), and it permeated certain body fluids, such as the cerebro-spinal fluid, and tissues better than the preceding antibiotics.

Aureomycin, discovered at Lederle by a retired academician,* marked the entry of the pharmaceutical industry into the antibiotic discovery process--a role in which many drug firms have

---

*Seventy-year-old Benjamin Duggar, Professor of Plant Physiology and Botanical Economics at the University of Wisconsin, retired from an academic career that included providing the basis for the United States becoming independent of imports of edible fungi, especially mushrooms, and joined the staff of Lederle Laboratories in 1944.

persevered ever since. More than 3,000 isolations were made by Duggar and his over 100 assistants from 800 samples of soil. Charles Pfizer & Co.--a pioneer in industrial fermentation, especially the mass production of citric acid--had been one of the first three producers of penicillin, a wartime effort. Its search for a novel antibiotic involved 116,000 soil samples.

After an enormous amount of tedious labor, including the reisolation of all the then-known antibiotics, A. C. Findlay and his associates found a new compound which elegant chemical detective work showed to be remarkably similar to Aureomycin. The Pfizer product, Terramycin, which came to be called oxytetracycline, and Lederle's Aureomycin (chlortetracycline) were members of the same complex, four-ring structural family of compounds--the tetracyclines--produced by different members of the genus *Streptomyces*, soil fungi. Their therapeutic properties were practically identical, and thus was launched competition in the antibiotic field.* By the end of the 40s, there were half a dozen competitors in the American market.

Both Lederle and Pfizer soon produced the "parent" compound, tetracycline, Lederle naming it Achromycin and Pfizer, Tetracyn. Initially it was produced by chemical removal of a chlorine atom from Aureomycin, but then a strain of *Streptomyces aureofaciens* was developed, by depriving the organism of chlorine, which had lost the ability to make Aureomycin and yielded, instead, its metabolic precursor, tetracycline. That step was accomplished by a German firm, Farbwerke Hoechst, and it provided the starting material for preparation of a derivative that would dissolve in water and be suitable for injection (rolitetracycline).

---

*Like Pfizer, Lederle possessed fermentation know-how (from production of the vitamin riboflavin) upon which to base rapid expansion.

The tetracyclines were identified at just the right time, for the phenomenon of penicillin resistance was then emerging. First one, then another, then 34 of 50 strains of staphylococcus, the microorganism upon which Fleming had first observed the antibiotic's action, were found to be unaffected by penicillin. Aureomycin proved effective against the 34 strains of staphylococci that resisted penicillin and streptomycin.*

Subsequently, novel drugs active against such resistant strains were prepared by making chemical derivatives of penicillin also. The resistant strains produce enzymes that deactivate most compounds containing the structural features, called a β-lactam, which is critical to their antimicrobial action. Synthetically-modified penicillins were eventually devised which are stable in the presence of the most common of these enzymes, but a few microbes make β-lactamases that attack not only the penicillins but related compounds called cephalosporins. The contest goes on between researchers and germs, a new strain always emerging not long after the invention of a less vulnerable antibiotic, proliferating while the antibiotic eliminates its normal competition, and sometimes even passing the resistance gene on to other species by the sort of natural conjugation and plasmid transfer described earlier in this report.

One class of pathogen not impeded by penicillin or tetracycline was that which grows only in the absence of air (anaerobes). These organisms, such as the Clostridia, that cause tetanus and gangrene, proved vulnerable to another discovery of the 40s. Parke-Davis' chloramphenical not only filled that gap, but it conquered typhus and other diseases caused by rickettsiae,

*Only in hospitals, where the germs encountered antibiotics all the time, did such resistant strains develop.

very tiny, non-viral organisms that live inside human cells.* It proved to be a real life-saver in typhoid fever and rapidly came to be widely used abroad. Sales zoomed from $9 million in 1949 to $52 million in 1952. Yet serious, but rare, side effects later precipitated a contraction of usage that has become a common sequel to success in this field, e.g. ampicillin, the oral penicillin, replaced chloramphenicol for treatment of typhoid.

Other aminoglycosides were developed which were superior to the first, streptomycin, for all but tuberculosis, and affected most anaerobes. Erythromycin, a macrolide, proves a useful substitute where there seems to be a danger of allergic hypersensitivity to penicillin, and its penetration of certain tissues is superior to that of penicillin. In many ways the cephalosporins, which resist deactivation by penicillinase, replaced penicillins for several purposes. Furthermore, cephalosporins affect some Gram-negative organisms that are resistant to penicillin. Thus, they rose to prominence in the '70s.

Other antibiotics, such as the antifungals and the lincomycins, also found niches--all subject to challenge by new agents or of adverse effect allegations. Dynamic evolution characterizes the antibiotic field to this day and seems likely to persist into the future, with new agents challenging old and physicians challenged to choose between familiar agents with certain limitations and new agents, less certain but possessed of definite advantages. In that respect, antibiotics represent a microcosm of the host of medicinal agents from which today's doctor must select the proper therapy for a given patient.

---

*Although originally found as a product of a streptomycete, chloramphanicol became the only antibiotic to be produced commercially by chemical synthesis rather than fermentation.

6.13 Other Drugs and Vitamins will not be treated in the manner that biologics and antibiotics have been discussed. As Table 27 set forth, we are talking about $7.4 billion worth of products and that scope exceeds our capacity or purposes. Most of those products are derived by synthetic chemistry from bulk chemicals produced by the petrochemical industry. However, plants provide the basic raw materials for a number--not just the familiar digitalis and opiates derived from flowers, but, for example, steroids (such as contraceptives) made from yams or the host of so-called alkaloids that have been incorporated into modern medicine from folk remedies. Vitamins, those co-factors essential in small amounts for the maintenance of normal health, are also produced primarily by chemical manufacture--although it is becoming more common in recent years to use microorganisms to perform a specific transformation or to prepare a key starting material rather than eschew biochemistry altogether. Finally, we must reiterate the role played by a few animal extracts, such as insulin and other hormones or the enzymes mentioned earlier.

The Physician's Desk Reference, which provides information on only the pharmaceutical products most used in the U.S.A.--perhaps half the total number available--contains about 2,500 entries. According to the PMA, there are approximately 2,400 different pharmaceutical entities on the market--1000 of which have been introduced since 1940--represented by a great range of dosage forms and brand names.

The categories used by the PMA (Table 27) in the presentation of sales data provide little aid in relating the products of the pharmaceutical industry to the needs of patients as interpreted by their physicians. Biochemical tests supplied by the industry assist the physician in a small but growing way to diagnose. However, classical observation of signs and symptoms, plus response to medication, more often guide the choice of therapy. The expanding pharmacopaeia enables the physician to shift such key physiological variables as blood volume, heart rate and output per beat, and the resistance of the vascular system in dealing with cardio vascular problems, for example. Much of medicine remains empirical because so little is known about any one patient's underlying disorder(s). The mechanisms of action of many agents are known, but at least as often they remain mysterious. Together, the homeostatic mechanisms inherent in the human body and the responses provoked by the medications restore health more often than not, but a great deal remains to be learned, as we all know.

Just as specialties have developed in medical practice and research, pharmaceutical companies show a tendency to specialize in disease areas for which their experience and current expertise is particularly relevant. Often they market to specific kinds of physicians and provide a range of agents for related conditions. However, the largest firms are quite diversified as a result of their historical evolution and the multiple uses of so many pharmacologic entities.

6.14 International and multinational are certainly terms that apply to the pharmaceutical industry. U.S.-based firms have been far more aggressive in penetrating foreign markets for drugs and other health products than for many other American products. Conversely, there is a long tradition of licensing pharmaceutical products from inventors abroad, and in the past two decades, increasing entry of foreign-based firms with their own marketing, manufacturing and research establishments in this country.

Although disease is universal, there are considerable differences in the practice of medicine, government regulation and other factors relating to the delivery and payment for healthcare from country to country. For example, pharmaceutical expenditures as a percent of total healthcare expenditures vary from 12 percent in the U.S.A. to 33 percent in Japan. (Schaumann, 1976.)

The marked differences in medical priorities and preferences in seven representative countries are apparent from the data presented in Tables 35 and 36.

Table 35: Leading Therapeutic Classes by Sales Through Retail Pharmacies in Selected Developed and Developing Markets

Country/Class	Market Share (%)
Brazil	
systemic antibiotics	14
cough and cold preparations	6
vitamins	6
antispasmodics	4
sex hormones	4
analgesics	3
antirheumatics	3
psycholeptics	3
psychoanaleptics	3
cholagogues/hepatic protectors	3
Pakistan	
systemic antibiotics	25
vitamins	13
cough and cold preparations	5
analgesics	5
nutrients	3
antianemics	3
antidiarrheals	3
antacids	3
tuberculostatics	3
antispasmodics	3
Philippines	
systemic antibiotics	19
cough and cold preparations	12
vitamins	8
analgesics	6
tuberculostatics	5
nutrients	5
antiasthmatics	3
antidiarrheals	3
topical steroids	2
antacids	2
Venezuela	
systemic antibiotics	14
vitamins	8
cough and cold preparations	7
analgesics	4
sex hormones	4
topical steroids	3
Venezuela (con't.)	
antianemics	3
psycholeptics	3
systemic steroids	3
antirheumatics	3
Japan*	
systemic antibiotics	26
vitamins	6
antirheumatics	4
antacids	4
hematologicals	4
hospital solutions	4
cardiac therapy	3
cytostatics (anti-cancer)	3
psycholeptics	3
cholagogues and hepatic protectors	3
USA	
psycholeptics (tranquilizers/sedatives)	9
analgesics	8
systemic antibiotics	7
cough and cold preparations	6
vitamins	5
hypotensives	4
diuretics	4
sex hormones	4
antirheumatics	3
psychoanaleptics	3
West Germany	
cardiac therapy	11
psycholeptics	6
peripheral vasodilators	5
cough and cold preparations	5
analgesics	4
vasoprotectives	4
systemic antibiotics	4
hypotensives	4
sex hormones	3
antidiabetics	3

*Measured by sales through wholesalers.

From "The Pharmaceutical Industry and the Third World," a pamphlet published by the Pharmaceutical Manufacturers Association, 1979, p. 36.

Table 36: Comparative Ranking of the Leading Ten Therapeutic Classes in Selected Developed and Developing Countries

Therapeutic Class	Brazil	Pakistan	Philippines	Venezuela	Japan	USA	West Germany
systemic antibiotics	1	1	1	1	1	3	7
cough and cold preparations	2	3	2	3	—	4	4
vitamins	3	2	3	2	2	5	—
analgesics	6	4	4	4	—	2	5
psycholeptics	8	—	—	8	9	1	2
sex hormones	5	—	—	5	—	8	9
antirheumatics	7	—	—	10	3	9	—
antacids	—	8	10	—	4	—	—
hypotensives	—	—	—	—	—	6	8
cardiac therapy	—	—	—	—	7	—	1
psychoanaleptics	9	—	—	—	—	10	—
cholagogues	10	—	—	—	10	—	—
antispamodics	4	10	—	—	—	—	—
nutrients	—	5	6	—	—	—	—
antianemics	—	6	—	7	—	—	—
antidiarrheals	—	7	8	—	—	—	—
tuberculostatics	—	9	5	—	—	—	—
topical steroids	—	—	9	6	—	—	—
diuretics	—	—	—	—	—	7	—
hematologicals	—	—	—	—	5	—	—
hospital solutions	—	—	—	—	6	—	—
cytostatics	—	—	—	—	8	—	—
peripheral vasodilators	—	—	—	—	—	—	3
vasoprotectives	—	—	—	—	—	—	6
antidiabetics	—	—	—	—	—	—	10
antiasthmatics	—	—	7	—	—	—	—
systemic steroids	—	—	—	9	—	—	—

From "The Pharmaceutical Industry and the Third World," a pamphlet published by the Pharmaceutical Manufacturers Association, 1979, p. 37.

One measure of international markets is the foreign sales of U.S.-based firms. Table 37 presents the relative shares of the therapeutic sectors in 1978 sales for domestic and foreign markets. A view of where in the world the sales are made is provided by Table 38, which shows the predominance of Western European markets in the overseas efforts of U.S. firms.

Foreign sales rose 22% in 1978, compared to a 12% increase in domestic sales, and foreign sales now account for 42% of volume for U.S.-based firms--up from 28% a decade ago.

Table 37. Relative Share of Sales in 1978 by Product Class, Dosage Form Ethical Drugs (Prescription and Over-the-Counter Ethicals) for Human Use

	Domestic	Foreign
Central Nervous System	23.9%	14.5%
Anti-infectives	15.1	25.1
Gastrointest. & Genitourinary	12.0	10.8
Neoplasms & Endocrine	9.8	10.2
Vitamins & Nutrients	9.8	5.5
Cardiovasculars	9.5	9.2
Respiratory System	7.9	12.2
Dermatologicals	2.9	2.8
Biologicals	2.8	1.2
Diagnostic Agents (In Vivo)	1.0	0.1
Other	5.3	8.4
	100.0%	100.0%

Table 38. Distribution of Foreign Sales by U. S. firms, 1978

	Percentage of Total Foreign Sales
Western Europe	47.3%
Latin America	19.9
Far East & Pacific	18.4
Middle East & Africa	6.8
Canada	6.6
Socialist Bloc	1.0
	100.0%

Table 39 summarizes the dollar values of domestic and foreign sales of human and veterinary products in bulk and dosage form, as compiled by the PMA. The $17 billion sales in 1978 amounted to a 15.9% increase over the preceding year--the rate of increase has averaged about 10% the past 25 years. In that year, the U.S. pharmaceutical industry contributed $954 million to the U.S. balance of trade. In 1979, exports increased 13% and imports decreased slightly compared to 1978. While a substantial deficit was again reported for U.S. foreign trade in general, the year-end data on exports and imports of medicinal and pharmaceutical products for 1979 indicate that the industry made a positive contribution of $1.15 billion to the balance of trade in 1979, (up 21%).

Table 39: Ethical Pharmaceutical Sales[a], Domestic and Foreign: Shares of End-Use Totals, 1978 (millions of dollars)

	Domestic U.S. Sales			Foreign Sales			Total	Sales for Export
	To Private Sector	To Govern-ments	Total Domestic Sales	Exports to Other Firms	Sales Abroad[b]	Total Foreign Sales		Intra-Firm Trans-actions[c]
Dosage Form Human Use	$8,418.8 (54.6%)	$736.7 (4.8%)	$9,155.5 (59.4%)	$220.0 (1.4%)	$6,050.0 (39.2%)	$6,270.0 (40.6%)	$15,425.5 (100.0%)	$ 252.3
Dosage Form Veterinary Use ..	272.1 (44.3%)	.9 (.1%)	273.0 (44.4%)	14.2 (2.3%)	327.1 (53.3%)	341.3 (55.6%)	614.3 (100.0%)	53.0
Bulk Human Use	253.0 (45.9%)	1.6 (.3%)	254.6 (46.2%)	87.4 (15.9%)	209.1 (37.9%)	296.5 (53.8%)	551.1 (100.0%)	385.9
Bulk Veterinary Use ..	233.8 (56.0%)		233.8 (56.0%)	7.4 (1.8%)	175.8 (42.2%)	183.2 (44.0%)	417.0 (100.0%)	44.5
TOTAL	$9,177.7 (54.0%)	$739.2 (4.3%)	$9,916.9 (58.3%)	$329.0 (1.9%)	$6,762.0 (39.8%)	$7,091.0 (41.7%)	$17,007.9 (100.0%)	$ 735.7

[a]Sales are before deducting cash discounts and other marketing expenses, but after returns and allowances (domestic returns and allowances totaled $253 million in 1978).
Export sales are f.o.b. U.S. port. A majority of the firms reported most domestic U.S. sales were made f.o.b. purchaser's location or equivalent. The above domestic U.S. dosage form sales to the private sector are gross at invoice price. For "f.o.b. manufacturer's plant" totals, deduct $292 million ($146 million transportation-out and $146 million company branch or field warehousing).
[b]Sales abroad refers to sales in a foreign area by subsidiaries or other corporate operations and excludes U.S. export sales to other firms. Excluded also are sales outside the U.S. by foreign-owned firms that have subsidiaries in the U.S. as PMA member firms.
[c]Sales for Export Intra-Firm Transactions are sales to own International Division or for export to subsidiary abroad and are included as part of cost in Sales Abroad. Addition of the two export columns will provide aggregate ethical pharmaceutical exports.

6.2 Companies

Generally the participants in the pharmaceutical industry can be described as research-based, integrated, multinational businesses. Scarcely any significant firm today can afford to operate with an R&D budget less than 10 % of pharmaceutical sales. That is substantially greater than any other segment of industry, according to a National Science Foundation survey. Today almost every company manufactures and markets its own products; although there is a vestigial custom-manufacturing-

and-packaging trade from former days when it was common for half of the products sold by a marketing force to be produced by another company--often one with no field sales force. Historically many firms derive from chemical, especially dyestuff, producers and fermenters, but today the ties to specific raw materials or process technology are apparent only in a few instances. The multinational characteristic follows from the R&D basis and the fact that highly technical solutions to problems are not readily imitated without violation of patent laws in the major market countries. Furthermore, approval for marketing requires the submission of clinical data that takes time to amass, and promotion makes use of publications on a specific product in the international scientific literature. On the other hand, the different national characteristics of medical practice in various countries are such that experienced local representatives are essential to marketing success. Also critical is familiarity with the regulatory and healthcare reimbursement systems provisions of the local government. Hence, marketing agreements with local firms, joint ventures and acquisitions are all common mechanisms for penetrating foreign markets in this business. The export of finished drug products represents approximately 13% of the international business of U.S. manufacturers. The largest proportion of the business is conducted by overseas subsidiaries established by direct investment from the parent companies. Forty-two percent of the 307,500 persons employed by U.S. based firms were in foreign locations in 1978. [PMA, 1980]

6.21 National Origin

The status and trends that characterize the nationality of the pharmaceutical market in 25 important countries are revealed by the estimates shown in Table 40. In only eight countries was more than half the market controlled by domestic companies in

1975, and the expectation was that by 1985 the majority of pharmaceutical business would remain in the hands of locally based firms in only five countries. Part and parcel of that forecast was the assumption that European and Japanese pharmaceutical firms will gain at the expense of U.S. companies. [Schaumann, 1976]

Table 40. Pharmaceutical Market Shares Held By The Domestic Industry in 25 Selected Countries, 1975 and 1985.

Country	Estimated 1975	Expected 1985	Real Change
Argentina	30%	32%	+ 2%
Australia	15	20	+ 5
Belgium	10	12	+ 2
Brazil	15	20	+ 5
Canada	15	18	+ 3
France	55	45	-10
India	25	30	+ 5
Indonesia	15	20	+ 5
Iran	25	32	+ 7
Italy	40	45	+ 5
Japan	87	77	-10
Mexico	18	20	+ 2
Netherlands	40	40	0
Nigeria	3	10	+ 7
Philippines	35	35	0
Saudi Arabia	0	10	+10
South Africa	40	40	0
Spain	55	45	-10
Sweden	50	43	- 7
Switzerland	72	68	- 4
U.K.	40	45	+ 5
U.S.	85	73	-12
U.S.S.R.	100	100	0
Venezuela	12	17	+ 5
West Germany	65	60	- 5

Note: The "domestic" share accounts for sales (at manufacturer's level) invoiced by more than nationally 50%-owned manufacturing companies; products under license by these companies from foreign companies are included in domestic sales. By contrast, the foreign share includes both sales by companies that are over 50%-owned by foreign companies (including companies that have only a pro forma domestic majority shareholding) and sales by domestic companies that function as contract marketing, distribution, and in rare cases, manufacturing agents for foreign companies. In short, "foreign" companies are those whose policies, in the final analysis, are dictated by foreign interests. Sources: Internal SRI data and projections. Industry sources.

From: Pharmaceutical Industry Dynamics and Outlook to 1985, Leif Schaumann, Health Industries Research Dept., Stanford Research Institute, July, 1976.

Table 41 lists the 51 largest pharmaceutical companies in the world and the countries in which their headquarters are located.

Table 41. Leading Pharmaceutical Companies of the World (according to R.S. Aries, IDO, Dynachim Div.)

Ranking	Company	Origin	1977 Sales*
1	Hoffman-La Roche	Switz	2,000
2	Hoechst Roussel	Ger	1,700
3	Merck & Co.	USA	1,600
4	American Home Prod.	USA	1,500
5	Bayer	Ger	1,300
6	Ciba-Geigy	Switz	1,200
7	Warner-Lambert	USA	1,100
8	Pfizer	USA	1,000
9	Bristol-Myers	USA	1,000
10	Sandoz	Switz	1,000
11	Eli Lilly	USA	900
12	Rhone-Poulenc	France	800
13	Upjohn	USA	800
14	Boehringer (Ing)	Ger	800
15	Squibb	USA	700
16	Schering-Plough	USA	700
17	Sterling Drug	USA	700
18	Richardson-Merrell	USA	700
19	Johnson & Johnson	USA	700
20	Glaxo	UK	700
21	Abbott	USA	600
22	Dow-Lepetit	USA	500
23	Takeda	Japan	500
24	Beecham	UK	500
25	Cyanamid	USA	500
26	Akzo	Neth	500
27	Smith Kline	USA	500
28	Medimpex	Hung	500
29	Searle	USA	400
30	Wellcome	UK	400
31	Schering AG	Ger	400
32	Sanofi	France	400
33	Shionogi	Japan	400
34	ICI	UK	400
35	Boehringer (Man)	Ger	400
36	Montedison	Italy	400
37	Philips-Duphar	Neth	300
38	Taisho	Japan	300
39	Tanabe	Japan	300
40	Fujisawa	Japan	300
41	Syntex	USA	300
42	Sankyo	Japan	300
43	E Merck	Ger	300
44	Boots	UK	300
45	Gist-Brocades	Neth	250
46	A H Robins	USA	250
47	Astra	Sweden	250
48	Eisai	Japan	250
49	Daiichi	Japan	200
50	C M Industries	France	200
51	BASF Pharm	Ger	200

*Estimated world human and veterinary drug sales

Roughly 40% of them are U.S. based; another 20% are active marketers in the U.S.A. and belong to the Pharmaceutical Manufacturers' Association (PMA).* There are reported to be more than 10,000 companies around the world that consider themselves pharmaceutical manufacturers. Of the 2-3 thousand companies that can actually be considered broadly competent, dosage-form pharmaceutical operations, less than 100 probably supply 90% of total world shipments and the leading 25 have been estimated to account for almost 50% of manufacturers' sales. [Schaumann, 1976]

A very enlightening profile of the 14 companies whose pharmaceutical sales exceeded $500 million in 1974 (accounting for about 30% of worldwide sales) is provided by Table 42. The extent of diversification in production and research, as well as marketing, is extraordinary, and the analyst who prepared the table, Dr. Schaumann of SRI, concludes the pharmaceutical industry occupies a unique position among business organizations in the scope and depth of its participation in international markets.

6.22 Market Share Concentration

The share of the national pharmaceutical market held by the leading companies in the U.S.A., Germany and France is depicted in Figure 11 for representative years. While the concentration is generally somewhat higher in the U.S. pharmaceutical industry, the Department of Commerce found that the degree of concentration in the pharmaceutical industry falls into the lower third of the 450 industries examined. Since the industry does not

---

*A non-profit trade association, the 143 members of which are the major producers of ethical pharmaceuticals in the United States. Many also manufacture medical devices and diagnostic products. Much of the statistical material in this segment of the report is based upon the annual surveys conducted by the PMA and analyses published by the PMA.

Table 42. Pharmaceutical Operations of First-Tier Companies, by Type and Number, 1974

	Marketing Organization		Production Capacity		Research Facilities	
Company	Own	Agent	Manufacturing	Packaging	Major	Minor
Hoffmann-La Roche	32	76	16	30	6	21
Hoechst*	130	0	14	37	7	2
Ciba-Geigy	50	10	14	41	5	15
American Home Prod.†	80	50	5	25	3	10
Merck & Co.	60	100	28	28	2	8
Sandoz	46	61	8	22	3	5
Bayer	48	45	12	19	3	7
Warner Lambert	50	90	10	35	3	10
Eli Lilly	28	117	5	12	2	13
Pfizer†	75	75	10	45	3	10
Bristol-Myers	60	55	5	20	4	10
Boehringer-Ingelheim†	46	40	5	10	1	5
Takeda	12	40	4	3	1	1
Schering-Plough†	60	60	5	25	2	6

Note: "Pharmaceutical" includes only ethical and proprietary specialties, in dose form, for human use; proprietary products included are only those with a recognized therapeutic activity.

†Data estimated by SRI.

*Excluding Roussel-Uclaf.

From: Pharmaceutical Industry Dynamics and Outlook to 1985. Leif Schaumann, Health Industries Research Dept., Stanford Research Institute, July, 1976.

serve a single, broad market, the Department of Commerce looked at concentration in major submarkets. Their data, based on sales by the four largest firms in each of nine submarkets, disclosed only one which had a four-firm concentration rate in excess of 50% as of 1972 (Table 43). Most rapidly expanding, high-technology industries exhibit comparable levels of concentration, according to the PMA study on competition from which these comparisons are drawn. [PMA, 1977]

Figure 11: Concentration Ratios in the Pharmaceutical Industries of the United States, West Germany and France

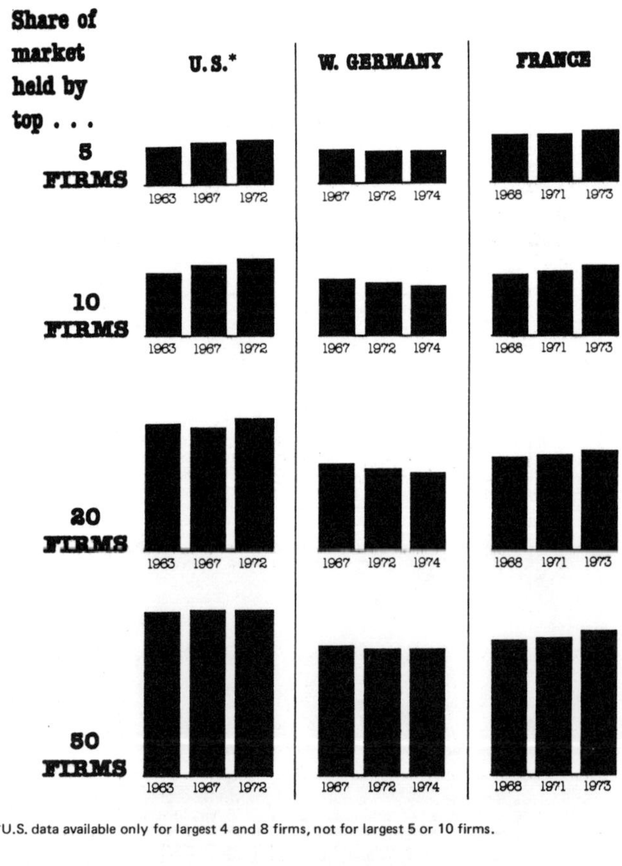

*U.S. data available only for largest 4 and 8 firms, not for largest 5 or 10 firms.

Sources:

United States: Bureau of Census, "Concentration Ratios in Manufacturing," MC 72(SR)-2, p. 94.

West Germany: Bundesverband der Pharmazeutischen Industrie, "PharmaDaten 76," p. 11.

France: Syndicat National de l'Industrie Pharmaceutique, "L'Industrie Pharmaceutique, Ses Réalités," p. 49, and "L'Industrie Pharmaceutique, Ses Réalités, Mise à Jour," p. 49.

Table 43: Concentration Ratios in the U.S. Pharmaceutical Industry (percentage of total sales by the four largest firms)

Category	1963	1972
Pharmaceutical Preparations (all)	22%	25%
Sub-markets:		
Neoplasms	53	48
Central nervous system	44	43
Cardiovasculars	52	57
Respiratory	39	48
Digestive system	30	35
Dermatologicals	33	35
Vitamins	28	27
Anti-infectives	47	45
Veterinary	37	39

Source: United States: Bureau of Census, "Concentration Ratios in Manufacturing," MC 72(SR)-2, pp 94-95.

The data of Table 43 and Figure 11 do not suggest dangerous concentrations in the pharmaceutical industry--even though the number of firms in the industry has declined. Furthermore, as Table 44 illustrates, many submarkets appear quite dynamic when examined in detail.

Table 44: Change in Market-Share Rank of Leading Products in Ten Pharmaceutical Markets, 1962-71

	Absolute Point-Change in Rank, 1962-71[1]									
Rank of Product, 1962	Anti-infective	Analgesic and anti-inflammatory	Psycho-pharmaceutical	Cough and cold	Anti-hypertensive and diuretic	Vitamin and hematinic	Oral contra-ceptive	Anticholinergic/antispasmodic	Anti-obesity	Diabetic therapy
1	-8	0	-1	-4	- 2	0	-9	0	-1	0
2	-5	0	-2	+1	0	-1	-1[3]	-1	-2	0
3	-1	-1	-2	0	- 7	-4	-6[4]	-1	-2	-4
4	-4	-8	-5	-3	- 7	+2	-7[2]	+2	+1	+1
5	-5	-5	-2	-6	- 4	+3	-3[2]	-1	-5	-3
6	+4	+1	0	-3	- 1	-3	+4[6]	+1	+5	-3
7	-4	-2	-3	-3	- 6	-3	+6[7]	0	-2	-3
8	-4	-3	+5	+6	- 4	-3	+1[7]	-1	0	0[4]
9	+8	+1	+8[3]	+1	+ 3	+5	+4[7]	-1	+2	+5[5]
10	+7	+4	+2[2]	+6	+ 2	+4	+5[8]	+2	+4	—
11	+6	+6[2]	—	+5[4]	+ 6[3]	+6	+5[9]	—	—	—
12	+6	+5	—	—	+ 7[2]	—	—	—	—	—
13	—	—	—	—	+12[5]	—	—	—	—	—

[1] + indicates increase in market share. - indicates decrease. 0 indicates no net change.
[2] Change from 1965, when product entered market.
[3] Change from 1963, when product entered market.
[4] Change from 1964, when product entered market.
[5] Change from 1966, when product entered market.
[6] Change from 1967, when product entered market.
[7] Change from 1968, when product entered market.
[8] Change from 1970, when product entered market.
[9] Attained such rank in year of introduction, 1971.

Source: Douglas Cocks, "Product Innovation and the Dynamic Elements of Competition in the Ethical Pharmaceutical Industry," in *Drug Development and Marketing,* R.B. Heims, ed., Washington American Enterprise Institute, 1975, p. 251.

This is most the case where novel products are introduced, such as the antihypertensive and diuretic market and oral contraceptive market in the period shown. If the same sort of study were to be made of more recent years, the anticholinergic/antispasmodic sector, which was quiescent in the earlier period, would reveal the radical upset caused by the introduction of the first H_2-inhibitor, Tagamet, for the treatment of gastric ulcers. What these data speak to is the role of R&D in the pharmaceutical industry.

6.23 Industry Growth

The growth of sales of U.S.-based firms since the founding of the PMA is set forth in Table 45 along with the R&D expenditures of those firms.

Table 45: U.S. Headquartered Firms' Sales and R & D Expenditures, 1954-1979 (millions of dollars)

Year	U.S. Sales[a] Human Dosage	Global Sales Human and Veterinary[b] Dosage Only	Global Sales Human and Veterinary[b] Dosage & Bulk	Global R & D[c] Human & Veterinary
1979[d]	10,154	na	19,202	1,600
1978	9,156	16,040	17,008	1,404
1977	8,233	13,896	14,674	1,276
1976	7,669	12,832	13,572	1,164
1975	6,895	11,543	12,183	1,062
1974	6,083	10,120	10,726	942
1973	5,507	8,755	9,191	825
1972	5,018	7,827	8,181	726
1971	4,667	7,020	7,838	684
1970	4,322	6,442	6,853	619
1969	4,008	5,832	6,208	550
1968	3,655	5,280	5,665	495
1967	3,226	4,707	5,102	461
1966	3,011	4,256	4,660	416
1965	2,779	3,841	4,219	365
1964	2,479	3,405	3,717	310
1963	2,317	3,152	3,469	292
1962	2,199	2,932	3,236	259
1961	1,954	2,685	2,992	245
1960	1,905	3,600	na	212
1959	1,850	2,500	na	197
1958	1,802	2,400	na	170
1957	1,742	2,200	na	127
1956	1,676	1,900	na	105
1955	1,457	1,650	na	91
1954	1,252	1,500	na	78

[a] U.S. sales of finished prescription and "otc ethical dosage forms."
[b] Global sales are defined to include total ethical pharmaceutical sales within the United States, exports to nonaffiliated firms and sales abroad of U.S. affiliates.
[c] Global R & D includes all ethical pharmaceutical research and development financed or conducted by U.S. pharmaceutical companies in the U.S. and abroad, and in the U.S. only by U.S. based subsidiaries of foreign owned firms.
[d] PMA estimates based on member company budgeted figures or estimates.

Source: PMA Annual Surveys.

The $1.6 billion budgeted for R&D in 1979 was a 14% increase over 1978. At 8.3% of global sales (dosage and bulk) the ratio of R&D to sales in the past two years is down somewhat from the levels close to 9% common from 1966 to 1973--due more to success abroad than decreased R&D commitment.

In 1939 the total sales of the U.S. pharmaceutical industry were $150 million at the manufacturers' level. Gerhardt Domagk, at I.G. Farben in Germany, had discovered the first sulfa drug, Prontosil, in 1932, setting off worldwide study of sulfanilamide and related compounds. World War II and penicillin enabled America to snatch the lead and move out strongly in the post-war period.* "Prior to 1939...the majority of the large industrialized countries were self-sufficient and the (export) trade which did take place was dominated by Germany." [James, 1977] Germany held a 39% share of the $130 million export market in 1938; the U.S. share was 13%; France and England each had 12% shares, and Switzerland had 7%. By 1955, when the immediate post-war imbalances had been overcome, the U.S. share was 34% of a $660 million export market, and the United Kingdom, Switzerland, France, and Germany had shares of 16%, 14%, 12% and 10% respectively. Ten years later, <u>exports</u> of Germany and the U.S.A. were equal at 16% of $1.286 billion; Switzerland had a 14% share, the U.K. 13%, France 11%, and Holland and Italy had risen to 5% shares.

It should be understood that the apparent decline in U.S. share of world markets was due, in large part, to the establishment of overseas affilitates which require export only of bulk active

*Although penicillin was a British discovery, war-torn Britain lacked the resources to pursue it. After demonstrating clinical efficacy, in 1941 Florey turned to the United States, where a collaboration was established between Merck, Squibb, Pfizer, and the Committee on Medical Research of the Office of Scientific R&D, headed by A.N. Richards.

ingredients or may be supplied entirely from plants abroad. [Ibid.] Thus, U.S. exports in 1965 may have been only $200 million, but total non-domestic sales of U.S.-headquartered firms were $1,440 million. By 1979, that figure is estimated to have grown to $9,048 million. [PMA, 1980]

6.24 Research Basis

Research and Development expenditures for human and veterinary pharmaceutical R&D by U.S. pharmaceutical companies (in the U.S. and abroad) and in the U.S. only by U.S.-based subsidiaries of foreign owned firms who are members of PMA are presented in Table 46. Human-use pharmaceuticals accounted for 94% of 1978 R&D expenditures--an 11% increase in human-use research and a 2% increase in veterinary-use research over the preceding year.

Table 46: R & D Expenditures for Ethical Pharmaceuticals: 1977, 1978 Actual and 1979 Budgeted PMA Member Firms (millions of dollars)

	1977 Actual	1978 Actual	1979 Budgeted
Company-Financed R & D Expenditures for Human-Use Pharmaceuticals:			
Amount spent in United States	$ 984.1	$1,089.2	$1,225.2
Within firm	862.7	929.3	1,028.7
Outside firm	121.4	159.9	196.5
Amount spent in foreign countries	197.7	222.0	269.9
Within firm	166.3	198.0	239.1
Outside firm	31.4	24.0	30.8
TOTAL Human-Use R & D	$1,181.8	$1,311.2	$1,495.1
Company-Financed R & D Expenditures for Veterinary-Use Pharmaceuticals:			
Amount spent in United States	$ 68.9	$ 69.9	$ 78.1
Within firm	61.4	62.3	70.2
Outside firm	7.5	7.6	7.9
Amount spent in foreign countries	15.4	15.9	19.6
Within firm	14.5	15.2	18.6
Outside firm	.9	.7	1.0
TOTAL Veterinary R & D	$ 84.3	$ 85.8	$ 97.7
TOTAL Company-Financed R & D Pharmaceuticals	$1,266.1	$1,397.0	$1,592.8
U.S. Government Grants and Contracts for Company-Conducted Pharmaceutical R & D	$ 10.0	$ 7.0	$ 6.8
TOTAL Expenditures for Pharmaceutical R & D	$1,276.1	$1,404.0	$1,599.6

Extra-mural expenditures for human-use drug R&D have risen considerably in recent years in this country. Abroad, the trend has been to expand R&D within the firm. Government contracts account for less than 1/2% of R&D expenditures and are declining.*

R&D spending abroad represents 17% of total expenditures--up from 8.7% a decade earlier. The largest proportion (68.6%) is for R&D in Western Europe, 18.5% for R&D in the Far East and Pacific, 6.3% and 5.1% for Latin America and Canada, respectively. That those proportions reflect the markets can be seen from the tabulation of estimated sales shown in Table 47.

The percentage of R&D expenditures directed toward new products is 76% abroad (up substantially from 68% in 1977) and 81% in the U.S.A. Most of the work abroad is on anti-infectives (24.1%) and cardiovasculars (21.4%), reflecting the prominence of those therapeutic classes internationally. U.S.-based firms employ 22,675 people in R&D in the U.S.A. (81%) and 5,200 abroad (19%). An NSF study completed early in 1979 noted that the greatest growth in overseas expenditures occurred between 1974 and 1975, when the foreign R&D budgets of U.S. drug companies almost doubled in response to new FDA regulations proposed in 1973 and passed in 1975.

Overall, the R&D expenditures are of the same order of magnitude for the categories of: anti-infectives; central nervous system and sense organs; cardiovasculars; neoplasms; endocrine system and metabolic disease (Table 48), which together have accounted for over 2/3 of the R&D budget for many years. These categories contribute ~60% of sales.

---

*In 1966 U.S. government grants and contracts for company-conducted R&D accounted for 3.36% of the $416 million spent on pharmaceutical R&D in industry.

Table 47. The World Market for Ethical Pharmaceuticals (Based upon estimates of IMS World Publications, as presented in PMA Newsletter, December 26, 1977, p. 5)

	Ex-manufacturer Sales in $ Billions		Percentage of World Market	
Western Europe	14.6		30.0	
West Germany		3.640		7.5
France		3.100		6.4
Italy		2.200		4.5
Spain		1.480		3.0
United Kingdom		1.200		2.5
Belgium		0.505		1.1
Sweden		0.430		0.9
Netherlands		0.422		0.9
Switzerland		0.365		0.8
Other		1.258		2.6
North America	9.6		19.7	
United States		7.800		16.0
Mexico		0.806		1.7
Canada		0.695		1.5
Eastern Asia	8.7		17.9	
Japan		5.400		11.1
South Korea		0.480		1.0
China & Other		2.820		5.8
Eastern Europe	7.8		16.0	
South America	3.4		7.0	
Brazil		1.500		3.1
Argentina		0.960		2.0
Venezuela		0.320		0.7
Other		0.620		1.3
Central & Southeastern Asia	1.7		3.5	
Africa	1.3		2.7	
Middle East	1.0		2.0	
Iran		0.400		0.8
Oceania	0.53		1.1	
Australia		0.450		0.9
Worldwide Total	48.7			

About one-third of R&D expenditures is for synthesis, extraction, biological screening and pharmacological testing; roughly another third goes into toxicology and clinical studies, and the remainder in product/process development, regulatory and other kinds of R&D activities as shown in Table 49. One can see from the comparison of two recent years that increases in other expenses have slightly decreased the share allocated to the discovery phases.

Table 48. R&D Expenditures by Product Class, 1978

	Percentage Share of R&D Dollar	
	In U.S.A.	Abroad
Anti-infectives	18.9%	24.1%
Cardiovasculars	16.6	21.4
Central Nervous System & Sense Organs	16.8	13.2
Neoplasms, Endocrine System & Metabolic Diseases	16.4	9.0
Gastrointestinal & Genitourinary System	6.2	10.5
Respiratory	4.0	2.9
Biologicals	3.2	1.5
Dermatologicals	3.1	2.4
Vitamins	2.4	0.8
Diagnostic (In Vivo)	0.5	<0.1
Veterinary	6.3	7.8
Other	5.6	6.4

Table 49. Distribution of U.S. R&D Expenditures for Ethical Pharmaceuticals by Function (Data from PMA Survey Reports for 1978-79 and 1976-77)

	1978	1976
Biological Screening & Pharmacological Testing	18.8%	19.4%
Clinical Evaluation: Phases I, II and III	16.5	20.2
Synthesis and Extraction	16.3	16.6
Pharmaceutical Dosage Formulation & Stability Testing	10.1	9.1
Toxicology and Safety Testing	10.0	8.9
Process Development for Manufacturing and Quality Control	9.2	9.5
Clinical Evaluation: Phase IV	4.8	3.3
Regulatory, IND & NDA Preparation, Submission and Processing	3.6	3.3
Bioavailability Studies	2.2	2.2
Other	8.5	7.5

The percentage of the domestic pharmaceutical sales dollar applied to domestic R&D has remained in the range of 11.5% since 1975--down somewhat from the beginning of the decade when the range was 11.9-12.1%.

Firms with the highest R&D to sales ratio (14.9%) were those whose domestic R&D expenditures fall between $10 and $30 million (average $16.8 million R&D with domestic sales of $112.5 million), but such firms only accounted for 15.3% of total R&D expenditures. Most (77.8%) of global R&D is performed by firms with R&D budgets over $30 million, and those firms average 8.4% on global R&D to sales ($64 million R&D on $758 million sales, average). However, firms whose domestic R&D exceeds $30 million show an R&D to sales ratio of 13.0%.

An industry analyst has shown recently that the ratio of R&D to sales has actually declined steadily since the mid-1960s in terms of constant (1967) dollars. [Lee, 1979] This study, the data for which are given in Table 50, related the cost escalation of biomedical research (as expressed in the Biomedical R&D Deflator used by NIH) to industry prices (as expressed in the BLS Producer Price Index for Ethical Pharmaceuticals), and reached the conclusion that the reinvestment rate in 1967 would have bought half again as much R&D as the 1978 rate.

The cost of development of a novel pharmaceutical product has risen steeply. A survey and analysis conducted for the National Science Foundation by a U. Rochester economist, concluded that the average development cost of a new chemical entity was $54 million in 1976 dollars, including the cost of associated failures* and allowing a discount rate of 8% for capitalizing

*Hansen found that only about 12% of the drugs which enter the human testing process actually reach the market.

Table 50: R&D to Sales Ratios, Constant (1967) Dollars

Year	R&D (millions)	Deflator (NIH, Bio-Medical R&D)	Deflated R&D	Sales (millions)	Deflator (BLS, PPI)	Deflated Sales	R&D/Sales Ratio, Not Deflated	R&D/Sales Ratio, Deflated
1960	206	81.9	252	1,855			11.1	
1961	227	83.6	272	1,954	108.2	1,806	11.6	15.1
1962	225	85.5	263	2,199	105.6	2,082	10.2	12.6
1963	248	87.9	282	2,317	104.2	2,224	10.7	12.7
1964	258	89.8	287	2,479	103.9	2,386	10.4	12.0
1965	304	92.5	329	2,779	103.2	2,693	10.9	12.2
1966	344	95.8	359	3,011	102.6	2,935	11.4	12.2
1967	377	100.0	377	3,226	100.0	3,226	11.7	11.7
1968	410	104.7	392	3,655	99.0	3,692	11.2	10.6
1969	464	110.4	420	4,008	99.5	4,028	11.6	10.4
1970	519	117.3	442	4,322	99.3	4,352	12.0	10.2
1971	577	124.1	465	4,667	99.0	4,714	12.4	9.9
1972	601	130.3	461	5,018	99.1	5,069	12.0	9.1
1973	644	136.5	472	5,507	99.9	5,513	11.7	8.6
1974	726	145.2	500	6,082	104.2	5,837	11.9	8.6
1975	829	160.7	516	6,895	113.2	6,091	12.0	8.5
1976	903	172.7	523	7,669	120.3	6,375	11.8	8.2
1977	984	186.4	528	8,233	125.4	6,565	12.0	8.0
1978	1,114*	200.3	556	9,344**	131.9	7,084	11.9	7.8

*Budgeted **Estimate

Sources of Data:

R&D and Sales Dollars: from PMA Annual Surveys of PMA members.

R&D Expenditures: in the United States, for human use ethical pharmaceuticals.

Sales: in the United States, human use, dosage form ethical pharmaceuticals.

R&D Deflator: Biomedical R&D deflator used by NIH, DHEW; 1967=100.

Sales Deflator: Producer Price Index for Ethical Pharmaceuticals (line item 0635), Bureau of Labor Statistics; 1967=100.

Pharmaceutical Manufacturers Association - R&P October, 1979 the R&D outlays from point of discovery up to the point of marketing. [Hansen, 1977] A plot of R&D spending and number of new chemical entities approved by the FDA, offset by five years to allow somewhat for the development period, suggests that the out-of-pocket cost was approximately $40 million each for the products approved in 1977 (Figure 12). [Blee, 1978]

Figure 12. R&D Spending and Resultant New Chemical Entities

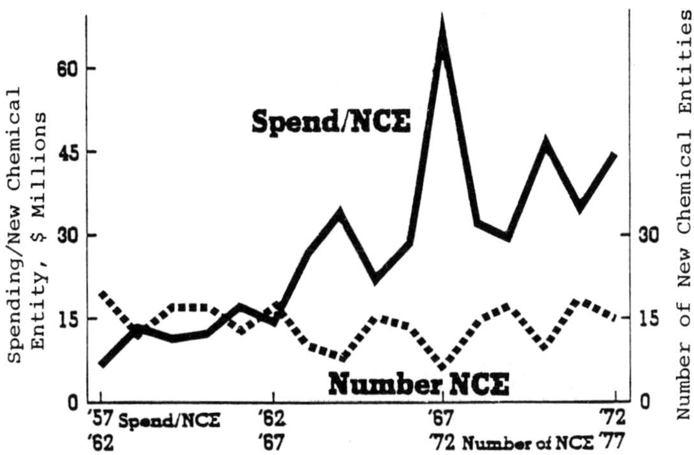

The five-year interval between discovery and approval implied by that plot is definitely on the short side. Francis Blee, corporate economist of SKF--which developed the altogether novel H_2-receptor inhibitor Tagamet during the late 60s and 70s--suggests a span of 6.5 to 8 years from identification of a compound to completion of clinical trials. To that must be added the time necessary to obtain regulatory approval, country-by-country, and at the other end of the cycle, for the research program which yields promising compounds. Dr. Blee estimates a $1 million/year research program can be expected to yield on average a single new compound suitable for testing through the clinical stage. Thus, he comes up with a 9.5-11 year and

$19.5-23.5 million estimate, aside from regulatory delays and costs, for the development of a new chemical entity in the 1970s. A number of other estimates place the cost in the $25-50 million range. Clearly, such investments are made in anticipation of major markets.

Yet, an analysis of 194 new chemical entities introduced between 1962 and 1976 found that 70% had annual sales of $5 million or less, and fewer than 9% had achieved U.S. sales in excess of $20 million/year. Only two products had sales of over $100 million/year, but that 1% of the total number of products accounted for 17% of the average annual sales overall.

These figures suggest the critical role of international markets, with their variety of therapeutic tastes, in the anticipation of pharmaceutical industry investors.

6.25 Profitability

On the face of it, the pharmaceutical industry shows a higher return on net worth than any other. In part, this is misleading in that R&D investment is treated by accounting practices as a current expense unlike investment in plant and equipment,* which results in overstatement of profits relative to the assets upon which the business apparently depends. A number of analysts have drawn attention to this anomaly. [Friedman, 1973; Schwartzmann, 1975; Clarkson, 1977] However, even when steps are taken to adjust for R&D investment and revalue it into constant dollars, the comparable return on real investment still handsomely exceeds that of other manufacturing businesses in

*The above-average risk associated with the pharmaceutical industry is symbolized by this practice, in that conservative accounting principles encourage the write-off of an investment which has an unpredictable future value.

the U.S.A. Blee has pointed out that a decline in pretax margins in the 1970s has been offset through a declining tax rate--derived from plants in Puerto Rico and other temporary tax havens--and predicts that by 1983 the pharmaceutical industry's will be equal to the U.S. average, even as measured by the accountant's standards. [Blee, 1978]

Nevertheless, the pharmaceutical industry remains a relatively untarnished growth industry in the eyes of investors, especially with the advent of a new wave of products expected to result from applied molecular genetics.

6.26 Regulation has strongly influenced new product introduction. Until 1962, the FDA required only evidence of safety prior to marketing of a new drug. Since that time several pieces of legislation have increased the role of FDA before and after marketing. From 1941 to 1959 there was a four-fold increase in new chemical entities; from 1960 to 1975 there was a three-fold decrease. The effect on new product introductions overall was much greater, as shown in Figure 13. Two-thirds of the new entities introduced by the modern pharmaceutical industry appeared before 1960. However, as the PMA Factbook expresses it: "Even without the increased regulatory requirements for approvals, a decline in new product introductions would probably have occurred as industry researchers developed more sophisticated instruments and techniques for measuring safety and effectiveness, and addressed more complex problems, including therapy for more intractable diseases."

Current regulations require that substantial evidence from well-controlled clinical investigations demonstrating the effectiveness of a new drug as well as its safety be submitted prior to marketing, and they make provisions to safeguard

Figure 13: Number of NDAs and New Molecular Entities Approved by Calendar Year from 1950 Through 1978

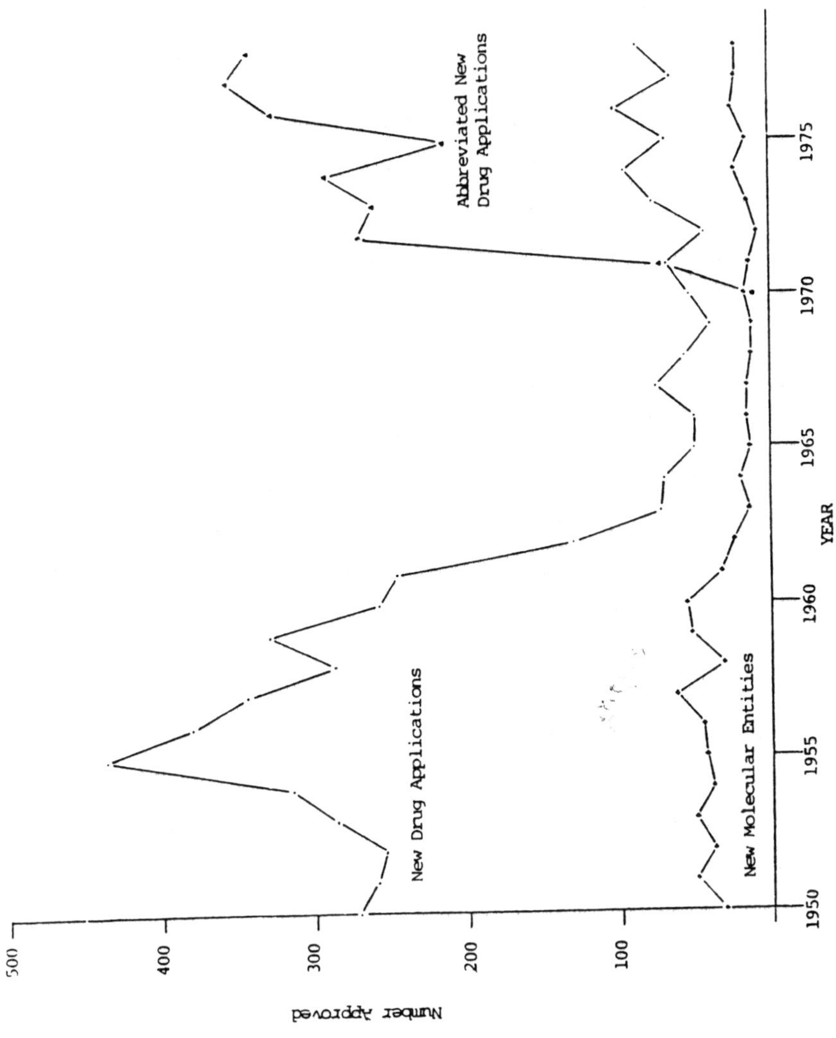

research subjects receiving a new drug during its period of clinical investigation. Sponsors of investigations of new drugs or new uses of approved drugs must file a Notice of Claimed Investigational Exemption for a New Drug (IND) before beginning human experimentation. FDA evaluates new IND submissions and monitors ongoing clinical investigations.

FDA approval of a New Drug Application (NDA) is required prior to marketing. The NDA contains: 1) full reports of investigations, including clinicals, that show whether or not the drug is safe and effective; 2) a full list of articles used as components of the drug; 3) a full statement of the drug's composition; 4) a full description of the methods used in, and the facilities and controls used for, the manufacturing, processing, and packing of the drug; 5) samples of the drug and components as may be required; and 6) specimens of the proposed labeling. Holders of approved NDAs are required to submit periodic reports on adverse reactions, published and unpublished medical literature, and other marketing data on the drug.

The NDA holder must notify FDA and the medical profession immediately if serious adverse reactions occur. Standards have been established that characterize Good Laboratory Practices, which are especially influential during the R&D phase, and Good Manufacturing Practices, which influence production and quality control.

Only a 30 day delay is involved between the submission of an IND and initiation of acceptable clinical studies, but the period required for approval of the average NDA is nearly three years. An average of 35.2 months was clocked for NDAs approved in 1979--up from 33 and 26 month averages in 1978 and 1977, respectively.

Nine NDAs considered by the FDA to involve important or modest therapeutic gains were fast-tracked and took only 19.4 months on average to gain approval. However, only five of these actually involved new chemical entities. Since 1973, the FDA has ended each fiscal year with about 170 NDAs pending, and each year about 125 original NDAs have been submitted. The output (Figure 14) has typically involved 25-35% new entities. Of the 5400 active INDs as of September 30, 1979, 1893 were held by commercial sponsors, and only 199 were considered by FDA to involve the potential of important or modest therapeutic gain. Only 110 involve new chemical entities in those classes.

Half of the 22 active INDs for new molecular entities which have been rated by FDA as promising important therapeutic gains are in the Metabolic and Endocrine Division. That will be the site of review of applications on such peptide hormones as insulin and growth hormone derived from molecular genetic developments.

7. Industrial Bioproduction and Genetics

In general the limited use of fermentation as a production technology has made only empirical use of genetics. The major field of experience in antibiotics has already been described (pp. 315-346). However, a few additional examples can be provided to broaden the view of the context in which developments based upon molecular genetics will be commercialized.

7.1 L-Asparaginase is presently the only pharmaceutical product obtained by fermentation of *E. coli*. Apparently this enzyme exerts its antitumor activity by removing from the blood stream an essential nutrient for certain leukemia cells which lack the ability, possessed by normal cells, to make their

Figure 14

Number of NDAs and New Molecular Entities Approved by Year

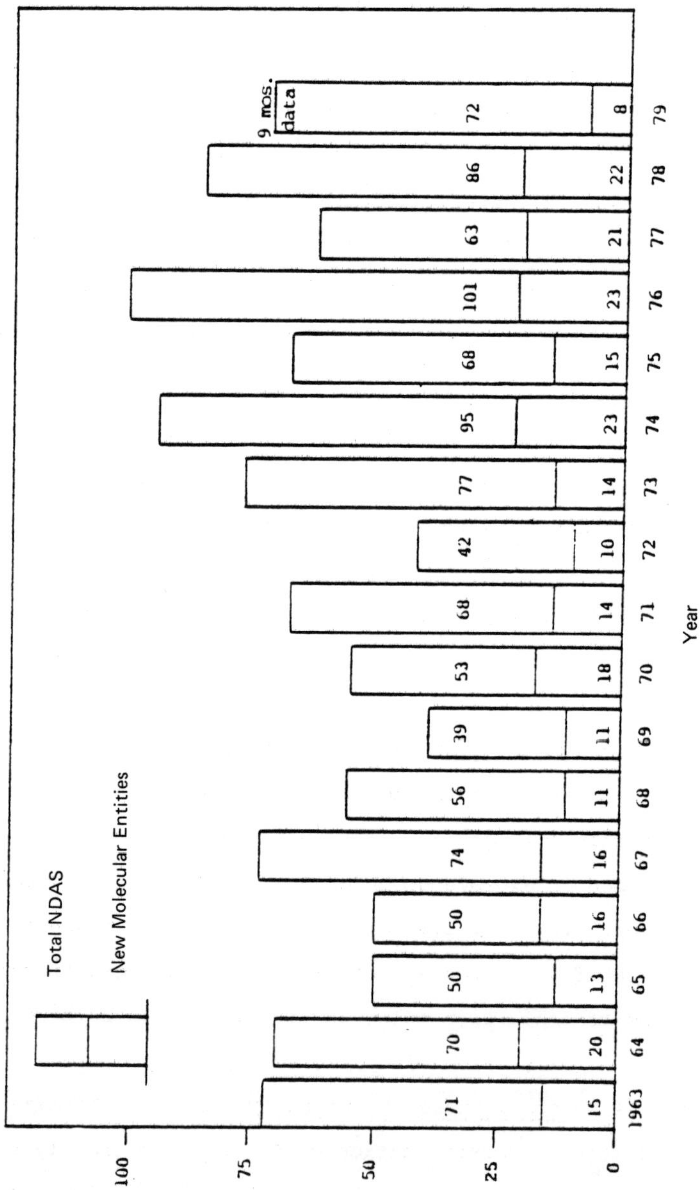

own asparagine. Clinical researchers at Wadley Research Institute approached Merck and Co. in 1967, and Merck undertook the development of a practical process, understanding the market would be small at best and primarily as a public service.

Glitterman and Wallick at Merck employed chemical mutagenesis and an automated selection technique to isolate a superior mutant that produced about 100 times as much enzyme as the strain received from Wadley. Scale-up went smoothly from 250 ml shake flask through 4000 l pilot stage, and within a year Merck was making 20,000 gallon fermenter batches. Factory titers are only 70% of those achieved in small fermenters and yields are quite sensitive to media composition. Nevertheless, Jim Lago, who directed the scale-up, says that the fermentation is an engineer's dream:

- Agitation-aeration demands are low, albeit critical to yield.
- Fermentation cycle is much shorter than for more common industrial organisms (~18 hours)
- Contamination with extraneous organisms is no problem because *E.coli* grows so rapidly.
- Principal nutrient, cornsteep liquor, is cheap and readily available.
- Mechanical refrigeration is not necessary.
- Cell mass is compact without voluminous mycelium.

The isolation-purification involves:

- Centrifugal separation at 7gal/min, which increases cell concentration from 6g/l to 120g/l reducing volume by 97%.
- Sterilizing the spent broth in a tubular continuous sterilizer.
- Rupturing the intact cells by passage of the liquid through an orifice at 8000 psig (twice) into a continuous homogenizer, with subsequent cooling to overcome the 30°C temperature rise.
- Incremental precipitation with methanol.
- Filtration of 2000-3000 gal, first to remove cellular debris. Then, after increasing methanol concentration and lowering pH, to collect crude asparaginase on diatomaceous earth.

- o Reprecipitation from aqueous solution with methanol.
- o Redissolving the filtercake in a 100 gal vessel and removal of diatomaceous earth filter-aid by centrifugation (specific activity at this stage is 250-270 units/mg).
- o Ultrafiltration (with 97% recovery) to reduce volume several fold.
- o Ion exchange purification - which is the rate and capacity limiting step.
- o Crystallization and collection by solid bowl centrifugation.
- o Sterile finishing by dissolving the crystals in buffer, treating with carbon, adding mannitol, sterile filtering, and freeze-drying in vials.

The Merck workers say the key step was the identification of a high yielding strain, for that made the task of isolation and purification much easier. As a result they note the project was actually completed for much less than had been budgeted. Furthermore, the cost of a course of therapy was slashed from as high as $15,000 to approximately $300. The specific activity of Elspar, Merck's brand of L-asparaginase, is typically 300 units per mg. Dr. Lago, who kindly provided the detailed description outlined above, says the concentration of the fermentation broth is 150 units/ml, or the equivalent of 500 mg/liter.

Although Merck is reluctant to disclose its costs, one can get an idea of the costs from those incurred at the Wadley Institute with a less productive strain. Under factory conditions this strain produces 20 units/ml according to W.R. Barnes, and it costs $1500 to make a thousand liter run.

The work-up has an overall yield of only 30%, that results in pure asparaginase at a cost of 25-30¢/1000 units, which corresponds to approximately 10¢/mg.

Barnes reports the preparation of a new strain, R-69, by mating a good asparaginase producer (marked with a deficiency gene that makes it auxotrophic) with a strain of *E.coli* (Hfr) known to conjugate more readily than other strains. By selecting for prototrophic recombinants (able to grow on a medium that would not support the growth of the deficient parent strains), he determined that the frequency of recombination was about 0.05%. Most of the recombinants proved inferior with respect to asparaginase production, but about 10% were better producers than their parents. The strain R-69 consistently produces 50% more than its parent, A-1, which is the standard, outside Merck, for production. [Barnes, 1978] Actually, the purpose of the initial matings was to get an asparaginase producer with the Hfr characteristic so that breeding could be done in which both parents were asparaginase producers. But the incidental improvement of productivity achieved in R-69 is sufficient (30units/ml) to cause Wadley Institute to consider re-entering the production of asparaginase.* The important point is that *E.coli* has a mating system that made improvement on the basis of genetic analysis possible. Most commercial organisms do not possess a practical mating system to permit *in vivo* recombination.

That is why *in vitro* recombination opens new possibilities for improvement. There simply has been no alternative approach!

7.2 *Dextrans* are high molecular weight polymers of glucose, that are commonly synthesized from sucrose by members of the bacterial family Lactobacilleae. The common oral bacteria *Streptococcus mutans* produces such polymers as part of its role in the development of dental caries [Montville, 1978], but the organism of commercial importance is *Leuconostoc mesenteroides* [Lawford, 1979]. *L. mesenteroides* is not a pathogen. The significance of dextrans in the present context is twofold:

*There is a reference, without details, in the literature to a mutant (A-1-3) that produces 50-70units/ml. [Nakamura, 1972]

- No other bacterial product has ever been infused into humans in as great an amount as dextrans. (The enzymes streptokinase and asparaginase have been infused in doses of a few milligrams.) When dextran is administered as a plasma expander, 30-100 grams may be infused in a short time.
- Dextrans are actually produced outside the bacteria through the action of an extracellular enzyme secreted into the medium. In fact, some producers use the isolated enzyme to prepare the polymer, but Pharmacia - the primary U.S. supplier - combines live bacteria with sucrose to obtain its high molecular weight starting material.

Apparently there has been little need or interest in the genetics of dextran production, or rather of the production of dextransucrase, the enzyme responsible for transfering glucosyl groups from sucrose to the growing dextran chain, where they are linked by α-(1-6) linkages. A second enzyme introduces the occasional attachments at carbons 2,3 or 4 which give rise to branching. The strain, B512 (F), used by Pharmacia produces a relatively low(5%) level of branching. At least 80% of the branches produced by this strain contain only one glucose unit.

Antigenicity of dextrans is directly related to their degree of branching.

The crude dextran obtained from the bacterial action on sucrose is polydisperse with a molecular weight range up to $35\text{-}125 \times 10^6$. Acid hydrolysis is used to reduce the molecular weight, and alcohol precipitations are employed to fractionate and purify the products. The final materials are felt to be free from bacterial debris. However, dextran cross-reacts with bacterial antibodies. Hence, it has not been proven that no material from the bacteria is carried along with the product, but neither have clinical problems been attributed to this possibility.

Two dextran fractions for intravascular use are produced by Pharmacia of Sweden for marketing in this country.* Macrodex

* Abbott, Mcgaw, Baxter and Cutter also market these materials - purchased primarily from Pharmacia.

(dextran 70) is a 6% solution, in either normal saline or 5% dextrose, of a fraction having an average molecular weight of 70,000 (90% between 20,000 and 115,000) which has colloidal properties similar to human serum albumin. It is administered by intravenous infusion, for the same purposes described earlier for albumin, at dosages not recommended to exceed 1.2 g dextran/kg body weight per day. Thirty grams of dextran is a common first dose.

Rheomacrodex (dextran 40) is composed of a 10% solution of 40,000 m.w. average dextran (90% between 15,000 and 75,000).* Its viscosity is lower and oncotic activity is greater than the higher molecular weight dextran, and it causes a rapid expansion of plasma volume by up to two times the volume infused. The FDA-approved labelling says that total dosage during the first twenty-four hours should not exceed 20ml/kg (2g dextran/kg), and subsequent therapy should involve no more than 10ml/kg/day for a maximum of five days - that is a maximum dose of about 500g dextran for an adult. However, dextran is not under the sort of lifetime dose constraint that is imposed on polyvinyl pyrolidone (500 ml/patient/lifetime). Pharmacia reports dextran infusion is much more common and extensive in Europe than in this country. Problems encountered seem to relate primarily to the effects of dextran on coagulation rather than its bacterial origin, and its record with regard to anaphylaxis is quite favorable.# [Ring & Messmer, 1977] Dextran molecules below 50,000 m.w. are eliminated by renal excretion. Dextranases present in the body break down the higher molecular weight dextran at a rate of 70-90mg/kg/day.

---

*Rheomacrodex was originally developed as a blood flow improver and to reverse red cell clumping. One of its major applications is as a priming fluid for pump-oxygenators in cardiovascular surgery.

#Pharmacia, and others, make many other products from dextrans, including medicinals, but they all represent lesser toxicological challenges.

Approximately 40% of Macrodex appears in the urine within the first 24 hours. Rheomacrodex is lost even more rapidly. As a result their effects do not have the duration of human serum albumin.

A spokesman for Pharmacia, Dr. Shinago, says that genetics have not played an important part to date, but they have a large department in Sweden actively interested in the potential of new developments through which genetics will influence the utilization of microorganisms for production.

7.3 Vitamins are primarily produced today by chemical synthesis, but microorganisms have been used for some purposes and may be more widely used in the future. Riboflavin - which is more properly viewed as a food and feed additive than a pharmaceutical product - consumption in the USA was of the order of 1.3 million pounds in 1977, with an estimated value of $26 million. Merck and Roche are the domestic manufacturers, but 40-50% is imported. In this country Merck is the only one using fermentation to produce riboflavin. They are in their third year of doing so, and characterize the experience as a"hell of a struggle but worth it". Like Roche, Merck had produced riboflavin by synthesis for 40 years. The organisms employed are probably yeast-like. Dr. Hendlin, head of Merck's Developmental Microbiology, says that yields of 10g/l are reported in the literature.

Vitamin B_{12} is a much smaller volume item with unit value about 100 times as great as riboflavin. Animal feed represents about two-thirds of the market with the balance going into food and pharmaceutical products. It is not practical to synthesize B_{12} chemically. Merck and Roche again are the domestic manufacturers, and imports account for about 35% of the market. Merck has been producing B_{12} by fermentation since the early '50s, and publications and patents in the '60s indicate a concentration on Pseudomonas dentrificans. The maximum yields reported in the literature, from mutants selected for ability to produce high levels of B_{12}, are in the 30-50mg/l range. Beyond these mutation-selection efforts, genetic knowledge has had little influence to date. Given that the market is in the $15 million range and the unit value around $5/g, one can assume that companies like Merck and Roche will attempt to apply such knowledge, as they and others develop it.

Little is evident about the role played by genetics, but Roche certainly makes use of fermentation in production of biotin, and in the transformation of sorbitol into L-sorbose, the precursor of vitamin C, its largest selling vitamin by far - Roche's world capacity for vitamin C exceeds 20,000 metric tons per year. Fermentation processess are available for β-carotene and pro-vitamin D, but chemical synthesis is presently being used. Technical or economic developments obviously could change those situations.

7.4 Microbial transformations such as the sorbitol to sorbose conversion are becoming increasingly common in pharmaceutical processes. This is especially true to the realm of steroids. [Charney & Herzog, 1967]

An enormous variety of highly specific structural modifications have been catalogued. These are often used in research and, when a compound merits scaling-up, are incorporated into the production scheme. A host of different genera among the bacteria, fungi and actinomycetes have proved useful for such purposes.

Commercial examples are:

- 11α- and 16α- hydroxylations of progesterone by Rhizopus nigricans and streptomyces argenteolus, respectively, involved in glucocorticoid production.

- o 11β- hydroxylation by *Curvularia lunata* to produce hydrocortisone.
- o 1-dehydrogenation by *Corynebacterium simplex* to produce prednisolone.

Essentially nothing is known about the genetics of these organisms. Developments based upon genetic techniques that might increase the production and secretion of key enzymes could substantially improve the economics of some presently very inefficient processes, by enabling the use of reaction conditions under which substrate compounds are more soluble, for example. However, this seems likely to be a derived rather than a direct benefit.

7.5 *Claviceps fermentations* represent a special category of bio-production. For one thing, the compounds - which include the infamous LSD - are among the most potent pharmacological agents known to man. Secondly, until the 1960s the production of these ergot alkaloids was based entirely upon the cultivation of the fungus, *Claviceps purpurea*, on rye growing in (carefully guarded) fields. The exploitation of the manifold pharmacological activities* of these compounds has been greatly facilitated by the development of fermentation conditions and strains of *C. purpurea*, *C. paspali* and *C. fusiformis* suitable for production under more controlled conditions than in grain fields. No doubt the genetic knowledge of these organisms is even less than that for antibiotic producing fungi, but they are worthy of mention here, particularly for two reasons:

- o Sir Ernst Chain, who was the biochemist who made it possible for Florey to get enough penicillin to set the field of antibiotics in motion, was the one who brought ergot production into the laboratory and fermenter.
- o Production of ergot alkaloids under fermentation conditions was simpler, safer, and less expensive - as the molecular clone products should be - and it enabled experimentation that is currently (>15 years later) resulting in novel products of significant utility.

---

*Used to treat migraine, senility, pituitary disorders, and to induce lactation or abortion, for example.

That is to say, academic researchers (Chain was then a professor at the Instituto Superiore di Sanita in Rome) can provide practical solutions to problems with far reaching implications, and the implications may take a decade or two to be expressed in pharmaceutical innovations!

The Business of Molecular Genetics

8. New Enterprises

Only a few major firms seem to have been at the cutting edge when molecular genetic manipulation became feasible. Hoffman-LaRoche, with its Institute for Molecular Biology, was fully cognizant of emerging developments; nevertheless, its collaboration with Genentech acknowledges the lead achieved by small firms. Merck, which was represented at the Asilomar conference in 1975, and Lilly, both had relevant R&D programs (in cancer, fermentation and immunology). Others must also have sensed the pregnancy of molecular biology in the mid-70's, but it was entrepreneurs and academicians who first took noticeable action. The following summarizes commercial activities as of Spring, 1980.

8.1 Cetus Corporation was founded in 1971 by two M.B.A.'s trained in biochemistry and medicine, a Nobel Prize-winning physicist turned molecular biologist and two other scientists. Their first clients were pharmaceutical companies who asked Cetus to improve the yields of existing commercial antibiotic cultures (as described earlier, pp. 320-322). They now have over 200 employees, 80,000 sq. ft. of space,* and have raised over $36 million in capital.# Cetus is presently working in six areas, of which pharmaceuticals is but one, on about 15 projects. Peter Farley, M.D., the President,

* 276 employees, 45 PhDs, and 110,000 ft^2 of space as of December 1980.

\# $120 million raised by public offering in March 1981.

says they will only work in areas in which they believe Cetus can achieve a significant share of a market that is worth over a billion dollars annually.

In the pharmaceuticals area, the synthesis of interferon is Cetus' largest project. Farley says, "Cetus believes that interferon will be at least as important, dollar-wise, as the antibiotics market - $5 to $10 billion - by the late 1980s." [Staver, 1980] Cetus has 28 people working on interferon as of April 1980. A large "P-3" laboratory was established in 1978. Since June 1979, their work in recombinant DNA has accelerated from 4 PhDs (12 people) to 14 PhDs (45 people). Vaccines created by recombinant techniques are the object of a Cetus-funded project, and Cetus' first such product has been undergoing animal testing in the facilities of a major pharmaceutical company. Cetus has an interest in monoclonal antibodies, and has generated evidence that it may be able to develop " an early/fast, general detection method for rapidly identifying pathogens."

Cetus established a pattern by recruiting a prestigious Board of Scientific Advisors - which includes Stanley Cohen, Joshua Lederberg, and a host of prominent academicians as formidable as any assembled in this field - that is intimately involved in the direction of R&D. Self-funded research accounts for approximately 40% of all R&D expenditures. The primary corporate goal is long-term growth and profit; it is anticipated that Cetus may have a negative cash flow for the next several years.

8.2 Genentech, Inc. was founded in 1976 by a chemist/MBA and one of the two principal inventors of gene-splicing technology, Dr. Herbert Boyer. Robert Swanson, President of Genentech, perceived in the work of Boyer and Cohen the basis for a business that would exploit the novel capacity to incorporate genetic characteristics by design. Boyer concurred, and they established research agreements with the University of California at San Francisco (where Boyer is a professor) and with the City of Hope National Medical Center at Duarte (where Dr. Keiichi Itakura, a specialist in DNA synthesis, and Dr. Arthur Riggs, who specializes in the control mechanisms of bacterial genes, are located). These affiliations formed the basis for the series of pioneering developments described earlier in the section on peptide hormones (pp. 163-167, 174-176, and 193-195). Genentech has developed an in-house scientific staff, including over 30 PhDs, in a total roster of about 70 employees at a 30,000 sq. ft. facility in South San Francisco.* Venture capital was crucial to the initiation of Genentech - in large part through limited partnership tax shelter firms, which have been the primary source of start-up capital since 1976. Genentech raised three increments of capital ($1.9 million total) from these sources before obtaining major financing. In 1979 Lubrizol Enterprizes purchased a 15% share for $10 million, and it bought an additional 10% share from Genentech and one of the earlier investors (Inco) for $15 million early in 1980.# Genentech's cash flow has been positive (revenues greater than expenses) since the second quarter of 1978.

Genentech has about 12 projects underway, almost all of which are in the pharmaceutical field. Just this year it has set in motion programs to capitalize on the technology in fields outside of medicine, but that activity is only in the formative stages. In terms of published developments and requests for permission from the NIH Recombinant-DNA Advisory Committee (RAC) to scale-up, Genentech appears to have established a broad lead in the area of polypeptide production. It has demonstrated the practicality of synthesizing genes and arranging translation-controlling elements to produce peptides of known structure. For example, Genentech reports capacity to produce multi-gram

*112 employees, 40 PhDs, and 114,000 ft^2 of space as of August 1980.

$35 million raised by public offering in August 1980.

quantities of somatostatin and growth hormone, by this means. In June 1980, Genentech disclosed results of a collaboration with Hoffman La Roche. They had produced interferon by fermentation with an *E. coli* into which they had placed cDNA derived from mRNA isolated from human leukocytes. They demonstrated the product's antiviral activity in monkeys.

In August 1978 Genentech disclosed a contract with AB Kabi, the Swedish-based world's largest commercial supplier of human growth hormone, to begin work on a method for producing that hormone by recombinant-DNA * - in exchange for cash, royalties and a share of marketing rights to Genentech. A month later it signed a contract with Eli Lilly & Co., the largest producer of insulin in the United States, on the basis of Genentech's synthesis of the A and B chains of human insulin.

This agreement provides for multi-million dollar payments to Genentech and continuing royalty rights, in exchange for Lilly's use of Genentech processes in the manufacture of human insulin.

Early in 1979 an unidentified large European firm signed a contract with Genentech for a long-term development concerning the production of a vaccine through recombinant-DNA technology. In March Genentech requested and received permission from RAC to begin a collaboration with the USDA on hoof and mouth vaccine. Letters of intent are presently reported to be in hand for collaboration with two other companies on undisclosed new projects.

Genentech's initial dependence on outside scientists - based on its fundings and agreement to share income with the non-profit institutions from patentable inventions arising from the efforts

*This goal was achieved, two months ahead of contract schedule, by June, 1979. Kabi is said to have quickly decided to fund process development and scale-up.

of their staffs - has largely been eliminated. However, it maintains extensive relationships with academic and institutional scientists. Mr. Swanson says that, "Genentech's success in attracting scientists of the highest caliber from around the world is traceable to its sense of being a part of the scientific community with responsibilities to both its own scientists and to science at large." It is corporate policy to publish in the scientific literature, and it has published an average of better than one new application a year of *in vitro* recombination:

- Somatostatin, August 1977
- Human Insulin, August 1978
- Human Growth Hormone, June 1979
- Thymosin-α_1, August 1979 (disclosed March 1980)
- Proinsulin, March 1980

Genentech is planning to sell some materials, such as somatostatin, for research (including clinical studies) in the near-term, and its middle-term plans call for a staff of 1,000 and revenues of $100 million by the late-1980s. Genentech's business goal is: "To make significant new products by genetically engineered microorganisms....to make such products safely and to make them available in the shortest time to the people who need them." [Swanson, 1979]

8.3 Biogen S.A. was originally incorporated in Luxemburg May 5, 1978, but it is now headquartered in Geneva, Switzerland. The founder, Chairman of its Scientific Board and Co-chairman of its Board of Directors is Dr. Walter Gilbert, a professor at Harvard. With one other American (Dr. Phillip Sharp of MIT) and six European scientists*, he organized a company "to develop commercial products in the pharmaceutical, chemical, energy and other fields using advanced microbiological science, principally DNA recombinant technology and advanced mutational technology." A pres-

*The board has since been expanded to ten members.

ident, Robert Cawthorn and a staff of about 30 (16 PhDs) were recruited by April 1980. The founding scientists were assisted in organizing by several venture capitalists, including people at Inco who had been involved in the Genentech start-up. Four European and two U.S. investors took shares for about $2 million and the Schering-Plough Corporation put in $8 million, in exchange for a 16% interest and rights to three pharmaceutical products.

Analogous to Genentech's strategy, Biogen placed R&D contracts with academic researchers. Two of these have come to light. The work performed at the University of Edinburg on the hepatitis antigen and at the University of Zurich on leukocyte interferon (discussed on pp. 166 and 176, respectively, in this report).

To date Biogen has hired only a fraction of the staff intended for the 8,000 sq. ft. facility leased from Battelle Institute in Geneva, but they advertised 18 post-doctoral positions recently and received 600 "very good quality" applications.

Mr. Cawthorn says that Biogen is planning greater links with European companies, particularly in chemical, petrochemical, and process engineering areas. They "already have strong European shareholding: 36%, with 41% in the U.S. and 23% in Canada; but they are not looking for shareholders any more."* "Rather, Biogen wants to learn from relevant industries what might be the most profitable products to develop with genetic engineering". [Walgate, 1980] Dr. Gilbert admits that Biogen was "driven off-shore" by concerns raised by the moratorium on recombinant-DNA in the mid-1970's, but says they have always operated consistently with governmental guidelines. A major

* Monsanto made a $20million dollar investment, and Schering-Plough and Inco both increased their investments to provide Biogen with an additional $8.8 million in October 1980. Plans to establish a second laboratory, in Cambridge, Massachusetts, were disclosed at that time.

advantage of Biogen in his eyes is its positioning to tap European know-how in molecular biology.

Dr. Gilbert says that they anticipate the biggest economic impacts will be non-pharmaceutical. Nevertheless, Biogen has programs related to vaccines for diseases, such as malaria, for which the organism cannot be cultivated, and human proteins other than enzymes and hormones, as well as the hepatitis and interferon projects underway. Richard Bennett, President of Schering-Plough, says that Schering has designated two undisclosed pharmaceutical projects upon which Biogen will be obligated to give Schering a first look and to make an offer with respect to development.

Schering-Plough has submitted plans to RAC for scaling-up synthesis of leukocyte interferon based on the process developed by Dr. Charles Weissman and his group in Zurich. Mr. Cawthorn told a Nature reporter recently (April, 1980) that Biogen was nine months from producing enough interferon for clinical trials.

He said a number of laboratories may undertake pilot plant production including Professor Brian Hartley's laboratory at Imperial College, London (a Biogen Scientific Board member). [Walgate, 1980] Schering has responsibility for much of the further development, including scale-up and toxicology testing, and it has the distribution rights.

Mr. Bennett, of Schering, told a stock analysts meeting (3/11/80) that "under the most optimistic schedule, with everything going without problems, with FDA giving it a fast-track, and with all the approvals coming through on that side, the most optimistic thing I can think of is three to four years" before commercial availability of interferon from this work.

8.4 Genex Corporation, too, got started by contracting research in universities. Robert Johnson, a venture capitalist, and Dr. Leslie Glick, president, established the firm in mid-1977. They obtained financing from Koppers and from Innoven Capital. (The latter, a venture capital firm set up by Monsanto and Emerson Electric, also participated in the early funding of Genentech.) In early 1980 Genex completed installation of laboratories, including one "P-3" and several "P-2" facilities, in its 18,000 sq. ft. building at Rockville, Maryland.*

Genex does not plan to undertake projects to develop novel pharmaceuticals on its own account, for such products take too long to reach the market. However, they are quite interested in becoming a source of intermediates for pharmaceuticals, and other specialty chemicals, for which they feel there is a substantial potential.

OTA has obtained a study in which Genex presents its arguments for a genetically engineered microbial conversion of aniline to acetaminophen(APAP), a very widely-used aspirin substitute. As they note therein the market for the raw compound is in the $50 million range and the retail value is 10 times that or more. We are not in a position to critique Genex's estimates of their or competitors production alternatives, but feel the study provides good support of the strategy that recognizes the potential for large volume, relatively inexpensive chemicals that could possibly be produced by fermentation.

Presently, Genex is conducting work for a Japanese client to fabricate a microorganism that will produce extremely high levels of an amino acid, and expects to have that process in pilot plant in early 1981.

*Additional funding from Koppers, expansion to 70,000 ft^2, and collaboration with Bristol-Myers on interferon were announced in August 1980.

Genex is entertaining, and soliciting, possibilities for collaboration with pharmaceutical companies. Most radical is its proposal to attempt commercial development of the oft-discussed possibility of treating sickle cell anemia by inserting the appropriate hemoglobin gene into marrow cells. Dr. Glick feels that pharmaceutical companies: "have asked themselves whether they want to be seen to be interested in gene manipulation" and decided not to touch it. He says the economics look good - based on the estimation that the average annual treatment costs $15,000 and the average life expectancy is between 20 and 25. He thinks the recombinant-DNA approach (which would involve withdrawing hematopoietic bone marrow or circulating cells, treating them to insert the gene and then replacing them in the patient, probably three or four times a year), would have an annual cost aroud $4,000, and would also improve longevity. [Scrip 11/7/79]

Genex is reportedly discussing other, more conventional, product development programs with pharmaceutical companies from both the U.S. and abroad. Dr. Glick said last winter that he "does not expect to see the commercial production of pharmaceuticals based on recombinant-DNA technology for another four to five years, with the possible exception of insulin. Of the drug companies in the field, he thinks Upjohn is the most advanced. [Ibid.]

8.5 New England BioLabs, founded in 1974 by Dr. Donald Comb, has the distinction of being the first company to bring to market a product made by recombinant-DNA methods. Included in the line of restriction enzymes it offers to researchers are four that are gene products resulting from gene amplification and expression, achieved by academicians or Comb's associates (22 people). He says they intend to stay small ($3 million sales according to Nicolas Wade) and family owned. BioLabs merits mention in this context because it has started to recruit molecular biologists to do cloning work "for production of proteins applicable to world health problems," such as malaria. Comb says they are not going to get into the interferon race.

8.6 Bethesda Research Laboratory (BRL) is also in the restriction enyzme (and other supplies for molecular geneticists) business. It was founded in 1976 by Stephen Turner; now employs 150 (35 PhDs) and has sales of more than $2.5 million. Half of BRL sales are restriction enzymes, but it is diversifying into other fields, chiefly recombinant-DNA and hybridomatechnologies. While BRL is not actively engaged in medical product developments, no doubt its products, along with those of New England BioLabs and Collaborative Research, are involved in many such projects. Such firms could easily be nudged into the mainstream.

8.7 Collaborative Genetics was created in November 1979 as a majority-owned subsidiary of Collaborative Research of Waltham, Massachusetts. Chairman of the scientific advisory board is David Baltimore - a member of the RAC and Nobel Prize winner (for his discovery of the reverse transcriptase used to make DNA complementary to messenger RNA). The board includes three top authorities in the field of yeast genetics, and many feel that sophisticated descendants of baker's or brewer's yeast may ultimately become the workhorse microorganism of modern biotechnology. The initial focus of Collaborative Genetics will be on energy, enzyme, and chemical production, but Collaborative Research has been involved in interferon, urokinase, growth factor, and other work of medical interest, so that a collaboration with a pharmaceutical company seems very probable.

One interesting aspect of Collaborative Genetics' approach is its proposal that the risks associated with developing a new application might be shared among a group of sponsors by dividing the licensing of the technology on a geographic basis, i.e. a Japanese company might have exclusive license for Japan and the Far East, while other companies which participate in underwriting the developmental costs might be licensed for the UK, West Europe, East Europe, the Near East, USA, etc.

Collaborative Genetics is collaborating with National Patent Development Corp. on the production of interferon via rDNA-modified yeast, and Dow Corp. signed a $5 million contract with Collaborative Genetics in 1980.

8.8 Hybritech, Inc. was formed in 1978 with the same sort of perceptions and planning that led to the formation of Genentech. Brook Byers, a venture capitalist (with Kleiner, Perkins, Caufield & Byers - KPC&B - former associates of Bob Swanson and familiar with biotechnology as a result of the Genentech experience) recognized the opportunity represented by monoclonal antibodies, put together a small team of immunologists, sowed the seed money and set out to find a president. He and Howard (Ted) Greene found one another - Greene, who knew the immunodiagnostics market, had been contemplating just such a venture. Byers, as chairman, and Greene have now built a staff of about 50 and an expanding product line. Dr. Ivor Royston, of the University of California, San Diego (UCSD), nucleated the scientific planning/recruiting, and most of the staff, such as Dr. Gary David, were drawn from the ample pool of immunotechnologists at Salk Institute, Scripps Clinic and Research Foundation and UCSD. (All of which are within walking distance of Hybritech.) While the company already has three products on the market, for research purposes, Hybritech's management says that its main focus is, and will be for several years, the development of the technology base. An initial investment of $300,000 by KPC&B was soon followed by $1.6 million more and the involvement of two additional venture capital firms. Hybritech currently occupies 8,000 sq.ft. of laboratory and office space, and plans to expand to 40,000 sq. ft. during 1980 to have adequate facilities for manufacturing commercial quantities of clinically useful antibodies.

Hybritech is the leading firm working on the sort of applications described previously (pp. 303 - 315). Specifically the monoclonal antibodies it now has on the market or in advanced stages of development are the following:

Antibody to	Potential Clinical Significance
Hepatitis B surface antigen	Hepatitis
Prostatic acid phosphatase	Prostatic cancer
Creatine kinase isoenzymes	Myocardial infarction
Carcinoembryonic antigen	Colorectal cancer
Alpha-fetoprotein	Prenatal neural tube defects
Immunoglobulin E	Allergies
Immunoglobulin G	Gammapathies
Chorionic gonadotropin	Pregnancy
Lymphocyte surface antigens	Immune deficiencies and proliferations

An important part of Hybritech's first-phase strategy is to develop products rapidly with the hope of supplying them in bulk to other companies that will obtain the necessary approvals and then develop and market kits for diagnostic use. "We want to concentrate on the technology," says Greene, "not the channels of distribution". [Franson, 1979]* Phase two will involve marketing diagnostic kits, and phase-three will involve the therapeutic use of antibodies.

Enthusiasm for the therapeutic uses of antibodies is very high, although such applications are far from certain. Of the diagnostic

*As of April 1981, Hybritech is expecting momentary approval of the first monoclonal diagnostic products, and it is putting together a sales group to market those products.

utility for monoclonal antibodies, there is little doubt. What is most uncertain is the issue of patent protection. However, Greene points out that identification and rapid penetration of markets with satisfactory products has traditionally been far more important to success in high technology businesses than patents.

8.9 Centocor, Inc. was founded in 1979 by Edwin Allen, with funding by the Bank of Paris and Venroc, the Rockefeller family's venture capital firm. It is closely affiliated with the Wistar Institute, also of Philadelphia - the Institute's director, Dr. Hilary Koprowski, being chairman of Centocor's board of scientific advisers. Centocor has been assigned exclusive licenses for two important hybridoma patents which were recently granted to the Wistar Institute. The broad patent for monoclonal antibodies directed against tumor cells, granted in October 1979, has already been described (p. 312). A second patent broadly covering antibodies to viral antigens was awarded in April 1980. The validity of these patents will, almostly certainly, be contested in the courts.*

Centocor has assembled a "senior staff" of 20. Mr. Allen says Centocor is "interested in stretching the technologies, so most of our projects are long term in nature". They do not expect to launch their first product before the end of 1981. Centocor is reported to be interested primarily in applying the hybridoma and other technologies to four areas of diagnostics, those concerning tumors, liver, heart and viruses. How they intend to exploit the cancer therapy patent has not been disclosed.

---

*Ceasar Milstein – who with Georges Köhler first described how to make hybridomas in 1975, and never filed for a patent on the invention – says a patent might be justified for particular clones, but "essentially they are patenting our procedure".

8.10 Becton-Dickinson, the major healthcare product company, obtained a license from Stanford University for the fluorescense-activated cell sorter (FACS) developed by Stanford scientists, and during 1979 it launched a satellite business, marketing monoclonal antibodies - which the FACS makes it easy to develop. Whether this will develop into a major commitment, apart from its main lines of business, remains to be seen. Becton-Dickinson (B-D) certainly is quite active in the diagnostic testing and laboratory field, and this new business undoubtedly provides an excellent channel for incorporating or accomodating the new technology in connection with B-D's other product lines.

8.11 Monoclonal Antibodies, Inc. of Palo Alto, CA, has the name, a small laboratory and a staff of about 5 upon which to build. Its president, Tom Glaze, incorporated in 1979. The first products, introduced in April 1980 primarily to give the company visibility, did not appear to be distinctive from monoclonal antibodies already offered by others.

However, Mr. Glaze says that discussions are in progress with diagnostic kit manufacturers for whom the new company would produce antibodies initially.

8.12 Clonal Research, Ltd. was founded in October 1979. The founder David Hutton, organized the firm as a joint-venture with Monitor Science Corp, a diagnostic firm which will market some of Clonal Research's products. In particular, Monitor Science is interested in agents for monitoring the blood levels of aminoglycoside antibiotics. Clonal Research has only 3 full-time employees and 4-5 part-time ones. By autumn 1980 it expects to market its first product, an antibody against human chorionic gonadotropin. However, that, too, is not a novel product; although some firm undoubtedly will be the first to obtain approval on a pregnancy test utilizing such an HCG antibody. Mr. Hutton says that their major medical interests at this time concern polypeptide hormones and isoenzymes. He is also quite interested in cell fusions other than myelomas, including especially plants and agricultural applications of cell culture.

8.13 Other entrepreneurial firms with pharmaceutical products in mind as of Spring 1981 include: Applied **Molecular** Genetics (Newbury Park, CA) DNA Sciences (New York); DNAX, Ltd. (Palo Alto, CA); EnzoBiochem (New York); Genetic Systems (Seattle); and Kappa Laboratories (Escondido, CA).

9. Established Pharmaceutical Companies

Categorization of companies as U.S.-based or foreign may be a convenient, appropriate distinction for a report to a U.S. government audience, but the point should have been made in the profile of the industry, presented earlier, that pharmaceuticals is unquestionably a multinational business. Furthermore, it will become apparent that several foreign-based companies are taking full advantage of technology developed in the U.S.A.--by profit and non-profit institutions. There are cases, too, of U.S.-based firms locating their molecular genetic developments overseas. The precedents and possible implications of such cases shall be considered briefly. It is not the purpose of this report to publicize or criticize anyone's activities, but to facilitate appreciation of the potential impact of molecular genetics by presentation of current facts.

Leading U.S.-based pharmaceutical firms known to have acquired recombinant-DNA manipulating capability include: Abbott, Bristol, DuPont, Lilly, Merck, Pfizer, Schering-Plough, Searle, Smith Kline, Syntex and Upjohn.* Firms especially

*In alphabetical order.

active in monoclonal antibody development include the diagnostic divisions of Abbott and Johnson & Johnson (Ortho). Other firms whose research programs make it seem likely that they are or soon will be engaged in such work are: Lederle, Parke-Davis, Revlon and Wyeth.

The nature of some of these activities shall be described and, where possible, the current level of effort characterized to illustrate how such new developments relate to previous activities of a given firm and to environmental factors. The anecdotal quality of this section derives from both the sparcity of information (quite understandable in an industry with the sort of long lead-times that are involved in pharmaceutical product development) and from our desire to explore some implications rather than exhaustively catalog what everybody is doing at this particular moment.

9.1 Abbott Laboratories' announcement in April, 1980, of successful cloning and expression of a gene for the thrombolytic agent urokinase (p. 232-238) can be viewed as exemplary from several perspectives:

- It is the first case of production of a human enzyme by means of molecular cloning.
- A pharmaceutical company is attempting to use molecular cloning to produce a product that it already markets for which cost is a substantial factor and probable deterrant to use.
- This is an example of the evolution of production technologies in a situation where product patents are not available (i.e., Abbott utilized state-of-the-art cell culture first and now may devise a fermentation process based on a new "art.")
- A possible outcome is that a U.S.-based producer/marketer, utilizing molecular clones for production, will be competing with another firm (also U.S.-based) marketing the same product made via a conventional process by a foreign firm.*

The history of urokinase illustrates the slow progress of products

* As described on p. 233, urokinase is the first Japanese-made drug formulation ever to receive production and sales approval from FDA.

to market and narrowness of claims allowed for what would appear to be a more broadly useful agent. Hence, there is a powerful incentive to build upon what is established, rather than start all over again with a novel agent (e.g. "tissue activator"), if one can be assured of some advantage relative to other marketers of the same product who do not have to bear the risk/cost associated with pioneering clinical research. Cost of production would certainly be such an advantage in this case, where the unit cost of the product is presently very high.

The existence of a >$90 million market for urokinase in Japan* plus substantial use in some European countries, certainly makes it possible that a brand of urokinase with a price advantage could achieve worldwide sales of the order of $100 million--a level achieved by less than 2 percent of new drug products [Blee, 1978]. In fact, the cost and logistic advantages of fermentation versus collection and extraction of urine could result in near universal transformation of urokinase production through licensing of the patents and know-how on a non-exclusive basis.**

We understand that cell culture production of urokinase never possessed such a cost advantage, but was adopted, rather, to gain better control of the source of the products than could be achieved in the collection of human urine.

---

*Abbott is involved in a joint venture with Takeda Chemical Industries, Ltd., of Japan to develop and market Takeda drugs in the U.S.A. and Canada. Takeda, the largest pharmaceutical company in Japan, is not presently marketing urokinase, but their largest selling product is a bacterial enzyme preparation, serratiopeptidase, which is used as an anti-inflammatory agent. Abbott also has marketing joint ventures in Japan.

**It is noteworthy that Abbott has announced plans to make fibroblast interferon and conduct clinical tests independent of NCI and ACS trials. It will use tissue culture techniques initially but recombinant-DNA work on interferon would be a natural sequel to its urokinase success.

Abbott's recombinant-DNA activities began in 1977 and intensity of effort increased markedly in 1979. Eight people were identified with the initial urokinase announcement and probably at least as many are involved in other efforts, especially those related to antibiotics.* Furthermore, Abbott has been a leader in exploitation of immunotechnology, particularly for diagnostic purposes and almost certainly has a substantial effort in the monoclonal antibody field. Abbott management declines to say how many people are working on products to be derived from molecular genetics, but we would estimate that about thirty of Abbott's 1200-person R & D staff have been performing such work and the number can be expected to increase with the urokinase success.** Abbott management stresses that the value of genetic techniques and monoclonal antibodies for insight and understanding throughout its exploratory research may far exceed the products directly derived from those technologies.

9.2 Bristol Laboratories has not made the newspapers with its efforts in molecular genetics, but its interests in antibiotics, cancer therapy, fermentation and biotransformations give it many reasons to be prepared to make use of the latest developments in genetics.#

Two unusual arrangements figure in Bristol's exploratory efforts in genetics and other areas. (A third was footnoted on page 411.)

* The antibiotic of paramount importance to Abbott, erythromicin, is one for which conventional genetics has raised yields only into the 2-4 g/l range, one-tenth the level achieved for penicillin and tetracycline, for example. [Malik, 1979.]

**Abbott reports that approximately half of their $85.44 million R & D expenditures in 1979 was devoted to pharmaceuticals (10.55% of pharmaceutical sales in that year, or 2.8% of human health care products). Its $419 million pharmaceutical sales in 1979 derived from roughly 350 pharmaceutical products. Like many firms in the pharmaceutical business, Abbott is a diversified company, with total sales of $1.683 billion.

Like a number of other pharmaceutical companies, Bristol delayed commitment to recombinant-DNA research for over a year after the Asilomar Conference.

The older of the two is Bristol's relationship with Banyu Pharmaceutical Company of Japan. Since the initial marketing agreement (for Banyu to sell Bristol's long-acting form of penicillin G in Japan) in 1953, the relationship between Bristol and Banyu has spawned many offspring, one of which is the Bristol-Banyu Research Institute (BBRI). Bristol established a lab in cooperation with Banyu in 1958, acknowledging the high skill they perceived the Japanese to possess in the sort of fermentation research that is used to develop antibiotics. Bristol in Syracuse, N.Y., directs the work of BBRI, which consists of screening thousands of soil samples a year for microorganisms capable of producing novel antibiotics and antitumor drugs. BBRI also works on modifying existing antibiotics and antitumor agents to develop new and more effective derivatives.*

Genetic manipulation and recombination is becoming part of that effort--which really includes all aspects, other than clinical, of R & D connected with such agents. Only a small fraction of the 92-person staff of BBRI is involved in recombinant work to date. An interesting comment made by a Bristol executive alluded to the fact that inflation has erased the economic advantage that originally pertained to doing work in Japan--it now costs 10 percent more per man to do research in Japan--but the productivity and excellence of the work make it well worth while.

The other collaboration referred to above is with Baylor College of Medicine, in Houston--another multifold relationship. In addition to substantial grants for unrestricted cancer research at Baylor (and four other universities), Bristol has established a laboratory on the Baylor campus. It is there, because Baylor

* This approach resulted in the laboratory's biggest success--amikacin, an aminoglycoside effective against many strains resistant to gentamicin, synthesized in 1971.

has the "P-3" containment facility necessary for some genetic manipulations, that Bristol has set up its principal recombinant-DNA research group. About six people work there* on programs sponsored by both Bristol Laboratories (one pharmaceutical subsidiary of Bristol-Myers Company) and the Industrial Division (which is responsible for the company's worldwide production of antibiotics). Dr. Stanley Crooke,# Bristol's Associate Director of R & D and head of its cancer research program, continues as a faculty member at Baylor, although he is paid only by Bristol, and the head of Baylor's Pharmacology Department is a Senior Consultant to Bristol. The fact that molecular genetics is central to the cancer problem requires the Baylor Molecular Pharmacology Group to appreciate and utilize the latest insights into genetic transmission and translation processes.

Bristol Laboratories has the most intense cancer chemotherapy program of any pharmaceutical company--certainly in the U.S.A. and probably in the world. After a stepwise involvement with various anti-cancer agents since 1960, a corporate strategic decision was taken in 1976 that resulted in an almost immediate tenfold increase in R & D on cancer. Spending on cancer R & D has gone from ∿ $1 million in 1973 to ∿ $18 million in 1980.** In many ways the course of cancer chemotherapy could serve as a model of developments related to molecular genetics. The prime investor in cancer is the federal government, through the National Cancer Institute (NCI). Bristol and other industrial labs (particularly Parke-Davis and Upjohn) have conducted

* Under the direction of a molecular geneticist hired away from Abbott.

Dr. Crooke became Vice President, R&D, at SmithKline in September 1980.

** Roughly 25 percent of R & D; 50 percent is spent on antibiotics.

screening tests for NCI for twenty years.* Today, this still constitutes ∿ 15 percent of Bristol's activities related to cancer. Such screening includes isolation, characterization and testing in vitro of samples from many sources. When an agent merits tests in animals, they are donated by Bristol so that it can begin to establish a proprietary position upon which to base scale-up, clinical testing and commercialization, if the compound proves useful. In addition, Bristol has cast a large net to acquire rights to anti-cancer agents developed elsewhere. It has agreements with other pharmaceutical companies (Leo, Sandoz, Schering and Searle), with Dr. H. Umezawa's Institute of Microbial Chemistry in Japan (perhaps the most prolific source of antibiotics and anti-tumor agents anywhere), with the U.S.S.R. (for two active compounds), and with a processor of precious metals, Johnson-Matthey (a source of compounds like cis-platinium). Furthermore, Bristol aggressively pursues opportunities to license promising anti-tumor agents from any source. With eleven drugs marketed, Bristol-Myers offers more anti-cancer products than any other pharmaceutical company, and there are at least sixteen more in clinical or late preclinical development. Some are what would be called orphan drugs; for example, an NDA is being filed on teniposide, a mitotic inhibitor, for treating neuroblastoma, of which there are perhaps 40 cases a year. There is good will to be gained by such commitment, but only a company that has achieved critical mass in a field can afford such an activity.

Bristol has an immunotherapy department, responsible for development of immunomodulators that seems likely to make use of molecular clone products. They are active in human tumor cloning and feel that that method may facilitate determination

---

* Parke-Davis (Warner-Lambert) has recently made a strategic commitment also to the field of cancer therapy.

of the sensitivity of specific malignancies to various drugs. Such clones may also prove useful in the prescreening of potential antineoplastic agents.

Diagnostic agents are also a component of Bristol's cancer program. They have an intense interest, for example, in a nucleolar antigen which has been found by a Baylor team to be associated with a broad range of malignancies and not with benign tumors. [Busch, 1979.]

The recently successful antitumor agent, bleomycin, provides an example of how Bristol expects to be influenced by molecular genetics. Bleomycin is a glycopeptide produced by *Streptomyces verticillus* which produces single and double strand breaks in DNA at very specific sites in the DNA sequence. Dr. Crooke says, "We now know that by modifying a certain site in the molecule, we can change the cleavage sites in DNA. Thus it is conceivable that by using recombinant DNA techniques we can construct a new bleomycin that selectively cleaves specific genes, perhaps genes important to the expression of cancer. We also know that modification at another site in (the bleomycin) molecule alters the ability to produce breaks in both strands of DNA. Since this is the fatal lesion, it is now conceivable to develop a new analog which might produce double strand breaks, only in cells which are usually acidic, for example, and cancer cells often tend to be more acidic than normal cells." [Crooke,1980] That is to say, molecular genetics is one more tool for engineering pharmacologically useful compounds.

Genetics has proven to be an extremely valuable tool, as far as Bristol is concerned. For one compound they were interested in, they were able to progress from a yield of 20 units/ml production by a surface culture to > 20,000 units/ml in a deep tank fermentation by means of mutation and selection, and provision of the proper medium and precursors. They are the largest penicillin producer in the U.S.A. and, since they are running at capacity, even a small (10%) increase in yield is considered important. For some processes, such an improvement may occur only every four years, while for others they may change strain three to four times a year.* Protoplast fusion has become a routine technique for this sort of thing, since it permits combining the results of separate strain improvement efforts and, with the high frequency of recombination (thousands of times greater than for the mitotic recombination of parasexual genetics), biochemical markers are unnecessary. For the most part, Dr. Elander, Director of Fermentation Development, says they are using protoplast fusion between strains of aspergillus and penicillium developed by mutation and biochemical selection. He thinks yeast will, in the long run, become the predominant industrial host for plasmids assembled in vitro.

9.3 DuPont has never made a big impression in the pharmaceutical field. This is true despite the fact that for over 15 years theirs was the only systemic antiviral drug on the market.** DuPont acquired a small pharmaceutical company, Endo Laboratories, in the '60s, markets about 15 products through that outlet, and has persisted in biological research for a couple of decades.# Now it appears to be putting quite a bit of effort into interferon development and recombinant-DNA technology. An in-house

* FDA regulations permit such changes without requiring new toxicology.

** Amantidine, marketed as Symmetrel, is effective in the prevention and treatment of Typa A influenza.

\# Acquisition of New England Nuclear, which has a sustantial effort in monoclonal antibodies, in May 1981, is evidence of duPont's interest in this field.

team that has been working on interferons for more than ten years has been augmented recently by collaboration with Professor Lee Hood at Cal Tech and by a **new** group of molecular geneticists that DuPont has recruited in 1978-80. In addition to the interferons, Ralph Hardy, PhD., who directs research in life sciences, says they view recombinant-DNA as the key to all sorts of rare regulatory molecules that may prove to be either valuable pharmacologic agents or vital intermediaries in processes for which antagonistic or agonistic molecules may be desirable.*

Furthermore, DuPont is in a position to exploit many non-pharmaceutical uses: in chemicals, obviously, but also in plants, for which they have been studying nitrogen-fixation for more than twenty years.

9.4 Eli Lilly & Company almost certainly is the farthest advanced U.S.-based pharmaceutical company in applied genetics. Lilly scientists are reported to have generated 20 percent of the world's currently available antibiotics, more than any other firm [Lilly, 1980]. They are one of the largest practitioners of fermentation in the world and have long exploited the mutation and selection method of strain improvement. They point out, however, that none of the seven cephalosporins Lilly now markets could be produced by fermentation alone. In recent years, Lilly has published [Queener, 1979] and obtained patents on protoplast fusion techniques for antibiotic producing organisms.** They have also made at least one novel antibiotic by interspecies

* Endo has emphasized agents acting on the CNS, and they are among the many attempting to make use of the recent knowledge that the receptors which bind CNS-active drugs have a natural affinity for certain polypeptides.

** One patent (U.S. 4,159,226, 6/26/79) describes a method of obtaining Streptomyces protoplasts which are able to regenerate viable cells with high efficiency, explicitly claiming its application to 30 species.

recombination of Streptomycetes [Godfrey, 1978]. They were among the first to attempt cloning of Streptomycete DNA in E. coli [Johnson & Burnett, 1978].*

Like Bristol, Lilly has a strong interest in cancer chemotherapy; however, it has shunned government-supported contracts and advocates decentralization to increase the probability of identification of novel chemotherapy modalities. Their long-standing involvement in hormones gives an endocrinologic flavor to their research. Furthermore, they are convinced that "understanding of immunological mediators will lead to new therapeutic agents with great specificity of action which will be broadly applicable to the treatment of malignancy." [Johnson, 1979]. Irving Johnson, Ph.D., a vice president for research, has summarized Lilly's philosophy with respect to cancer as follows: "With modern day molecular biology, i.e., recombinant-DNA, etc., we will find tools and mechanisms for therapeutic intervention which, in fact, may lead to the control of malignant disease called for in the cancer legislation of 1971 (National Cancer Act). It will, however, be in spite of it (the legislation) rather than because of it." [Ibid.]

Of course, the collaboration between Lilly and Genentech on insulin is well known, and Lilly appears to be investing considerable effort, especially on such problems as the combination of the A and B chains into active insulin and other aspects of scale-up and purification. They also have an agreement with the University of California on growth hormone. Lilly has been involved in brain peptide research. They used to be a major vaccine producer but have withdrawn from that field,

* Lilly, too, has a close relationship with a Japanese company, Shionogi, and with other firms and institutions that may be sources of new pharmaceutical products.

except for rabies vaccine of which they have been the sole source. However, they have an interest in another antigen, myelin basic protein (MBP), an 18,750 m.w. substance extracted from porcine nervous tissue, which Dr. Jonas Salk and his associates have shown to suppress the symptoms of experimental allergic encephalomyelitis, an animal disease resembling multiple sclerosis. Early clinical trials are inconclusive with respect to efficacy, but they have provided evidence that patients given substantial quantities--in order to mop up antibodies that would otherwise attack the myelin component of the patients with multiple sclerosis--did not suffer major adverse effects. Of course, such peptide might be made via recombinant-DNA techniques if the approach proves beneficial. Dr. Johnson will not say whether Lilly is working on interferons, but he does view them as examples of the 40 or 50 cytokines that may prove useful.

Of its $2.2 billion sales, Lilly reported $600 million antibiotics sales and $403 million sales of other pharmaceuticals in 1979.* The $175 million invested in R & D would be 7.9 percent of total sales, and it would be reasonable to assume that the proportion of pharmaceutical sales that is spent on pharmaceutical R & D is substantially higher, but that information is not provided in Lilly's public financial reports. They do report that research expenses for the average new drug now exceed $60 million. [Lilly, 1980.] A rough estimate of the number of people Lilly may now have working within the field covered by this report would be less than 2 percent of its 2,600-person R & D staff.

*Lilly is the largest U.S. company with respect to domestic sales of prescription-legend pharmaceuticals; it has only begun to expand significantly in foreign markets within the last 10 years.

9.5 Merck & Co. emphasizes the fact that molecular genetics provides a series of techniques that can be used to more quickly and easily solve problems related to general microbiology and, in particular, strain improvement programs. They report 6-7 separate microbiology and biochemistry projects that fall into this category, as of October 1980. This is in addition to the molecular genetics laboratories, that use genetic engineering for production of specific proteins, according to Dr. Birnbaum, Vice President Microbiology and Veterinary Research. His original estimate was that Merck would have a staff of about eight in the genetic engineering group by the autumn of 1980, but he reported in October that a substantially greater effort is planned. He expresses confidence that "vaccines are a great near term medical target", but his colleague Maurice Hilleman, who is responsible for Merck's vaccine development program maintains a "show me" position.

9.6 Pfizer Inc. has been one of the foremost companies in development of fermentation technology and has the broadest range of fermentation- derived products of any major company, including organic acids, antibiotics and enzymes. Pfizer reports that it has been actively using modern genetic engineering methods for a number of years to supplement traditional mutation/selection methods for upgrading yeasts and streptomycetes for its fermentation processes. Chromosome mapping, and *in vivo* recombination have been employed in Pfizer laboratories both in the UK and USA, and protoplast fusion is used routinely for improvements across the whole spectrum of product groups. Pfizer considers itself among the most advanced of the fermentation companies in successful application of these techniques to improve its manufacturing processes. It is expanding its genetics base with a substantial commitment to recombinant DNA work. A broad range of projects are under study, including both fermentation process improvement, and the preparation of new medical materials.

9.7 Schering-Plough Corporation invested in Biogen in 1978 and obtained rights to at least three pharmaceutical products to be made by molecular cloning. Of the first, leukocyte interferon, much has been written. While it is certainly an important option, by no means did the Biogen invention confer a monopoly upon Schering. Two other pharmaceutical developments of Biogen remain to be claimed by Schering. Schering's $8 million equity in Biogen was evidence of its valuation of the approach, and it is probably pouring a good deal of additional funding into the development based on the initial cloning of the interferon gene in *E. coli*. They have also developed a small in-house capability for genetic manipulation by *in vitro* recombination.

In the segment on antibiotics, we described how Schering, lacking expertise in microbial genetics, established a relationship with tiny Cetus. Within five years, Schering claims it developed in-house skill and an automated screening capability superior to that of Cetus--equaled only by two other labs in the U.S.A. and one in Europe, according to Dr. Lawrason, Schering's Senior Vice President, Scientific Affairs. Biogen could represent a similar short-term relationship but the substantial equity invested (unique on the part of a pharmaceutical company with any of the new enterprises based on recombinant-DNA) makes that unlikely. Rather, it appears that Schering has simply elected to locate this R & D activity outside its own labs--in fact, outside the country.

9.8 G. D. Searle & Co., too, has gone outside the country, albeit to one of its own laboratories abroad. Searle took advantage of the molecular biologists and facilities at its laboratories in England when it established its molecular genetics program. These facilities had been set up for virology and tissue culture, so that it would not be necessary to invest in new containment provisions.* In fact, the laboratory had been used for processing hepatitus virus-contaminated blood from 1972 to 1978. [Richards, 1980.] With a rare (for industry) interest in vaccines, the High Wycombe research group under the direction of Dr. Arthur Hale, set out to clone, sequence and express in bacteria the gene for hemagglutinin (HA), one of the key types of protein in the antigenic profile of influenza viruses. Their accomplishment of that objective has been described (pp. 267-269). They achieved good yields of the polypeptide corresponding to fowl pox HA, although no evidence has been published with regard to the carbohydrate substituents that are normally attached to that backbone. The feat was hailed in the popular press as a basic advance that would enable the production of a more universal vaccine which might eliminate the need to produce a fresh vaccine every time the virus mutates. Certainly, the methodology makes possible rapid characterization of antigenic determinants and their production much more readily than by conventional techniques.

Whether Searle will commercialize novel flu vaccines itself, or will again try to produce a hepatitis vaccine, remains to be seen. They are moving ahead with scale-up in an Intermediate Development Laboratory, which will provide a level of containment normally reserved for virulent pathogens--such as the hepatitis virus it was designed for--to all stages of processing before evidence is assembled that the process is safe enough to put into a Biological Development Pilot Plant, where the emphasis will be (as it usually is in pharmaceutical production)

*Searle's $68 million R & D budget for 1979, primarily for pharmaceuticals, amounted to over 15 percent of its $450 million pharmaceutical sales, and it is doing everything reasonable to stretch for the new products it needs to stay in competition. Pharmaceutical/consumer products account for only 53 percent of Searle's sales but 94 percent of profits and half those sales are international.

more on protecting the product than protecting the environment. [Ibid.] Of course, all this work is being conducted under supervision of the British Genetic Manipulation Advisory Group.

Early in 1980, Searle disclosed plans to build a $10-12 million pilot plant at High Wycombe, initially to produce fibroblast interferon by cell culture. The 18,000-square-foot plant, which is expected to be operational by mid-1981, will also be used for the development of other biologicals, including "drugs for the treatment of gastrointestinal and cardiovascular diseases," as well as interferon, influenza and other vaccines made by recombinant-DNA methods--according to Dr. Daniel Azarnoff, Senior Vice President, R & D. Searle has developed new technology which involves a high yield stack-plate system that has a production capacity sixty times that of the previously used 20-liter roller bottle system for cell culture. The pilot plant capacity is reported to be intended to produce 9.6 billion interferon units per month. [Scrip 3/8/80.]

Clinical trials are under way in Japan and at M. D. Anderson Hospital in Houston. The Texas trials involve three types of cancer: advanced breast tumors, multiple myeloma, and a form of lymphoma. "Ten patients in each category will receive three million units of interferon per day for eight weeks by intramuscular injections. If they respond, they will then be treated with three million units three times a week for the next four months. The dosage will then be stepped up to ten million units a day for an eight-week period* and will then return to the three million units three times per week regimen."

*To this point, the protocol would require 872 million units per patient, worth $22,000 at Searle's estimated $25/million units.

The Japanese trials are related to an agreement Searle also disclosed this spring, under which Searle and Mochida will jointly develop and market interferon in Japan. Searle has agreed to supply Mochida with technical know-how and information developed on fibroblast and lymphoblastoid interferon and make Mochida its sole licensee in Japan. In addition to cash and other benefits, Searle also obtained the right of first refusal on a new pharmaceutical product developed by Mochida and Mochida's agreement to manufacture interferon for Searle, if requested. Both companies intend to concentrate on fibroblast interferon. [Ibid.] With all of the people cloning interferon genes today, one can readily imagine how such a collaboration might shift to bacterially-derived interferon, since the purification technology that needs to be put in place should be useful for material from either source.

9.9 Smith Kline Corporation is currently enjoying--and so are ulcer patients around the world--the fruits of a successful product developed at its laboratory in the United Kingdom. However, its activities in the recombinant-DNA field are concentrated in its Pennsylvania research labs, and they are an outgrowth of its antibiotic R & D program, rather than something altogether novel.* SKF expects to complete its "P-3" facility toward the end of the summer, 1980, and it had to wait until the new Guidelines were issued (about 3 months delay) to be able to do what it wanted to under "P-1" conditions. Overall, Claude Nash, Ph.D., who heads up the program, estimates SKF lost about a year because of the uncertainties imposed by the evolution of the Guidelines. Nash anticipates substantial

*SKF's $60 million worldwide sales of antibiotics in 1979 were only 10 percent of those of Lilly. Overall, SKF pharmaceutical sales went from $33 million to $680 million in the '70s, led by Tagamet (cimetidine) the H_2-receptor antagonist treatment for gastrointestinal ulcers that accounted for sales of about $450 million in 1979.

impact from the new genetic techniques on antibiotic innovation and production. He feels that molecular cloning can be used to improve the production of primary metabolites that serve as starting material for antibiotics. He also points out the potential of such methods to enhance the resistance of an antibiotic producer to the toxic product it produces, and to enable the organism to get the antibiotic out of its cell and into the medium--all important limitations on productivity and the practicality of some candidate antibiotics. Furthermore, he is an advocate of genetic techniques to modify the structure of antibiotics in ways not readily achieved by the chemists. They have identified such a novel antibiotic made by protoplast fusion, and now they are preparing a plasmid to facilitate synthesis of altered aminoglycosides. Generally, Dr. Nash supports the viewpoint we have expressed (pp. 315-346) about the potential of molecular genetics with respect to antibiotics. In his presentation on protoplast fusion, plasmids, and recombinant-DNA applied to antibiotic discovery at the ICAAC/ICC meeting in October, 1979, Nash concluded, "In addition to the construction of new antibiotics, microbial genetics provides the tools for rapidly identifying groups of antibiotics, isolating diverse species, increasing genetic diversity, and developing unique testing procedures. The latter approach could be the most crucial to the discovery of new antibiotics." [Nash, 1980.]

Bryce Douglas, Ph.D., Vice President, R & D Pharmaceutical Products, says, "We are now focusing on the most basic functions of the human system to unlock the secrets of the body's own defenses. The rapid evolution of DNA technology, leading to explosive growth in the field of immunology, is but one of our new capabilities." He also points out, "Our partnership with the academic world is growing, with new collaborations between the university and industry that permit rapid transfer of basic knowledge into practical health care applications." The 1979 annual report states, "Having developed our core (research) facilities in the United Kingdom and the United States to what we believe to be an optimum size, we are now planning a major new facility in France...Well-staffed and broad-based research programs--in the U.S., U.K., Belgium, Germany, and planned in France*--combined with strong development capabilities serving the U.S. and international markets, give us the capability to maintain our history of growth through discovery." [SKF, 1980.] No better example of the benefits of international research could be cited than that of the H_2-receptor antagonists, a field in which the basic discoveries were made by the SKF Research Institute in the U.K. The first product from this research, Tagamet, which achieved greater first-year sales than any other pharmaceutical in history when it was introduced in 1977, is reputed to have the potential for saving some $650 million/year--a 29 percent reduction in the $2.2 billion/year cost of duodenal ulcer disease in the U.S.A.**

9.10 The Upjohn Company management expects molecular genetics to make significant contributions and has supported recombinant-DNA research for five years. Nevertheless, apart from the special cases of human insulin and human growth hormone, which may be marketed within 2 to 5 years, Dr. W. N. Hubbard, Jr. said in January, 1980, that it is unlikely that other pharmaceutical products resulting from recombinant-DNA technology will be marketed within the next decade. Yet, Upjohn is the only

*Europe accounts for 32 percent of SKF sales.

**Assuming Tagamet was used to treat 80 percent of all U.S. duodenal ulcer patients in 1977, it is calculated direct hospital costs would have declined $258 million, losses from morbidity would have been reduced by $329 million and $30 million in surgeons' fees would have been saved, for a rise in drug cost of only $34 million. [Ibid.]

pharmaceutical company with two "P-3" labs, and they have sizeable groups working in two distinct areas: antibiotics and human proteins. Upjohn considers itself a pioneer in fermentation technology and is determined to retain that leadership position when the new era of fermentation comes to pass.

As early as 1977, Joseph Grady, Ph.D., Infectious Diseases Research Manager, told a senate subcommittee that a modified antibiotic, produced through a simple enzymatic procedure otherwise impossible to accomplish, is likely to be Upjohn's first use of molecular cloning. Malik, of Upjohn, points out that, "So far, it has not been possible to map any locus that controls a well characterized enzymic reaction involved in the biosynthetic pathway of a secondary metabolite." Obviously, Upjohn is struggling to break that barrier to improvement or novelty by design.

Dr. Malik distinguishes five classes of genes governing antibiotic production:

1. Structural genes that code for the enzymes involved in the biogenesis of the antibiotic molecule.
2. Regulatory genes that determine the onset and extent of expression of the structural genes for antibiotic biosynthesis.
3. Resistance genes that determine immunity of the producing organism to the produced antibiotic.
4. Cellular-permeability genes that control entry, exclusion, and excretion of antibiotic.
5. Regulatory genes that control primary pathways. These genes influence the level of precursors and cofactors needed for antibiotic biogenesis.

He suggests that classes 1, 2 and 3 may be clustered like operons in other microorganisms and that they may even be plasmid-borne, while classes 4 and 5 are likely to be scattered all over the chromosome. [Malik, 1979.] Certainly Upjohn has been searching for and dissecting plasmids in Streptomycetes (e.g. *S. venezuelae*, *S. espinosus* and *S. fraidiae*). Since Upjohn produces about ten different kinds of antibiotics now and has a broad research program, it is impossible for an outsider to predict where the first utilization will occur or what its impact may be.

Upjohn is also one of the leading practitioners of microbial transformation, especially in connection with their extensive line of steroids. Not only finished products, but intermediates supplied in bulk to other manufacturers, flow from such fermentations and chemical processes operated by Upjohn. As is the case in many pharmaceutical companies, the production of intermediates and drugs takes place in a separate chemical division which, among other things, provides incentives to make the fruits of new processes available to companies other than Upjohn.

The other area of Upjohn activity which we have discussed in some detail earlier (pp. 250-255), concerns the production of large proteins such as albumin. In a presentation where he said, "It would be reasonable to assume we are working on human albumin," Dr. Hubbard emphasized the differences between Upjohn's approach to this opportunity and the more widely publicized activities. He said Upjohn is concentrating on near-anaerobic yeasts as production organisms, exploring the promotion of growth by low pressure evaporation of the ethanol produced by their metabolism. He anticipates major problems with government regulatory agencies, especially for complex proteins that defy the sort of tests of identity one can apply to insulin. He said that it will be a long time before the questions of patentability of the basic methodology and of the production of the specific microorganisms are resolved, and that, as is the case in the chemical industry where improvements in process are rarely patented, he felt companies would rely on secrecy rather than seek patents on their donors, carriers or recombined hosts. He thinks most of the time firms assume "the disadvantage of general knowledge is greater than the advantage of patent protection." [Scrip 2/?/80.] It is of interest that Upjohn sacrificed a patent appeal with regard to its clindamycin in order to avoid clouding the issue of patentability of recombinant organisms.

It would appear to this observer that Upjohn has about 10 Ph.D.s working in molecular genetics at this time. They have close ties with academic leaders, such as Professor Bert O'Malley at Baylor, and high expectations for the long term. [Fraser, 1980.] However, they have been quite conservative in their estimates of the rate of progress and have not made public any activity with regard to interferon--the promising possible product that has attracted so much attention.

9.11 Other American companies (e.g. J & J, Revlon and Syntex) have been involved in developments related to monoclonal antibodies for some time, but only now are they beginning recombinant-DNA activities. Some who are heavily committed to fermentation (e.g. Lederle, Parke-Davis, Pfizer and Wyeth) may have adopted such techniques as protoplast fusion for strain improvement, but have yet to see more than a minor role for molecular genetics in their R & D programs. The small groups of molecular geneticists at some major pharmaceutical firms appear to operate more as intelligence gatherers and opportunity evaluators than implementors of management's plans. Everyone in diagnostic immunology who hasn't already been mentioned must be looking at monoclonal antibodies as opportunities or potential competition. Almost all research-based firms will utilize the new genetic techniques to achieve insight into the molecular biology of disease. In many cases, of which Squibb could provide a timely example, such insights prove critical to the definition of a new product (e.g. the antihypertensive, captopril) although the ultimate product may be produced by conventional chemistry.

10. Foreign-based Firms

10.1 Switzerland

Hoffman-La Roche has pharmaceutical sales in the United States which are surpassed only by those of Merck and American Home (the parent of Ayerst, Ives and Wyeth). Roche also has one of the largest research establishments in this country, as well as a number of labs in Switzerland and elsewhere. Part of that establishment is the Roche Institute of Molecular Biology (RIMB), a prestigious academic-affiliated bastion of pure science, with a staff of about 200, including about 35 very bright lights from academia, and especially the NIH. In Basle, Switzerland, there is a similar facility concerned with basic research in immunology that has been the source of some fundamental breakthroughs in that revolutionary field which is so closely tied to molecular genetics. In terms of longevity (RIMB was founded in 1967) and scale, these institutions constitute the largest investment by any pharmaceutical company in molecular genetics. RIMB scientists began to work with recombinant-DNA in the spring of 1976. Because Roche is a broad-based company with products over the full spectrum from immunodiagnostics to anti-infectives, besides the psychopharmaceuticals that have become household words,* it

*Roughly 80 percent of Roche's pharmaceutical sales are derived from benzodiazepine compounds, of which Librium and Valium are the best known.

is in a position to exploit the pioneering developments of which such institutes make it a part. Roche invests over a quarter of a billion dollars a year in R & D worldwide.* Never-the-less, Roche has gone outside its own labs many times and licensed products for its formidable marketing force to sell.

Recent developments with respect to interferon will suffice to illustrate a Roche modus operandi. The RIMB was the source of some of the earliest work (1975) on messenger RNA that serves as the template for interferon synthesis. In the latter part of 1978, scientists at RIMB were the first to succeed in producing pure interferon from human leukocytes in amounts sufficient for physical and chemical analysis. About the same time, the collaboration, only recently revealed, was established with fledgling Genentech, which, early in 1980, resulted in successful production by molecular clones of materials with the chemical and biological properties of fibroblast or leukocyte interferons. The yield reported at the initial disclosure (June 4, 1980), 250 million units per liter, was substantially greater than that by previous workers and this announcement also provided the first evidence of efficacy of bacterially-produced interferons against viral infections in whole animals.

In addition to its activities related to interferon, of which we shall hear much more, RIMB and Roche's commercially-oriented labs have polypeptide research programs, including work in conjunction with NCI on immunomodulators such as thymosin-α_1 and neuroactive peptides that may be advanced--if not brought to fruition--by molecular cloning. Perusal of the scientific annual reports of RIMB creates the impression that there are

* U.S. R & D in 1978 cost $90 million, according to Mr. Lerner, Roche's new president.

few areas of molecular biology, especially genetic regulation and expression, in which Roche does not have a dozen researchers working. Not being a publicly traded company subject to SEC requirements, Roche releases relatively little information on its activities, but it is obvious that few new drug products have materialized from the Roche R & D investment in recent years. No one would appear to be better positioned, or more patient about pharmaceutical product gestation, than Roche to take advantage of the commercial implications of molecular genetics.*

The other "Swiss" multinationals, Ciba-Geigy and Sandoz, have much more modest commitments to molecular biology. Ciba-Geigy has a substantial interest in immunology and polypeptides--its scientists were the first to synthesize human insulin--and can be expected to utilize molecular genetics ultimately, but they have maintained a very low profile. Ciba-Geigy was not doing recombinant-DNA work in the U.S.A. as of March, 1980, but they had some activity in Switzerland. Both Sandoz and Ciba-Geigy have major research laboratories in the U.S.A., and both utilize fermentation to produce a small quantity of their total output.

In 1977, Ciba-Geigy acquired a controlling interest in Alza Corporation, a pioneering U.S. firm developing systems for administering pharmacologic agents in prescribed patterns under conditions that will minimize the need for medical attendance or side-effects. To the extent that molecular genetic development makes the use of physiologic substances desirable, it may be necessary to resort to such delivery systems to make practical

*Roche has much greater experience in chemical synthesis, but it has limited experience in fermentation, and a record of willingness to have others manufacture its products (e.g. Beecham is the source of amoxycillin). The agreement with Genentech on interferon permits both parties to manufacture.

the therapeutic potential of such agents. Access to these patented delivery systems could provide Ciba-Geigy with an important edge at the experimental stage and when it comes down to definition of feasible dosage forms of these frequently evanescent molecules.

10.2 Germany

German companies are heavy spenders on R & D and biotechnology has recently been receiving strong emphasis. Hoechst is reputed to be second only to Roche in worldwide pharmaceutical sales--having surpassed Merck and American Home by acquiring Roussel of France. It spent a total of over $650 million on R & D in 1978.* Recently, pharmaceuticals is reported to be receiving the largest share, about 40 percent, of the R & D budget, although only 16 percent of Hoechst sales results from pharamceuticals. [C & EN, 7/24/78 p.8]. Nearly 20 percent of that quarter billion dollars per year is spent on pharmaceutical R & D in the U.S.A.

In addition to the Hoechst-Roussel Pharmaceutical organization in New Jersey, Hoechst acquired a window on U.S. molecular biology in 1977 by taking over Calbiochem of La Jolla, California--then a $15 million/year (sales) manufacturer and marketer of biochemicals, including immunochemical and diagnostic reagents.

Hoechst is reported to be the most enthusiastic leading German firm with respect to the prospects for genetic engineering. [O'Sullivan, 1980]

*Hoechst has averaged an R & D level of 188 percent of net profits for the past 10 years. "Support for R & D by the West German chemical industry--of which pharmaceuticals is a part--greatly outstrips R & D funding by U.S. chemical companies as a percentage of sales or profits...One factor in the high West German percentage (4.9 percent versus 2.4 percent of sales for W.G. and U.S. respectively) may be relatively more drug research than in the U.S." [O'Sullivan, 1980].

Hoechst has long been involved in insulin production and views molecular cloning as a natural outgrowth of that activity. They established a synthetic peptides group about 20 years ago, according to Dr. Günther Seidl, head of Hoechst's pharmaceutical research, and recently have prepared a strain of E. coli which expresses a gene that yields a product immunologically identical to human insulin. They are also studying interferon production, says Seidl. [Ibid.] Hoechst is a leading producer of vaccines and blood products, two other categories of products that could be cloned. Hoechst is a major antibiotic producer and has significant growth prospects in that field, some of which may well hinge on the economics of production of competitive cephalosporins. "In the past, biotechnology studies at Hoechst centered on classical fermentation techniques to produce antibiotics," says Dr. Weissermel--overall R & D director--and, "now we have switched to new fermentation methods. We started with single-cell protein. We will explore what can be done with naturally occuring resources, how they might be converted to chemicals in general."

Several Hoechst scientists from Frankfurt are presently on sabbatical visits to U.S. laboratories, including NIH and the genetic engineering groups at UCSF--where the American efforts on bacterial production of insulin originated.

Bayer is a very rapidly growing member of the top five pharmaceutical companies in the world, which made two major acquisitions during the '70s of U.S. firms in the field--Miles, Inc. and Cutter Laboratories. It is difficult to accept as coincidental the fact that Cutter's experience in vaccines and blood products, when combined with Miles' involvement in diagnostics, peptide hormones, enzymes, fermentation and pharmaceuticals, complement the resources of Bayer for exploitation of the full range of biotechnological opportunities. Professor Düchal, Director of Research at Bayer, says, "We are doing some work (in genetic engineering) at Miles, and we are setting up a unit in the biochemistry department of our pharmaceutical division (in Leverkusen)." [Ibid.] There are 45 people in the genetics group at Miles, according to Dr. Blake Ingle, v.p. R & D in Miles' Industrial Products group--which produces enzyme preparations for a host of purposes ranging from gene splicing through food processing to sewage treatment. He reports it took several years to get management interested in applications of molecular genetics, but the pace of things has picked up considerably. Miles is now struggling to be a contender in a field in which it was once a leader. Since Bayer spends about as much as Hoechst on R & D and is committed to becoming a major factor in U.S. markets, it can be expected to be a major competitor in global developments based upon molecular genetics. In 1979 Miles entered into a joint-venture with Scripps Clinic in the field of immunology that could place it at the forefront of monoclonal antibody exploitation.

BASF, the third descendant of I.G. Farben, is much less a factor in pharmaceuticals than the preceding, but it is rumored to have set up a large group to study many applications of genetic engineering. Of the other four leading West German firms, (Boehringer Ingelheim, Boehringer Mannheim, Schering A.G., and E. Merck), Boehringer Mannheim is the most intensively involved in molecular genetics and immunology--being a leading participant in the market for diagnostic reagents and biochemicals in the U.S.A. and abroad. Schering A.G. is also reported to be performing genetic engineering research. [Chem. Week, 6/27/79, p. 32.]

10.3 Great Britain

Four British companies rank among the 50 largest pharmaceutical companies in the world. The two with the largest global pharmaceutical sales, Glaxo and Beecham, are primarily antibiotic producers and neither has yet achieved a substantial presence in the U.S.A. However, the third U.K.-based firm ranks about 15th in the U.S. market. Burroughs-Wellcome, the North Carolina-based subsidiary of the Wellcome Foundation* is the only non-Swiss foreign company among the top 20 competitors in the U.S.A.**

The U.S. research program under Pedro Cuatrecassas, M.D., who had established a reputation in the field of cell-surface receptors for hormones at Johns Hopkins, George Washington and NIH before becoming Vice President for Research at Burroughs-Wellcome in 1975, has been strongly influenced by modern molecular biology. Similarly, Dr. John Vane, a distinguished molecular pharmacologist, has guided developments in the U.K. into such diverse areas as anti-inflammatory biochemistry and interferon research. Wellcome has been involved in vaccines, immunodiagnostics and cancer for some time. Thus, it can be expected to utilize molecular genetics in many medical applications.

The management of Imperial Chemical Industries (ICI), on the other hand, is quite skeptical about the utility of genetic engineering for medicinal products. The skepticism is certainly not derived from ignorance. ICI has been doing recombinant-DNA research in the corporate laboratory in the U.K. In fact, a recently disclosed bit of genetic engineering by ICI may well be the first commercial application in Western Europe.

---

* The Wellcome Foundation, Ltd., was founded by two American pharmacists, for whom the American subsidiary is named, to distribute American drugs from England to other parts of the world. There are no stockholders; all profits are used for medical research on a worldwide scale.

**Roche and Ciba-Geigy are in the top ten and Sandoz ranks about 17th with respect to U.S. sales. Hoechst, Bayer and ICI would be included in the top 25.

However, the product involved is single-cell protein (SCP) not a pharmaceutical.*

Nevertheless, Dr. Henry Freedman, Vice President and Director of Biomedical Research at the ICI Americas headquarters, is much more optimistic about drug design based on identified receptors and chemical synthesis than he is about the molecular cloning approach. Given the opportunities ICI perceives for new products based on synthetic chemistry and limited resources, Dr. Freedman says ICI's pharmaceutical divisions have decided to take a wait-and-see position on cloning. He denies that is because fermentation has only been used a small amount by ICI. Of all the company representatives approached, Dr. Freedman was the least enthusiastic about molecular genetics--in a context of general confidence about the feasibility of devising novel solutions to commercially and medically important health problems.

The English are going through a great deal of self-criticism with respect to the so-called "biotechnology gold rush."

---

*Methylophilus methylotrophus is the organism used to produce SCP from a medium in which methanol is the carbon source. Nitrogen assimilation is inefficient with naturally occurring M. methylotrophus because that organism assimilates nitrogen via a pathway utilizing glutamate synthetase. Recognizing that glutamate dehydrogenase might enable the use of a more efficient pathway, Dr. John Windass mutated the glutamate synthetase-producing gene, thereby inactivating it. Then he took the glutamate dehydrogenase gene from E. coli and inserted it into M. methylotropus, and got an organism with quite a significant improvement in its nitrogen metabolism efficiency. "Carbon conversion efficiency is improved without increasing energy input. The company's agricultural division currently is putting its continuous fermentation SCP plant through startup trials...when fully operating, it will have an annual output of up to 155 million lb. of product, based on unaltered M. methylotrophus. The modified bacteria (appears) identical to the naturally occurring organism, apart from its improved metabolism. Consequently, chances are it will be used eventually in the existing SCP plant after further study, and it could be the basis for production of larger SCP units if the company decides to build them." [C & EN, 5/5/80, p. 30]

Articles with headlines like "California set to cash in on British discovery" are common in both their financial and scientific news publications. Sera-Lab, in Crawley Down, was the first firm to sell monoclonal antibodies--which had been invented at the MRC in Cambridge--but the surge to commercialize typified by Hybritech and Genentech in this country has not developed in the U.K. Non-profit institutions in the U.K. (e.g. MRC, John Innes Institute) have been the source of many basic discoveries and inventions that underlie the new biotechnologies--after all, Isaacs and Lindenmann were at the National Institute for Medical Research in London when they discovered interferon. In response to the perception of missing the boat, "there are moves afoot to set up a British version of the American (sic) Biogen Company." [Nature, (London), 1/10/80, p. 126] A government working party recommended that the National Enterprise Board fund a 50 person, £2 million/year company, code named "Greenfields". [Nature 4/10/80, p.502]

10.4 France is generally considered to be trailing in the biotechnology race. A report presented in November, 1979, to Giscard d'Estang stated that the French biological research effort, especially in the field of recombinant-DNA, needs increased government financing and cooperation between the government, universities and private industry to make it competitive with that in the U.S., Japan and West Germany. [Gros, Jacob & Royer, 1979] Three programs were rapidly approved, two of them related to pharmacology and biological technology for drug production. Rhone-Poulenc, France's largest pharmaceutical company, reacted strongly to the criticism of industry by the report authors. The chairman of the firm emphasized that Rhone-Poulenc ranks fourth in the world for fermentation products output and second for immunological products. The company took out a full-page advertisement in several French newspapers to explain its major biological research effort, especially with respect to fermentation, immunology and recombinant-DNA research. The prestigious Institut Pasteur Foundation (IPF) is seeking a merger with equally prestigious Institut Merieux (51 percent owned by Rhone-Poulenc) and Choay (another biologicals producer, one-third owned by a Sanofi, which, in turn, is government controlled) to form a biologicals firm second only to Hoechst's Behringwerke in size.* Such a move would eliminate needless duplication of efforts and counter-productive competition in foreign markets, IPF argues.

Other EEC Country pharmaceutical companies reported to have activities in genetic engineering include the Dutch firms: Gist-Brocades and Phillips-DuPhar; Italy's Montedison** and Scandanavian firms such as Kabi and Pharmacia, whose specific interests have been referred to previously. All have the capacity to be significant competitors in a new flowering of biotechnology. For example, Gist-Brocades, which has enzyme production plants in The Netherlands, Belgium, France and the U.S.A., is estimated to have the second largest fermentation capacity in the world.# [Scrip, 7/25/79, p. 6] Several

* The composite sales would be over $240 million annually, greater than the biological product sales of either Merck or Wellcome, the next in rank. Part of the rationale presented by IPF is that a company engaged in modern biological research, including genetic engineering, must be able to spend at least Fr 100 million (∿$24 million) a year for this purpose.

** It is not clear whether some firms, like SNAM-Progetti (Italy) and Unilever (Holland), which have major biotechnology and genetic engineering programs are working on medical applications in addition to their food, chemical and energy interests.

Gist-Brocades NV of The Netherlands purchased the Wallerstein Division of Baxter-Travenol in 1977 and became one of the two largest enzyme producers in the USA. (Miles, owned by Bayer, is the other.) Although GB Fermentation Industries, Inc., maintains an R & D laboratory in the U.S.A., all strain improvement and development of novel microbiological processes is conducted by the over 600 R & D personnel Gist-Brocades has working in Holland.

Scandanavian firms are major processors of human hormones and blood products.*

10.5 "Japan has taken an incontestable lead in biotechnology," says Robert Walgate of Nature. [Walgate, 1980.] It may be true that "Japan has led the way with a profitable fermentation industry producing amino acids, nucleotides, and organic acids," as Professor Brian Hartley points out, "but this relies largely on conventional microbiology, genetics and fermentation feedstocks," and, "it can certainly be challenged by new biotechnology." [Hartley, 1980.] "There is a feeling that Japan is not as advanced as some other countries in fundamental biology, so international competition, particularly through genetically engineered organisms, is not closed."

One area on which Japanese firms do seem to have a stronghold is the production of amino acids, which are used in nutritional fluids for infusion, and also in animal feed supplements. Around 90 percent of worldwide sales of medical amino acids are produced by Japanese companies, including Ajinomoto, Tanabe Seryaku, and Kyowa Hakko Kogyo. Ajinomoto is currently constructing a $37 million plant in Raleigh, N.C., to more efficiently supply the U.S. market. Genex, the Rockville, MD genetic engineering firm is doing work for an undisclosed Japanese client to construct a microorganism that will produce very high yields of a particular amino acid, and Genex management expects to be in a position to move that process into pilot plant scale-up early in 1981.

*Novo Industries of Denmark is the world's largest producer of enzymes, and two-thirds of the free world supply of insulin comes from Lilly or Novo. Novo is steadfastly defending the utility of animal insulin, but it would be surprising if there were no work in the area of molecular genetics going on in the R & D labs of the firm, 50 percent of whose business is based on fermentation.

The practice of medicine differs, and therapeutic use of parenteral nutrition has been more extensive in Japan. But the rate of growth in the U.S.A. is impressive, and the American firms (e.g. Abbott, American Hospital, Cutter and Travenol) all obtain their animo acids primarily from Japanese sources.

Many comparisons have been made in recent months, such as the stagnant R & D manpower situation in the U.S.A. relative to the 62 percent growth in Japan* over the past decade, and the origin of biotechnological patents since 1977: 124 originated in Japan, 39 in the U.S.A. (and less than ten each for Germany, France, Denmark and Great Britain). [Walgate, 1980.] However, our physical sciences manpower resources have been more harshly affected than the biological sciences--in 1975, the U.S. employed three times as many in pharmaceutical R & D as Japan (21,970 versus 7,300).** While the biotechnology patent figures are alarming (and we have not checked their basis), Japan and Germany each account for only about 10 percent of the U.S. patents issued on drugs and medicines (as of 1975, at least),# although that percentage is twice what prevailed around 1970. Without underestimating the potential of the Japanese and other foreign competition, it is important to retain a realistic

---

* Increase in technical manpower was 59 percent for West Germany, 55 percent for U.S.S.R. and 62 percent for Japan, while the U.S. stock of scientific manpower has been at a virtual standstill for nearly a decade, according to E. P. McGuire of the Conference Board. [C & EN, 3/31/80, p. 7.]

**The statistics provided by PMA, and through them from other national pharmaceutics industry associations, for R & D personnel employed by local pharmaceutical industry in 1975 are as follows: U.S.A. 21,970; West Germany 15,000; United Kingdom 9,500; Japan 7,300; France 6,000; Switzerland 5,000 and Italy 4,700. [PMA, 1977.]

The PMA "Factbook '76", a 1980 version of which will issue very shortly, contains tables on drug and medicines patents derived from OTA and other sources. Only 54 percent of patents in these categories originated in the U.S.A. in 1975, down from 64 percent in 1965 and 1970--the percentage originating from U.S.-owned firms in foreign countries had stayed pretty constant (6 to 8 percent) from 1963-1975.

appreciation of the strengths of U.S. industry at this early stage of application of molecular genetics, especially to pharmaceutical purposes.

Japan's fermentation industry has benefited from the prominant position of the traditional fermented food products: miso, shoyu and sake. The total annual production of the fermentation industries in Japan is approximately $10 billion, two to three percent of the GNP, which yields eight to nine percent of the nation's total tax income. [Arima, 1979.] Japan is one of the principal producers of antibiotics in the world and, as we have noted, many U.S. firms (e.g. Bristol, Lilly, Pfizer) have taken an if-you-can't-beat-them-join-them reaction to this expertise.* The Japanese government has a strong interest in biotechnology. For example, the Office for Life Science Promotion has set up a systematic program for R & D on enzyme technology and has made a detailed analysis of its impacts on society. [Ibid.] The Ministry of Health has set up a special study group to speed the development of interferon--a decision reported to have resulted from discussions between the Ministry and the two principal firms working in the field, Green Cross (working with leukocytes) and Torii (working with tissue culture), which concluded that "the separate approach being taken was costly both in terms of funds expended and time taken... It was agreed that, in future, all interferon studies will be coordinated within the study group which will have the Ministry's backing." [Scrip, 8/1/79, p. 9.] Since we understand that Toray Industries holds the rights to the exploitation of the interferon gene isolated by Tadatsugu Taniguchi last fall

*Antibiotics account for more than one-third of the 100 largest selling pharmaceuticals in Japan, in comparison to less than one-fifth in the U.S.A. (36 percent versus 19 percent in 1978). Forty percent of the antibiotics now marketed were discovered in Japan. [Arima, 1979.]

while he was working at the Japanese Foundation for Cancer Research, one can anticipate that "Japan Incorporated" will see to it that invention is utilized in keeping with national-industrial common interests.* As Daniel Thomas, of the University of Compiegne, France, wrote in his report to the EEC on the international race with respect to applied biology, "It is clear that in Japan there is no official or conceptual division between basic and applied research; there is a continuous exchange of skills and services between universities and industry." [Walgate, 1980.]

By mid-1979, Ajinomoto, Asahi, Kyowa Hakko Kogyo, Mitsubishi, Sumitomo and Takeda were reported to be planning genetic engineering programs. We have already mentioned Ajinomoto and Kyowa Hakko Kogyo in connection with amino acid production and Takeda is Japan's largest pharmaceutical company. Sumitomo owns Shionogi, the firm with the second largest pharmaceutical sales. Giant Mitsubishi is the DuPont of Japan. Asahi and Toray (mentioned in connection with Taniguchi's interferon work) are two of the several chemical firms diversifying into healthcare businesses.

10.6 The Soviet Union's commitment to biotechnology was proclaimed May 21, 1974, in a document that set Soviet science the task of "accelerating the development of molecular biology and molecular genetics" and the "wide use of their achievements in agriculture, medicine and industry." Several groups are reported to be working on applied genetic engineering and it has been rumored that "the industrial breakthrough point for certain medical products made in this way will be reached in the nearest future." [Rich, 1980.]

* Tanaguchi initially spent time in Weissman's lab in Zurich, where the first expression of an interferon gene in bacteria was achieved, and now is at Harvard trying to get expression of his human fibroblast interferon gene.

Polish specialists from the Institute of Biochemistry and Biophysics of the Academy of Sciences reportedly said that they expect the commercial production of unspecified hormones, vaccines and special microorganisms designed to clean up pollution, in the near future. [Ibid.] None of that sounds very original, but then one does not expect major impacts in pharmaceuticals to originate from the Eastern Bloc countries.

Hungary's international biological center at Szeged (built largely with UNDP funds) is reported to be still at the stage of establishing techniques and principles. [Ibid.] It might be noted that several Eastern Bloc nations have, in recent years, found the production of drugs for Western companies to be a good source of foreign exchange. Hungary is offering gentamicin, for example. In response to that proposal and possibly other competitors, Schering-Plough has agreed to supply gentamicin to Upjohn and Wyeth, since the patent expired in May, 1980.

Issues for Government Policy

11. ASPECTS GENERIC TO BIOTECHNOLOGY

11.1 Regulation and Safety Concerns

11.11 <u>RESEARCH & DEVELOPMENT</u> have come under extraordinary scrutiny in the case of recombinant-DNA manipulation through the mandatory requirements imposed upon institutions accepting government funding and voluntary compliance to the same guidelines by private industry. This topic has been treated adequately elsewhere in the OTA report, and here it is appropriate only to mention that the regulations themselves and the atmosphere of uncertainty surrounding their evolution deterred many large companies from undertaking R & D in this field for one to two years. It was a contributing factor to the initial decision of some firms to locate their molecular cloning research facilities outside the U.S.A. Our discussions with industry representatives suggest that concern about the effects upon corporate image and relationships with the communities in which their laboratories are located weighed heavier in the mind of top management than doubt that the ultimate regulations would be practical for them to abide by. Among leading pharmaceutical companies, there was definite shyness about drawing attention to themselves at a time when some alarmists were eager to tie the profit motive to allegations of irresponsible "tampering with the very stuff of life."

Knowing the long lead time of products in their industry, most established companies elected to let the dust settle before making much commitment. To a limited extent, the lead established by the entrepreneurial companies was the result of the hesitance of the establishment, but to a much greater extent it is the product of the entrepreneur's single-mindedness and proximity to the well-springs of the basic breakthroughs.

Uncertainty about the nature and, especially, costliness of the facilities that would be required to conform with what is coming to be accepted as appropriate precautions caused firms to delay establishing such R & D facilities. The possibility that they might have to expose their R & D plans to their competitors was extremely discouraging to an industry which places very high value on the simple recognition that something is feasible or that a novel correlation exists--and to companies which feel they have been harmed by the freedom of information treatment accorded some things long considered trade secrets. The mechanism that has been established under which the Recombinant Advisory Committee of the NIH will accept disclosures in confidence, along with the substantial relaxation of the guidelines that permits many exploratory activities to go forward without any disclosure, has made industry much more sanguine.

Industry generally seems eager to allay anxieties about molecular cloning and to avoid the suspicion of irresponsibility that would result in imposition of regulations likely to impede the utilization of one or more elements of molecular biology. Most representatives of the pharmaceutical industry view molecular genetics as a skeleton key capable of unlocking many doors that have resisted previous attempts. While opinions differ on the impact that molecular cloning will have with respect to the manufacturing of pharmaceutical products, there is essential unanimity about the value of the techniques for R & D in this industry. Since R & D extends into the clinic, scale-up beyond an arbitrary ten-liter limit is an essential component of pharmaceutical industry utilization of these tools, even for research.

11.12 COMMERCIAL OPERATIONS based upon molecular genetics do not represent the sort of departure from customary practices for the pharmaceutical industry that adoption of biotechnology will involve for other industries. Fermentation and coping with biohazards are part of the lore of the industry. Furthermore, regulatory monitoring of manufacturing practices has a longer history in this field than most others. Most of the major participants in the industry have some experience in handling real pathogens, at least in the laboratory, and so they feel confident in designing containment facilities for organisms that they expect to prove harmless in the long run.

As a general principle, so long as the powers-that-be require proof that the commercial processes will be safe, industry seems to prefer to satisfy those concerns early in the process by adequate characterization of the organisms they propose to utilize and agreement with respect to the type of precaution necessary for handling such a specified agent. What they wish to avoid is discriminatory policing and arbitrary regulations which might be imposed upon recombinant-DNA production that would not be required of alternative processes and, especially, would not be required of their foreign-based competitors.

Government must satisfy itself that due precautions are being exercised and monitoring is being performed, without handicapping American companies in international business, unnecessarily delaying the availability of valuable medical products, or adding unjustified cost to healthcare modalities. Industry has obligated itself to voluntary compliance with the NIH Guidelines and cooperated in the development of large-scale standards and procedures in the hope of persuading the public that it shares those values.

11.2 Communication and Technology Transfer

11.21 Patents are more important to the pharmaceutical industry than they are for many other industries. That is a reflection of the research intensity of this business. It also reflects the long lead-time of pharmaceutical products--a point to which we shall return later. Among diagnostic products, for which the development interval is generally much shorter, patent protection weighs less heavily in corporate evaluation of the attractiveness of opportunities. Companies are extremely reluctant to invest money and time necessary to prove the utility of a pharmaceutical product if they cannot be assured of a period of freedom from competition with someone else selling exactly the same thing. The absense of patent protection delayed the development of penicillin in the '30s and interferon in the '70s. In many potential applications of molecular cloning, the product itself, e.g. human insulin for treating diabetes, is probably too obvious for patenting.Thus, patent protection on the process for obtaining the product, including the modified organisms used to produce it, is practically a *sine qua non* for investment, other than the sort of defensive investment that may be made by a firm which already has a substantial franchise in a given market, e.g. Eli Lilly and insulin.

The decision of the Supreme Court on the Chakrabarty case was especially important for pharmaceutical developments based on molecular cloning. Patents on the production organisms may turn out to be critical elements in the establishment of the corporate sense of security necessary to unleash the flow of expenditures necessary to develop and make available certain products. From the government's point of view, the benefits could be multifold. Of course, the patents created will satisfy the basic premise of the patent system, i.e. encouraging disclosure of useful information. Moreover, because each such patent claim is inherently narrow, the granting of such patents will encourage the performance and reporting of more exploratory work, broaden the area of coverage and obtain protection on a variety of capable organisms including the optimal molecular clone--which will need to be characterized in increasingly sophisticated ways. No doubt, this will result in the development and reporting of new methods of broad utility. Firms may seek to exclude the use of competitive organisms through the medium of scientific publication, which by their existence precludes the competitor from obtaining patents on said organisms-- such a publication, of course, contributing at the same time to greater dissemination of the state-of-the-art.

The existence of patents will enable the marketing of specific multi-purpose organisms and licensing of the right to use them, which will facilitate the transfer of technology from firms expert in the development of engineered organisms to firms interested in accomplishing certain objectives, in pharmaceutical R & D as well as other fields, for which the engineered organisms constitute useful tools. That is to say, patents will promote the dissemination of the organisms as well as the principles and techniques which underlie their construction. A lesser proportion of useful methodology will be retained in the secrecy of proprietary know-how and more will be available for utilization as a result of the allowance of both composition of matter and process patents.

Pending the resolution of the Chakrabarty case, the patent office had suspended the examination of patent applications pertaining to plasmids--despite the fact that plasmids can be completely characterized and are totally inanimate chemical species, which would normally be patentable like any other novel, unobvious and useful chemical substance. Now, both these means by which the genetic capacity of organisms can be augmented and the resultant organisms containing the plasmids or those suitable for acting as hosts to them will, presumably, be accorded the same treatment under the patent law as any other useful, etc. composition of matter.

It was with the expectation that patent protection could be attained that small companies undertook innovative R & D in this field, and in some cases those patent guarantees may provide the leverage necessary for those entrepreneurs to extract from the established pharmaceutical companies, who will actually market many of the products, a fair return on the entrepreneur's investment and a basis for his expansion and extension of his technology and business, including competition with the establishment.

11.3 National Resources and International Competition

Relationships between business, academia and government are complex with regard to biotechnology. The United States government has made a far greater investment in basic bioscience research than any other government or institution, and practically all of the results are freely available for the use of anyone in the world.* Most of that basic science is actually performed in our universities. One of the beneficiaries is American industry. Today, we are witnessing or participating in the

*For fiscal 1980, NIH is funding 717 projects involving recombinant-DNA at the level of $91.5 million.

mobilization of resources, including portions of that basic knowledge, to wring some practical consequences from the perception that cellular production systems could perform new and important functions in several sectors of our economy.

Events of recent years have made us all more conscious of the importance of international trade to the vitality of the American economy. It is widely held that the ability of American firms to innovate and retain positions of leadership in the exploitation of new technologies is critical to the maintenance of a trade balance relative to our expenditures on oil and other imports. Of course, other nations have the same objective. Applications of modern biology, including molecular genetics, constitute just the sort of high technology upon which America's position is supposed to depend. Hence, governmental attention should be directed to the nurturing and encouragement of such enterprises. Furthermore, since the health of the nation literally depends upon the health of the people, special consideration is in order with respect to medical applications.

One aspect to ponder is whether there are, or ought to be, mechanisms by means of which advantages could be extended to American businesses in the race to exploit the basic science inventory that has been amassed at the expense of the American taxpayer. An indication that this is not presently the case is the fact that the biggest user of our National Technical Information Service is the Japanese government. We have cited statistics earlier that indicate Japanese patents on biotechnology have outnumbered U.S. patents in the same categories since 1977 by more than three to one. The Japanese have greater experience in fermentation than Americans, but at the moment we have the edge with respect to recombinant-DNA technology. Is there anything government can do to facilitate retention of that edge?

One part of the answer is to improve the conditions under which U.S. companies can collaborate with academic scientists and make use of the technology developed in whole or part at the taxpayers' expense in universities. Legislation, such as S.414 has been proposed to give small businesses and universities exclusive rights to control the marketing of products arising from federally funded research.* The administration and pharmaceutical industry seem to favor extending the legislation to large companies as well as the small ones and universities. Universities appear to be engaging in more collaboration with industry than was previously the case. Niels Reimers, of Stanford, for example, reports several companies are supporting areas of fundamental research on the understanding that the university will grant to them limited-term exclusive rights to practical results. He says that the fact that many academic investigators are recipients of federal funding limits the extent of the industrial sponsorship because the companies are afraid of their rights becoming entangled with those of the federal government. He says legislation like S.414 should alleviate such fear of contamination and lead to increased industry-university interaction.

A conference in October, 1979, at Midland, Michigan--attended by R & D executives of 27 major corporations, mainly chemical, and chairmen of 60 of the most prestigious chemistry and chemical engineering departments in the country--highlighted the divergence between industry and academia over the past 25 years, due in part to the big influx of federal money into academic research

*Currently, some federal agencies award contractors these exclusive rights, while others insist on nonexclusive licensing of inventions, allowing anyone to market them. Senators Bayh and Dole cite the fact that the government owns 28,000 patents derived from research contracts, less than 4 percent of which have been licensed, as evidence the present system doesn't lead to utilization.

starting in the mid-1950s. George Bugliarello, president of Polytechnic Institute of New York, pointed out that in the mid-1950s industry financed 15 to 20 percent of the research conducted at universities, but less than 4 percent of the $5.5 billion to be spent by universities in fiscal 1980 will come from industry. In the case of chemistry, the federal government will account for $139 million of a total $183 million and industry will provide about $15 million. There were lukewarm reactions to a proposal by Mac Pruitt, Vice President of R & D for Dow, to form a Chemical Research Institute to collect and distribute industrial funds to support basic research. Academicians seemed worried that such a scheme might syphon industry funding away from the researchers who currently receive it. David Baltimore, of MIT, complained that if industry wants better relations with academic chemistry, it has to be more specific about its needs--that industry seems to lack a sense of the problems that need to be worked on. [Heylin, 1979.] One senses that is less a problem with respect to interactions between pharmaceutical companies, many of whom were represented at the Midland meeting, and academic bioscientists, than for the chemical industry. However, Baltimore, as a leading bioscientist, ought to be talking to people who know what they don't know.

One major relationship has been established recently that is often cited as a model. As part of its long-range planning, Monsanto became aware of the commercial possibilities of biological developments and, in 1976, established a relationship with Harvard--in the broad, basic field of the biochemistry and biology of organogenesis--with a term of about 12 years and involving about $25 million.

> "Harvard provides the conceptual framework, identifies capable scientists, provides training, and controls relevant

> research at both Harvard and Monsanto during the research phase. Monsanto helps identify research needs, provides critical starting materials beyond the ability of Harvard to produce or buy, provides unusual and exotic analytical capabilities and controls the development phase, providing expertise in technological innovation, development and marketing. Harvard can receive royalties from either Harvard or Monsanto patents, but, in return, Harvard has waived royalties on initial sales. Monsanto controls its own inventions and receives exclusive license to all Harvard inventions in the area of cooperation for a specific period of time provided that they are progressing toward commercialization. There is no restriction on publication by either partner of its own results; the only obligation is that the other partner be informed prior to publication. The partners are in no way constrained from disclosing their own data, but are constrained from disclosing information gained from the other." [Prager & Omenn, 1980.]

The Monsanto-Harvard relationship, according to the Office of Science and Technology Policy analysts from whose study the preceding description was drawn, is unique in that it involves a partnership in a major new area of science of uncertain future applicability to Monsanto. Monsanto does not have or wish to develop basic research competence in the area of cooperation, although it has a highly developed research program in other areas of R & D. Those authors report potentially interested corporations and university research groups that might have similar capabilities are observing the Monsanto-Harvard experiment and their own opportunities with caution. The sort of university ties that we described earlier for Bristol in the realm of cancer, or that exist between Rochester University and Miles Labs, or which most pharmaceutical companies have with one or two major medical schools are usually much more modest in scope.

There are few examples in the field of medical products, of actual business ventures--such as the recently formed joint venture between Miles Laboratories and Scripps Clinic & Research Foundation in the field of immunology--and of very close collaboration related to exploitation of university inventions--such as the relationship between Becton-Dickinson and Stanford

University based upon the Fluorescence Activated Cell Sorter (FACS), which is used in the development of monoclonal antibodies. However, such institutional ties in the commercialization phase are rare, although consultation by individual academicians is common.

Some academic laboratories offer services to industry, e.g. the unique protein sequencing apparatus assembled by Lee Hood at Cal Tech is very much in demand by industrial and other clients who need a sequence for a minute quantity of peptide--a critical stage in many molecular cloning schemes. Hood says NIH and NSF both rejected his bids for financing. Now Cal Tech is enjoying some return on its investment and looking for a licensee to do for the sequenator what Becton-Dickinson did for Stanford's FACS. The University of California at San Francisco (UCSF) molecular geneticists have certainly found many ways to fund their efforts in the private sector. Besides Herbert Boyer's ties to Genentech, the Howard Hughes Medical Institute (the tax-exempt organization which owns all of Hughes Aircraft Company, a top defense contractor) supports a growing number of investigators, primarily in genetics, but the relationship is not without its drawbacks [Broad, 1979]. Furthermore, contracts with Lilly and Merck on specific molecular genetic developments seem to have become quite routine for UCSF workers.

Cancer is an area in which interaction is the rule. For example, Warner-Lambert (Parke-Davis) recently declared it will become a "major contender in the anti-cancer drug market." That decision, which is expressed in "a highly active acquisition effort and a new team of about 60 scientists," came after decades of contract work for the National Cancer Institute (NCI) on the screening and development of novel anti-cancer drugs. Furthermore, Warner-

Lambert has a long-standing relationship with the Memorial Sloan-Kettering Cancer Center, that has particularly focused on interferons in the past couple of years. Now Warner-Lambert has been awarded the NCI contract to produce 50 billion units of leukocyte interferon, using a variation that has been developed by Sloan-Kettering of the Finnish "Cantell process."

What role the government can or should play in facilitating industrial-academic interactions is far from agreed upon. Prager and Omenn, of OSTP, sum the controversy up as follows:

> "Opinions are divided over the proper federal role in the initiation and maintenance of formal university-industry research relationships. Some are convinced that the government has no role and should allow university and industry to develop relationships as appropriate. These people argue that government participation might limit the flexibility or diversity of industry-university arrangements. Others visualize a range of federal roles. At a minimum, they see the government contributing by attempting to remove disincentives to research and innovation in general. At the next level of involvement, the government could identify problems requiring research, help identify potential partners, and facilitate negotiations. A more active role would involve the government's providing start-up funds. Finally, the government could be a third partner, sharing costs with industry and the university. In this case, too large a government role could lead to federal intervention in activities that should be the responsibility of business and industry." [Prager and Omenn, 1980].

Certainly the government can and does facilitate communication, and in the health field the NIH is an effective stimulus to contacts and substantive interaction through its many symposia, conferences and task forces. Relatively small and intimate meetings such as the Gordon Conferences are generally agreed to be more fruitful than the massive national meetings. The recent two-day conference staged by the FDA on peptide hormones produced by recombinant-DNA was regarded by most participants as an auspicious beginning of three-way communication encompassing academic, industrial and FDA interests in this emerging field.

There is no quarrel about the importance of federal support of basic research in the universities, but there seems to be little interest on the part of industrial genetic engineers in the establishment of non-profit centers to develop and transfer generic technology in this field--even though such institutions seem to be prominent features in many foreign programs to exploit biotechnology. The message that was communicated at a conference sponsored by OTA in San Francisco (March 18, 1980) in connection with this study was for government to fund basic research and stay out of process development stages with respect to molecular genetics. There was strong sentiment in favor of finding ways to assure the return to basic research of some of the profits derived from it. Tax incentives were advocated. However, it seems fair to point out that Canada's use of tax credits is reported to have failed to produce the desired results. Prager and Omenn suggest that that is "because research is the cheapest end of the innovation process. Incentives at the later stages of the innovation process may be more effective." [Ibid.] Of course, measuring success in such a realm is extremely controversial. Certainly it would help consideration of multi-company sponsorship of R & D in institutes or universities to have anti-trust implications clarified as the Department of Justice has been asked to do. [Ibid.]

Dr. Lewis Thomas, President of Sloan Kettering, wrote recently,

> "There are signs that. . .the atmosphere for doing genuine basic research has undergone a gradual and dismaying transformation... Young, very bright and imaginative investigators feel constrained to write out great stacks of pages describing the work they propose to do in the next year or two, and they are convinced that if they do not achieve the results anticipated in the grant proposal, they will never get another grant... The sort of gambling on imaginative ideas that must, in the nature of things, carry a high probability of not turning out as predicted in the end, is no longer the driving force in the scientific enterprise that it was

> during the productive '50s and '60s... If things go on in this way, the country's basic science effort will come to a standstill, and the bank (of stored information in biomedical sciences) will go empty." [Thomas, 1980]

In a closely related talk (for a conference on interferon developments) which he gave a couple of months before that quotation was published, Thomas had suggested an answer:

> "I wish we could develop an auxiliary mechanism of patronage for pure science, in the interest of providing some degree of stability, and at the same time more freedom to make wild guesses, for this vital enterprise. I am not suggesting here any diminution in the governmental support, but I would like to see some other source of what might best be called venture capital, set alongside the federal funds.
>
> "What I have in mind is, of course, the private sector, but something different from private philanthropy or foundation support, which also tends to be more and more short-term. The time has come for the invention of a brand new model of investment in fundamental science by corporate America. One way to do this--or to try--might be to form genuine partnerships between industrial science and academic science. If we could discover better ways for linking the unique scientific resources now controlled by industry to their opposite numbers in the scientific departments of the universities and medical schools, with arrangements for protecting the vital interests of the individual scientist in both camps, including confidentiality and, most important of all, with opportunities for the interchange of investigators holding joint appointments at all levels, including training fellows, we might find ourselves with a new and powerful instrument for doing science and, at the same time, a way of preserving the quality of our best universities and insuring the future of our best industries."
>
> "I realize that partnership arrangements like this have already been tried, some of them with success, but on a relatively small scale considering the magnitude of the problem. I also realize that to talk about setting up partnerships between universities and corporations is to bring up a highly delicate matter, almost like using improper language in public. I am sensitive to this, and thus I have only raised the topic at the very end of this talk as a sort of footnote. But I have come to believe that this is a matter of great urgency and importance for the future of both the science and the technology of our culture, and I suggest that it is time to try some experiments, even with the risk, which holds for any good experiment, of being wrong most of the time. If we tried hard, and kept at it, my hunch is that it could eventually turn out to be a pleasant surprise. I am pretty sure that the academic scientists are ready for such linkages to industry, although this would not have been thinkable ten years ago. And I should

> think the corporate world would be ready to invest some venture capital on some experiments even with all the obvious difficulties of patent law and rights and their year's profits, if they are as worried as they ought to be about their own futures twenty years from now." [Thomas, 1979.]

Reference to government in this statement is conspicuous by its absence. Apparently, Dr. Thomas had no clear idea of what the government should do either - rare for a man who sees so many things clearly.

Dr. Clifford Grobestein, former dean, University of California, San Diego Medical School, said at the OTA conference in March:

> "In the past ten years, I think it is fair to say that the kinds of basic research from which this technology (molecular genetics) has arisen has shifted from log phase to lag phase of growth - to use a microbiological analogy, even as the growth of industrial application was shifting from lag phase to log phase. There is, at the present time, no direct feedback from the kind of economic value generation that is going to occur in this field to the universities that do the research which gave rise to it. The way in which we have been supporting basic research in the universities is essentially on the welfare side of the budget, and we all know what is happening to the welfare side of the budget at the present time. Until we recognize that the kind of research which went into these developments is part of the productivity budget and has to be reinforced when its reduction to practice is successful and value is generated, we are going to oscillate up and down in different fields and lose the kind of momentum that we have acquired in molecular genetics over the last ten-twenty years... I think the Congress ought to be giving its attention to how the government can see to it that budgeting for basic research in some fashion participates in value generated from that research when it goes into application."

12. ASPECTS SPECIFIC TO MEDICAL APPLICATION

12.1 Regulation of the Pharmaceutical Industry

The conference at the FDA (June 5-6, 1980) referred to above went a long way toward clarifying the issues with respect to regulation of hormones produced by microorganisms containing recombinant-DNA. The point was stressed repeatedly that the concern of the FDA should not focus upon the fact that these materials are made in a novel way, but rather upon their intrinsic safety and efficacy with evidence of identity to the natural hormone to serve as the initial basis for probable safety from the standpoint of immunologic reaction, for instance. The larger and more complex the molecule in question, the more difficult will it be to characterize in terms of identity to the natural hormone. Several methods of analysis were described and good progress seemed to be made toward preparing to address the concerns of the FDA raised by the peculiar way these products are to be produced and the possibilities implied by mutation, low-level incidence of misreading and other potential causes of immunogenic analogs.

However, even if recombinant products were treated by FDA just like any other drug product, the pharmaceutical industry would not be pleased. Dr. Robert Williams told an A.D.Little seminar, in March, 1980, that Lilly calculates the unnecessary delays in FDA approval of a new compound cost Lilly $6-8 million/year/compound, which could be productively employed if excessive regulatory expenditures were eliminated. He listed among the consequences of this cost: the increase in licensing of technology; greater use of off-shore preclinical and clinical testing facilities; new weight accorded to arguments "for staying in familiar therapeutic areas"; further diminished interest in products aimed at small markets, and a decrease in the real dollar resources allocated to corporate R & D. Williams pointed out that Bureau of Drugs director Crout asked recently for accurate data on how much regulation really costs. Previewing PMA's cost-of-regulation study, Williams noted that "to isolate such unnecessary regulatory costs, my company has joined with eight other firms under the auspices of PMA to conduct a study in this area. The results of this study should be released this year.

In this study, we hope to quantify such excessive regulatory costs as those associated with duplicative certification requirements for insulin and antibiotics." [F-D-C Reports, 3/31/80, pp.13-14]

One effect of U.S. regulation of clinical trials has been a decided shift to other lands of the initial tests of new chemical entities (NCEs) in man. Between 1963 and 1969, the proportion of NCEs generated by American companies first given to man abroad ranged from three percent to ten percent, with a mean of eight percent. From 1970 on, there was a rise in that proportion to 34 percent in 1973 and 47 percent in 1975.* This trend toward an increasing number of NCEs being first studied abroad has been particularly marked for the largest companies; 50 percent of all drugs of the top four U.S. companies were first studied abroad in 1972 and 1975.** [Wardell, 1978.]

Since 1964, there has been a mean time lapse of 19 months between the first foreign study and subsequent IND filing--a period long enough for a firm to determine a drug's general properties, pharmacokinetics, and probably even its pharmacologic efficacy in man, before filing an IND in the U.S. [Ibid.]

Since clinical pharmacology and drug development benefit from close interaction between the people synthesizing the compounds, testing them in animals, and testing them in man, such a trend implies a movement of all stages of drug research (not necessarily development) to the countries in which it is least encumbered. As science embarks upon new paths opened by molecular genetics, it is appropriate to consider whether changes can be made to improve the efficacy of regulation so as to enable research

* More than 30% of the first evaluations in man took place abroad for NCEs in two pharmacologic categories: gastrointestinal (58 percent) and endocrine (32 percent).

**These four companies accounted for one-third of all the NCEs tested in man by companies operating in the U.S.A. between 1963 and 1975.

without harm to U.S. citizens. Ample evidence has been provided for the international character of pharmaceutical developments and the declining dominance of the U.S.-based firms. Examples have been provided also of the steps being taken by foreign governments to foster the utilization of molecular genetics. It is our perception that the realization of the potentials of molecular genetics in the field of medicine will depend in very large part upon the development of disease models and screens that will facilitate the recognition of molecules for which it would be desirable to prepare producing clones. It would be a shame to drive the development of those models off-shore in a continuing migration of individuals involved in experimental therapeutics.

12.2 Patent Life Span and the Development Interval.

The regulatory mechanisms described previously prolong the period from invention to marketing. Since the life span of patent protection is fixed and commences at a time tied to the invention/filing/granting processes and is independent of the period necessary to prove safety & efficacy and obtain the requisite approval for marketing, prolongation of the interval between invention and marketing means decreasing the duration of patent protection. Dr. William Wardell, of the University of Rochester, has determined that the average effective patent life (after NDA approval) of drugs, in the U.S.A., fell from 13.8 years in 1966 to 8.9 years in 1977--the statutory period of patent life is 17 years from the date of issuance. Dr. Wardell finds that almost all of this difference can be accounted for in FDA-required clinical investigation time, although he points out that the time consumed by NDA review may be more amenable to shortening.* The mean IND duration for 1973-74 was 65 months,

*One-third to one-half the total IND-NDA time is occupied by the NDA review period, yet 12 percent or less of NDA submissions fail to be approved in the end, while almost 90 percent of drugs taken into the IND phase are dropped before they ever reach the NDA.

longer than the total time for NCE NDAs approved in those years. Dr. Wardell warns that unless there is a "dramatic shortening in the duration of the NDA phase or a decrease in the success rates of currently submitted NDAs, there will be a future increase in the duration of the total time for approved NDAs."

Dr. J. Richard Crout, Director of the FDA Bureau of Drugs, has commented in response to Dr. Wardell's findings, that he perceives a divergence in drug research and development trends in which academicians are increasingly focusing on drugs that could help smaller patient population while drug companies are concentrating on drugs with obvious mass-market potential. "Simultaneously, for a variety of economic reasons--costs going up, longer research--drug firms are facing a larger and larger threshold of cost for a marketed drug. It seems to be self-evident that as time goes on and those trends continue, we will not exploit a lot of good research ideas... My solution (to this problem) is to look at the patent system--there are any number of different ways. My personal view is that it's more in context with our society to find a way to reward the drug industry, help it get over that threshold cost... The options are things like beginning the patent at the time the drug is approved instead of when it's first patented as a molecule; another option is to develop a flexible patent system...that allows patents for specific health reasons and specific molecules--let them run 20, 30, 40, 50 years. The longer it runs, the more economically rewarding it is... and the more and more health rewards to a smaller and smaller group of people in our society."

A U.S. patent is reported to have less than a 50 percent chance of surviving a court challenge in the present state of the U.S. patent law, according to the Washington Food and Drug letter.

The PMA reports that there is little willingness in the industry to come out in favor of the increase in patent life because of the difficulties in contesting court challenges, and partly because of expected consumer reaction to such a stand. In the past, the NDA procedure, with its confidential documentation, has apparently been regarded as providing better product protection than patents. It also seems that any one company can have an ambivalent attitude to long patent protection, wanting it on the one hand for its own new drugs, but not wanting it, on the other, because of its own copying for generic activities.

The FDA has a task force working on proposals for patent law changes which would encourage the development of "orphan drugs." This task force is said to favor strengthening patents so that they would hold up in court against competitors and thus prevent generic companies from making copy products before a patent has expired. Patent reform would relieve the pressure on the FDA on the drug lag issue.

To our way of thinking, at the very least, the life span of patent protection ought to be related to the period in which a product is marketed. The inventor should not have diminished the monopoly--intended by the patent system as a reward for his risk, innovation and disclosure--simply because the invention involved is subject to government regulation. As such interventions become increasingly common in high technology businesses this would seem an equitable treatment to extend to all inventors so affected, not just a favor to the pharmaceutical industry. The fact that the welfare of patients is involved directly, to the extent such inequities affect the decision to allocate funds among alternative investments, makes the field of medicinal patents a particularly appropriate place to commence such a reform.

12.3 Government Participation in Product Development

12.31 Cancer is a field in which the government has been an active participant since the National Cancer Act in 1937. In fact, the NCI has taken as its mission to follow every lead for anti-cancer drugs and to develop drugs in which the industry would not be interested. It is interesting to note that NCI doesn't bother to follow the NDA route, since it is in a position to authorize the use of anti-cancer agents on an investigational basis indefinitely. By that means, it "supports" the "orphan drugs" that would not otherwise be claimed.

Presently, NCI is playing a leading role in the investigation of "Biological Response Modifiers," a term it uses to encompass a wide variety of agents that exert influences on bodily defense systems loosely included within the term "immune response." These range from such popular agents as interferon to vaguely characterized factors which manifest intriguing phenomena in cellular preparations *in vitro*. Since these are just the sort of rare molecules for which molecular genetics holds special promise, policies with regard to NCI programs are likely to strongly influence the immediate course of developments in medicine based upon molecular genetics.

12.31 Vaccines are another area in which government is commonly involved. General use vaccines, which are required for a substantial segment of the population, have traditionally been the province of private industry while special-use vaccines, needed only for limited populations, more commonly were developed and produced with the financial support of government. As profit margins declined, liabilities increased, and other alternatives brightened, even the general-use vaccines have lost favor with much of private industry, and there has been steady attrition of pharmaceutical manufacturers from the ranks of vaccine producers. Extensive analysis and case studies of this field have been assembled by OTA and others; it would serve no purpose to reiterate those facts and arguments here. Suffice it to say, fruits of the application of molecular genetics to the field of vaccines may turn out to be in direct proportion to the stimulus the government and other agencies, such as WHO, provide to increase the relative value of a vaccine program to a company skilled in the practice of recombinant-DNA technology.

The clinical study period is the costly stage and the one in which the federal government can do the most good. Either directly or by providing funds, the NIH or CDC could actively support clinical studies necessary to amass the volume of data required to support registration of such a product. NIAID almost always funds most of the basic research and the greatest proportion of developmental effort, including volunteer studies and early field trials. The government can exert a great deal of influence over the priorities of private firms, simply by making it known that it is prepared to support clinical studies on vaccines made by molecular cloning, for x, y, and z diseases. Since various levels of government, including state and local, are often the purchasers and distributors of vaccines, this seems all the more fitting case in which the federal government might take an active role in development.

12.4 Basic Bioscience of Disease and Health

Lastly, it seems appropriate to stress a theme that has been touched upon often in this volume. That is the perception that the identification of desirable biomolecules will be the primary determinant of how fast and in what directions the use of molecular genetics can extend in the field of medicine. Meaningful screens, cellular, animal and clinical models will be necessary to guide the exploitation of the almost infinite synthetic capacity that has been opened. The degree of real benefit derived from this technology will depend upon the quality of that identification process.

In turn, the identification process will be dependent for the most part upon the state of comprehension of the biology and biochemistry of disease processes and bodily defense systems. That is to say, the need for basic bioscience research aimed more at understanding than cure will be greater than ever. While we have developed remarkable understanding of biologic phenomena at the cellular level, the state of knowledge is nowhere near as satisfying with respect to disease processes which are almost invariably multicellular. Molecular genetics and other techniques provide powerful tools for dissection of this relationship, but the intellectual challenge is awesome and will take decades--more discoveries, new techniques and different perspectives will flow from it.

It is a vital policy issue to assure that we do not lose the basic bioscience momentum as we move into a phase of expanded application. A nation that would do so might easily find its promising leads rapidly exhausted and be reduced to witnessing institutions from other nations devising solutions to what are today insufficiently characterized problems. With the leadership position the U.S. holds today in basic bioscience, there is no reason, but myopia, why such an event should come to pass.

References

Abeles, J., "Antibiotics: Part I, A Quantitative Appraisal," Kidder, Peabody & Co., 30 July, 1975.

Alexander, M. R. et al., "Therapeutic Use of Albumin," JAMA 241: 2527-2529, 1979.

Alikhanian, S.I., "Induced Mutagenesis in the Selection of Microorganisms," Advances in Applied Microbiology 4: 1-50, 1969.

Anne, J. and J.F. Peberdy, "Induced Fusion of Fungal Protoplasts Following Treatment with Polyethylene Glycol," J. Gen. Microbiol 92: 413-417, 1976.

Arehart-Treichel, J., "The Human Sweetbread," Science News 117: 61, 1980.

Arima, K., "Recent Developments and Future Directions of Fermentation in Japan," Developments in Industrial Microbiology 18: 79-117, 1979.

Barker, L.F., "Can a National All Voluntary Blood Transfusion Service, by Adequate Blood Component Therapy, Cover Actual and Future Needs of Albumin," Vox Sang 31: 225-227, 1976.

Barlow, G.H., et al., "Production of Plasminogen Activity by Tissue Culture Techniques," in Thrombosis & Urokinase, New York: Academic Press, 1977.

Barnes, W.R., Vela, G.R., and Dorn, G.L., "Physiology of L-Asparaginase Synthesis in Recombinants of Escherichia Coli A-1," Applied and Environmental Microbiology 33: 766-770, 1978.

Benagiano, G., et al., "Sustained Release Hormonal Preparations for the Delivery of Fertility-regulating Agents," J. Polymer Sci.: Polymer Symposium 66, 129-148, 1979.

Bibb, M.J., et al., "Transformation of Plasmid DNA into Steptomyces at High Frequency," Nature 274: 398-400, 1978.

Bibb, M., Schottel, J.L., and Cohen, S.N., "A DNA Cloning System for Interspecies Gene Transfer in Antibiotic-producing Streptomyces," Nature 284: 526-531, 1980.

Bizzini, B., "Tetanus Toxin," Microbiological Reviews 43: 224-240, 1979.

Blalock, J.E. and J.D. Stanton, "Common Pathways of Interferon and Hormonal Action," Nature 283: 406-408, 1980.

Blee, F.J., "Economics of the Pharmaceutical Industry," Smithkline Corp., 1978.

Boldingh, J., "Genetic Engineering: A Key to Innovation in Industrial R&D," in Genetics of Industrial Microorganisms, edited by O.K. Sebek and A.I. Laskin, Washington: American Society of Microbio., 1979.

Broad, W.J., "Hughes Institute Reborn amid Strife," Science 206: 36-37, 1979.

Buchanan, T.M., "Vaccines: Classic Applications and New Prospects," Drug Therapy, 113, June 1979.

Burrel, C.J., et al., "Expression in Escherichia Coli of Hepatitis B Virus DNA Sequences Cloned in Plasmid pBR322," Nature 279: 43-48, 1979.

Busch, H., et al., "A Nucleolar Antigen Found in a Broad Range of Human Malignant Tumor Specimens," Cancer Research 39, 3024-3030, 1979.

Bushnell, L.S., "Respiratory Failure and Albumin," in Proc. Workshop on Albumin, Feb 12-13, 1975, Pub. No. 76-925, NHLI, 1975.

Cabasso, V., Personal Communication, 1980.

Chang, L.T. and R.P. Elander, "Rational Selection for Improved Cephalosporin C Productivity in Strains of Acremonium Chrysogenum" Developments in Industrial Microbiology 20: 367-379, 1979.

Chater, K.F., "Some Recent Developments in Streptomyces Genetics" in Genetics of Industrial Microorganisms, edited by O.K. Sebek and A.I. Laskin, Washington: American Soc. Microbio., 1979.

Clarkson, K., Intangible Capital and Rates of Return, Wash., D.C.: American Enterprise Institute, 1977.

Cline, M.J., et al., "Gene Transfer in Intact Animals," Nature 284: 422-425, 1980.

Cohen, S., Pick, E., Oppenheim, J.J. ed. Biology of the Lympokines, New York: Academic Press, 1979.

Crea, R., et al., "Chemical Synthesis of Genes for Human Insulin," Proc. Natl. Acad. Sci. U.S.A. 75: 5765-5769, 1980.

Crooke, S.T., "The Anticancer Therapy Development Program at Bristol Myers," presented at Financial Analysts Meeting, New York, Jan. 1980.

Danilenko, V.N., Lomovskaya, N.D. and Alikhanyan, Genetika 11: 67, 1975.

Demain, A.L., "Mutation and the Production of Secondary Metabolites," Advances Applied Microbiology 16: 177-202, 1973.

Demain, A.L., "Role of the Microbiologist in Obtaining New Antibiotics," abstracted in Proc. 11th Int. Cong. of Chemotherapy, p. 60, 1980.

DHEW Publication No. (NIH) 77-1274, Study to Evaluate the Supply-Demand Relationships for AHF and PTG through 1980, 1974.

DHEW Publication No. (NIH) 78-1588 and 78-1589, U.S. Department of Health, Education and Welfare. A Study of Insulin Supply and Demand, Appendix, 1978.

DHEW Publication No. (PHS) 79-1222, U.S. Department of Health, Education and Welfare. Facts of Life and Death.

Dowling, H.F., Fighting Infection, Cambridge, Mass., 1977.

Dulaney, E.L., and D.D. Dulaney, "Mutant Populations of Streptomyces Viridifaciens" Trans. N.Y. Acad. Sci 29: 782-799, 1967.

Elander, R.P., "Enhanced Pennicillin Biosythesis by Mutant and Recombinant Strains of Penicillin Chrysogenum," Induced Mutations and Their Utilization, edited by H. Stubbe, Berlin: Akademie-Verlag, 1967.

Elander, R.P., "Genetic Aspects of Cephalosporin- and Cephamycin-Producing Microorganisms," Developments in Industrial Microbiology 16: 356-374, 1975.

Elander, R.P., et al., "Genetics of Industrial Microorganisms," in Annual Reports of Fermentation Processes, Vol. 1, edited by D. Perlmand, New York: Academic Press, 1977.

Elander, R.P., et al., Genetics of Industrial Microorganisms: Proceedings of the Second International Symp. London: Academic Press, p. 253, 1976.

Elander, R.P., "Mutations Affecting Antibiotic Synthesis in Fungi Producing Beta-lactam Antibiotics," Genetics of Industrial Microorganisms: Third International Symp., edited by O.K. Sebek and A.I. Laskin, Washington: American Soc. Microbio., 1979.

Elander, R.P. and A.C. Demain, "Genetics of Microorganisms in Relation to Industrial Requirements," Handbook of Biotechnology I. Microbial Fundamentals, Weinheim, GFR: Verlagchemie, 1980.

Elander, R., Letter to T. McGurn, Office of Technology Assessment, July 2, 1979b.

Elander, R.P. and Chang, L.T., "Microbial Culture Selection," in Microbial Technology, New York: Academic Press, 1979c.

Elander, R.P., "Mutation to Increase Product Formation," in Microbiology-1976, edited by D. Schlessinger, Am. Soc. Microb., 517-521, 1976.

Elander, R.P., "New Genetic Approaches to Industrially Important Fungi," Biotechnology and Bioengineering 22 (Supp. 1), in press, 1980.

Fawcett, P.A., et al., "Formation and Properties of Protoplasts from Antibiotic-Producing Strains of Penicillin Chrysogenum and Cephalosporium Acremonium," J. Gen Microbiol. 79: 293-309, 1973.

Fleck, W.F., "Genetic Approaches to New Streotomycete Products," in Genetics of Industrial Microorganisms, edited by O.K. Sebek and A.I. Laskin, Washington: American Soc. Microbio., 1979.

Florey, H.W., "Antibiotic Products of a Versatile Fungus," Ann. Internal Med. 13: 480-490, 1955.

Fox, J.L., "Antibody Reagents Revolutionizing Immunology," Chem. Eng. News, p. 15, Jan. 1, 1979.

Fraser, T.H. and B.J. Brucè, "Chicken Ovalbumin is Synthesized and Secreted by Escherichia coli," Proc. Nat. Acad. Sci. 75: 5936-5940, 1978.

Fratantoni, J.C., "Urokinase as a Cell Product," in Cell Substrates, edited by J.C. Petricciani, et al., New York: Plenum, 1979.

Franson, Paul, "Medicine's New Pioneers," Venture, 28-31, Aug., 1979.

Friedman, J.J., "Economic Aspects of R&D Intensity in the Pharmaceutical Industry," 1973.

Gaffney, P.J. and S. Balkuv-Ulitin, eds., Fibrinolysis: Current Fundamental and Clinical Concepts, London: Academic Press, 1978.

Gargiulo, G.T., "Comprehensive Review of World Pharmaceutical Market," Wainwright Securities, Inc. (Summarized in Scripp pp. 14-15, May 28, 1977.)

Godfrey, O.W., Ford, L. and Huber, M.L.B., "Interspecies Matings of S. fradiae with S. bikiniensis by Conventional and Protoplast Fusion Techniques," Abstr. Ann. Mtg. Amer. Soc. Microbiol.: 123, 1978.

Goeddel, D.V., et al., "Expression in E. Coli of Chemically Synthesized Genes for Human Insulin," Proc. Natl. Acad. Sci., U.S.A. 76: 106-110, 1979a.

Goeddel, D.V., et al., "Direct Expression in E. Coli of a DNA Sequence Coding for Human Growth Hormone," Nature 281: 544-548, 1979b.

Gordon, A.S., "Erythropoietin," in Vitamins and Hormones 31: 105-162, 1973.

Gros, F., F. Jacob and P. Royer, "Sciences de la Vie et Societe," La Documentation francaise, 1979.

Haddon, J.W., et al., Immunopharmacology, New York: Plenum, 1978.

Hamlyn, P.F., and Ball, C., "Recombination Studies with Cephalosporium Acremonium," Genetics of Industrial Microorganisms: Proceedings from the Third International Symp. Washington, D.C.: American Society for Microbiology, p. 185, 1979.

Hansen, R.W., "The Pharmaceutical Development Process: Estimates of Current Development Costs and Times, and the Effects of Regulatory Changes," University of Rochester, Graduate School of Medicine, August, 1977.

Hansen, R.W., "The Pharmaceutical Development Process: Estimates of Current Development Costs and Times, and the Effects of Regulatory Changes," in Issues in Pharmaceutical Economics, ed. by R.I. Chien, Lexington: Health, 1979.

Hartley, B., "The Biology Business," Nature 283: 122, 1980.

Heylin, M., "Industry-Academic Ties in Chemistry Probed," Chem. Eng. News, 20-22, 29 October, 1979.

Holt, G.G., et al., "The Genetics of Mutants Impaired in the Biosynthesis of Penicillin," Second International Symposium on the Genetics of Industrial Microorganisms, edited by Macdonald, London & New York: Academic Press, 1976.

Hood, L.E., et al., Immunology, Benjamin, page 63, 1978.

Hopwood, D.A., "Genetic Recombination and Strain Improvement," Developments in Industrial Microbiology 18: 9-21, 1977.

Hopwood, D.A., "Genetics of Antibiotic Production in Streptomyces," in Microbiology - 1976, edited by D. Schlessinger, 558-562, 1976.

Hopwood, D.A., "Opening Address: The Many Faces of Recombination" in Genetics of Industrial Microorganisms, edited by O.K. Sebek and A.I. Laskin, Washington: American Soc. Microbio., 1979.

Hopwood, D.A., et al., "Recent Advances in Streptomyces coelicolor Genetics," Bacterial. Rev. 37: 371-405, 1973.

Hunkapiller, M.W. and L.E. Hood, "New Protein Sequenator with Increased Sensitivity," Science 207: 523-529, 1980.

Hughes, R.C., "Hexosaminidases and Phenylalanine Hydroxylase for Treatment of Storage Diseases and Phenylkentonurea," Nature 269: 288 (1977).

James, B.G., The Future of the Multinational Pharmaceutical Industry to 1990, New York: John Wiley & Sons, 1977.

Janicki, R., Personal Communication, 1980.

Johnson, A.J., et al., "Urinary Urokinase, Two Molecules or One?" In Thrombosis & Urokinase, ed. by R. Paolette, Rome: Academic Press, 1977.

Johnson, I.S., and J.R. Burnett, "Problems and Potential of Industrial Recombinant DNA Research," in Genetic Engineering, edited by H.W. Boyer and S. Nicosia, Amsterdam, Elsevier, North Holland Biomedical Press, 1978.

Kennett, R.H. and Gilbert, F., "Hybrid Myelomas Producing Antibodies Against a Human Neuroblastoma Antigen Present on Fetal Brain," Science 203: 1120-1121, 1979.

Kirby, R., et al., "Plasmid-determined Antibiotic Synthesis and Resistance in Streptomyces coelicolor," Nature 254: 265-276, 1975.

Lai, E.C., et al., "Ovalbumin is Synthesized in Mouse Cells Transformed with the Natural Chicken Ovalbumin Gene," Proc. Natl. Acad. Sci. 77: 244-248, 1980.

Laver, W.G., et al., "Amino Acid Sequence Changes in the Hemagglutinin of A/Hong Kong (H3N2) Influenza Virus During the Period 1968-77," Nature 283: 454-457, 1980.

Lawford, G.N., "Dextran Biosynthesis by Continuous Culture of Leuconostoc mesenteroides," Biotechnol. Bioeng. 21: 1121-1121, 1979.

Lee, A.M., "Pharmaceuticals and Health Care Costs: A Case Study of a Research-intensive Industry," presented to National Economists Club, Washington, D.C., 4 October 1979.

Lee, J., and J. Laycock, Essential Endocrinology, Oxford Press, 1978.

Lemke, P.A. and C.H. Nash, "Fungal Viruses," Bacteriol. Rev. 38: 29-56, 1974.

Lewin, R., "Modern Biology at the Industrial Threshold," New Scientist: 18-19, 5 October 1978.

Lilly, Annual Report 1980.

Lomovskaya, N.D., et al., "Genetic Relationship between Actinomycetes and Actinophages," in Genetics of Industrial Microorganisms, edited by O. K. Sebek and A.I. Laskin, Washington: American Soc. Microbio., 1979.

Malik, V.S., "Genetics of Applied Microbiology," Advances in Genetics, Vol. 20: Academic Press, Inc., 1979.

Malik, V.S., "Genetics of Applied Microbiology," Advan. Genet. 20: 37-126, 1976.

Martial, J.A., et al., "Human Growth Hormone: Complementary DNA Cloning and Expression in Bacteria,"Science 205: 602-606, 1979.

Marx, J.L., "Newly Made Proteins Zip Through the Cell," Science 207: 164-167, 1980.

Marx, J.L., "Vaccinating with Bacterial Pili," Science 209: 1103-1106, 1980.

Mauro, T., et al., "Production of Choriogonadatropin-like Factor by a Microorganism," Proc. Natl. Acad. Sci. U.S.A. 76: 6622-6626, 1979.

McCormick, J.R.D., et al., "A New Family of Antibodies: the Demethyltetracyclines," J. Am. Chem. Soc. 79: 4561-4563, 1957.

Meager, A., et al., "Involvement of a Gene on Chromosome 9 in Human Fibroblast Interferon Production," Nature 280, 493-5, 1979.

Mercereau-Puijalon, O. and P. Kourilsky, "Introns in the Chicken Ovalbumin Gene Prevent Ovalbumin Synthesis in E. coli K12," Nature 279: 647-9, 1979.

Miller, J.A., "The Cloning of an Antibody," Science News 114: 444-447, 1978.

Miyake, T., et al., "Purification of Human Erythropoietin," J. Biol. Chem. 252: 5558-5564, 1977.

Montville, T.J., C.L. Cooney and A.J. Sinskey, "Streptococcus mutans Dextransucrase: A Review," Advances in Applied Microbiology 24: 55-85, 1978.

Moore, F.D., et al., "Hemorrhage in Normal Man - I. Distribution and Dispersal of Saline Infusions Following Acute Blood loss, Clinical Kinetics of Blood Volume Support," Ann. Surg. 163: 485, 1966.

Murano, G., et al., "Inhibition of High and Low Molecular Weight Urokinase by Plasma Inhibitors," Circulation 58II: 136, 1978.

Muagh, T. H., "Hepatitis B Vaccine Passes First Major Test", Science 210: 760-762, 1980.

Nakamura, N., et al., "On the Productivity and Properties of L-Asparaginase from E. coli A-1-3," Agr. Biol. Chem. 36: 2251-2253, 1972.

Nakanishi, S., et al., "Nucleotide Sequence of Cloned cDNA for Bovine Corticotropin -B- Lipotropin Presursor," Nature 278; 423-427, 1979.

Nash, C.H., "Proplast Fusion, Plasmids, and Recombinant DNA Applied to Antibiotic Discovery," abstracted in Proc. 11th Int. Cong. of Chemotherapy, 59, 1980.

Neusch, J., et al., "Biosynthesis of Cephalosporin C," in Genetics of Industrial Microorganisms, Vol. 11, Actinomycetes and Fungi, edited by Vanek, E., et al., Amsterdam, Elseview, 1973.

Nossal, G., Antibodies and Immunity, New York: Basic Books, 1978.

O'Sullivan, Dermot A., "R&D Gets Heavy Emphasis in West Germany," C & EN, 15-21, 18 Feb., 1980.

Parish, P.S., "Regulation in the Public Interest," Food, Drug and Cosmetic Law J. 33: 164-172, 1978.

Peberdy, J.F., "New Approaches to Gene Transfer in Fungi," Genetics of Industrial Microorganisms: Third International Symp., Washington, D.C.: American Society for Microbiology, 1979.

Peberdy, J.F., "New Approaches to Gene Transfer in Fungi," in Genetics of Industrial Microorganisms, edited by O.K. Sebek and A.I. Laskin, Washington: American Soc. Microbio., 1979.

Peters, T.J., "Serum Albumin: Recent Progress in the Understanding of it's Structure and Biosynthesis," Clin. Chem. 23: 5-12, 1977.

Phillips, L.S., and R. Vassilopoulou-Sellin, "Somatomedins," New England J. Med. 302: 371-380, 438,446, 1980.

Porter, A.G., et al., "Complete Nucleotide Sequence of an Influenza Virus Haemagluttinin Gene from Cloned DNA," Nature 282, 5738, 1979.

PMA (Pharmaceutical Manufacturers Association), Annual Survey Report, Ethical Pharmaceutical Industry Operation, 1978-1979, 1980.

PMA, Competition in the International Pharmaceutical Industry, 1977.

Prager, D.J., and G.S. Omenn, "Research, Innovation, and University-Industry Linkages," Science 207: 379-384, 1980.

Pye, E.K., et al., "Purification of Urokinase by Affinity Chromatrography," in Thrombosis and Urokinase, edited by R. Paoletti, Rome: Academic Press, 1977.

Pye, E.K., T. Maciag and R. Iyengar, "Production of Enzymes and Proteins in Tissue Culture," in Biotechnical Applications of Proteins and Enzymes, edited by Z. Bohak and N. Sharon, New York: Academic Press, 1977.

Queener, S.W. and Baltz, R.H., Ann. Repts Ferm. Processes 3, New York: Academic Press, 1980.

Queener, S.W., et al., "Mutants Blocked in Antibiotic Synthesis," Ann. Rev. Microbiol. 32: 593-636, 1978.

Ramachandran, S., et al., "Bio-transformation of Thiolutin Production from Streptomyces Pimprina to Streptomyces Aureofaciens," Hind. Antibiot. Bull. 7: 197-202, 1965.

Reiff, T.R., "Colloid Osmotic Homeostasis in Humans - I. Theoretical Aspects and Background," J. Theor. Biol. 28: 1-14, 1970.

Rich, V., "Soviet Secrecy," Nature 283: 123-124, 1980.

Richards, B., "An Approach to Establishing Safety of Recombinant Processes and Products," presentation to Endocrinologic and Metabolic Drugs Advisory Committee, 4 June 1980.

Ring and Messmer, "Anaphylactoid Reactions of Various Plasmer Expanders," Lancet, 466-469, 1977.

Ritz, et. al., "A Monoclonal Antibody to Human Acute Lymphoblastic Leukemia Antigen," Nature 283: 583-585, 1980.

Roitt, I., Essential Immunology, 3rd Ed., Oxford: Blackwell Sci. Press, 1978.

Samama, M., et al., "Urokinase I and II Activity," Thrombos. Hemostas. (Stuttgart) 40: 578-580, 1978.

Schaumann, L., "Pharmaceutical Industry Dynamics and Outlook to 1985," SRI, 1976.

Schwartzman, D., The Expected Return from Pharmaceutical Research, Washington, D.C.: American Enterprise Institute for Public Policy Research, 1975.

Schwarz, R.T. and R. Datema, "Inhibitors of Protein Glycosylation," Trends in Biochem. Sci., 64-67, March 1980.

Sermonti, G.,"Mutation and Microbial Breeding," Genetics of Industrial Microorganisms: Third International Symp. Washington, D.C.: American Society for Microbiology, 10, 1979.

Sevastopoulos, et al., "Large Scale Automated Isolation of E. coli Mutants with Thermosensitive DNA Replication," Proc. Nat. Acad. Sci. U.S.A. 74: 3485-3489, 1977.

Sherman, I.W., "Biochemistry of Plasmodium (Malarial Parasites)," Microbiological Reviews 43: 453-495, 1979.

Sherry, S., "Therapeutic Thrombolysis: Past, Present and Future," in Thrombosis & Urokinase, edited by R. Paoletti and S. Sherry, Rome: Academic Press, 1977.

Shier, W.T., Rinehart, K.L., and Gottlieb, D., "Preparation for Four New Antibiotics from a Mutant of Streptomyces Fradiae," Proc. Natl. Acad. Sci. U.S.A. 63: 98-204, 1969.

SKF, Annual Report, 1980.

Smuzness, W., et al., in Viral Hepatitis, eds. Vygas, G.N., et al., Philadelphia: Franklin Institute Press, 1978.

Solomon, J.M., "A Federal View: an Examination of Five Years of FDA Licensure of Source Plasma Manufactures," presentation at the 1979 Plasma Forum, American Blood Resources Assn., Arlington, VA., 26 Feb., 1979.

Soria, M and L. Zeller, "Expectations and Priorities of Pharmaceutical Industry," in Genetic Engineering, edited by H.W. Byer and S. Nicosia, Amsterdam, Elsevier, North Holland Biomedical Press, 1978.

Staver, S., "Physician's Firm Applies Biological Advances to Industrial Problems," American Medical News, p. 8, 14 Mar 1980.

Steiner, D.F., "Peptide Hormone Precursors: Biosynthesis, Processing and Significance," in Peptide Hormones, edited by J. Parsons, University Press, 1976.

Swanson, R., Personal Communication, 1979.

Thomas, L., "Notes of a Biology Watcher, on Science Business," New England J. Med. 302: 157-158, 1980.

Thompson, W.L., "Rational Use of Albumin and Plasma Substitutes," John Hopkins Med. J. 136: 220-225, 1975.

Toufexis, A. and A. Jucius, "The Big IF in Cancer," Time, 60-66, 31 March, 1980.

Tullis, J.L., "Albumin: 1. Background and Use," JAMA 237: 355, 1977a.

Tullis, J.L., "Albumin: 2. Guidelines for Clinical Use," JAMA 237: 355, 1977b.

Wade, Nicholas, "Cloning Gold Rush Turns Basic Biology into Big Business," Science, Vol. 208: 688-692, 16 May 1980.

Walgate, R., "600 Apply to Biogen," Nature 284: 585, 1980.

Walgate, R., "Japanese Domination," Nature 283: 123, 1980.

Ward, C.B. and R.D. Bruner, "Method of Cloning Filamentous Organisms," U.S. Pat. No. 3, 950, 224, 1976.

Wardell, W.M., et al., "The Rate of Development of New Drugs in the United States, 1963 through 1975," Clin. Pharmacol. Ther. 24, 133-145, 1978.

World Health Organization, Proc. of the Internat. Conf. on the Role of the Individual and the Community in the Research Development and Use of Biologicals, Geneva, 1977.

Zahner, H., "New Ways to New Antibiotics," Current Chemotherapy and Infectious Disease, 59-60, 1980.

Part IV

Pharmaceutical Industry and Medical Research

The information in Part IV is from *Report to the Office of Technology Assessment on the Use of Applied Genetics in the Development of Pharmaceuticals and Medical Research Technology*, prepared for the Office of Technology Assessment of the Congress of the United States. This material was prepared at the Massachusetts Institute of Technology.

An Overview of the Use of Applied Genetics to Develop Pharmaceuticals and Medical Technologies

Applied genetics has been pivotal in the development of medically important therapeutic substances. High yields of commercially important antibiotics are the result of having applied a knowledge of genetics to the production of antibiotics by microorganisms.

The antibiotic activity of *Penicillium notatum* was reported in 1929 by Fleming. Ten years then passed before the isolation of a significant amount of bacteriostatic principle was accomplished. The demands of the World War II battlefields for an antiinfective agent constituted an important impetus for the development of large-scale antibiotic production.

In order to screen organisms for antibiotic production or to select from an existing culture an improved genetic strain, screening methods have been developed. Organisms can be tested for antibiotic production by plating out obtained colonies onto agar plates seeded with a suitable test organism. Colonies can be screened for changes in colony morphology or color that may indicate alterations in metabolism that arose spontaneously or after induced mutagenesis. A system has been developed whereby a computerized television monitor can examine up to 85 parameters of up to 10^8 colonies in order to isolate genetic mutants with desired characteristics.

UV light and chemicals are used to induce genetic

mutations in a microbial population. Mutants are then selected for those that resultantly produce either a novel antibiotic substance or increased amounts of an antibiotic that was being produced by the parent population. Genetic mutants may also be sought that exhibit altered growth rates, temperature sensitivities, feedback inhibition characteristics, growth and biosynthetic requirements, permeabilities, or altered enzyme activities that result in novel enzyme reaction products. In other words, genetic characteristics are selected for that might either expedite the production and isolation of a microbial product or result in a novel product.

Section IV. of this report is a case study of gentamicin, an aminocyclitol antibiotic of *Micromonospora pupurea*. The history, development, production, and importance of this important antibiotic are reviewed.

Commercially cell culture is used principally in the production of viral vaccines. Recently, however, the technology has gained renewed interest because of the potential to produce a number of other medically important substances. These substances include many hormones, enzymes such as urokinase, antibodies, and most notably the interferons. The interferons are a class of protein that are produced by cells in response to a viral challenge. They are presently being evaluated as therapeutic agents for the treatment of viral infections such as hepatitis, herpes, and rabies. The value of the interferons as beneficial agents in the treatment of cancer is likewise being evaluated.

There are presently three important cell sources of the interferons. These are leukocytes, lymphoblasts, and fibroblasts. Human leukocyte interferon is obtained from leukocytes that have been primed with interferon and then induced with Sendai virus. The cells themselves have been obtained from a blood bank of human blood donations. Lymphoblastoid cell lines are a second source of interferons. These cells can be grown in high density suspensions and can be induced by paromyxoviruses to produce both leukocyte and fibroblast interferons. Likewise fibroblasts can be grown in culture and induced to produce interferons by exposure to a synthetic polyribonucleotide and certain antimetabolites.

The use of human diploid cells grown in culture for the production of medically important substances will be examined in Section V. of this report as a case study of the human interferons.

Techniques utilizing recombinant DNA methodology are being used to develop products of medical importance. Gene sequences can be made from isolated mRNA sequences using the enzyme reverse transcriptase or they can be synthesized, or excised from a cell's chromosome. Excisions of DNA sequences are possible with the use of endonucleases, which are enzymes that make site-specific cuts in DNA. A synthetic or isolated gene can then be spliced into a section of bacterial DNA by enzymatic ligation. The hybrid DNA molecule or "plasmid" is then inserted into a host bacterium that has been treated to accept the plasmid. The host bacterium can then be induced to transcribe the plasmid, thereby producing the desired gene product.

The use of recombinant DNA technology in the commercial manufacture of human hormones is presently being pioneered. Genes coding for somatostatin and insulin have been synthesized, linked to the bacterial *lac* operon sequence of DNA, and inserted into the host bacterium,

E. coli. *E. coli* can then be induced to produce a hormone of therapeutic value.

Also this report will review the medically relevant products and technologies that may result from applied recombinant DNA methodologies.

Products and Technologies of Recombinant DNA: A Survey

I. INTRODUCTION

This report summarizes recent accomplishments of those using recombinant DNA technology to manipulate mammalian genomes. The potential for such technology to generate novel approaches to problems of medical prevention, diagnosis, prognosis, and treatment of disease as well as generating new tools of medical research are discussed. Problems confronting the implementation of recombinant DNA technology in clinical and basic medical research are presented.

To understand gene structure and function is to understand the processes by which biological information is ordered, stored, and implemented. Knowledge of molecular genetics has grown rapidly in the past few decades. This increased understanding of the regulation and processing of biological information may have an exponentially increasing impact on the development of medically important products and technologies. Much remains to be learned, however, before a molecule of predetermined biological activity can be designed by draftsmen and subsequently produced by economically feasible methods.

The use of rDNA technology to commercially produce medically important peptide products in the near future will be restricted to the production of simple peptides until the technology is developed to duplicate posttranslational modifications of proteins. A common posttranslational modification is the addition of covalently bound carbohydrate. This modification may increase the _in vivo_ half-life of a protein. Sialic acid residues covalently bound to several proteins such as haptoglobin and HCG have been shown to increase the serum half-lives of the peptides. Glycosylation of interferon renders it more biologically active than the unglycosylated form. Recombinant DNA techniques could be applied to produce commercially significant amounts of the enzymes that perform these posttranslational modifications.

Another important posttranslational modification is the alteration of the structures of amino acids that have been incorporated into the protein. For example, glutamate is cyclized and proline is amidated after the sythesis of Thyroid Releasing Hormone,the structure of which is pyroGly-His-Pro(NH_2). Such alterations of the amino acid moieties of peptides may alter the functioning of the peptides _in vivo_. Degradation of the protein may be impeded by such modifications,or the binding of the protein to its receptor may be facilitated by such modifications. Methods to alter the structures of amino acids that have been incorporated into protein will have to be developed should attempts be made to produce by recombinant DNA technology proteins characterized by such structural alterations.

The functioning of many enzymes depends on the proper positioning of specific metals within the tertiary structure of the protein sequence. For example, zinc and copper are constituents of Cu-Zn superoxide dismutase while iron and sulfur are constituents of aldehyde oxidase. The addition of metals to a protein sequence often takes place in a specialized compartment within a cell. Should metalloproteins be produced by recombinant DNA techniques, methods will have to be developed to properly position metals into the cloned peptide sequences.

The functioning of some proteins depends on the proper positioning of other nonproteinaceous moieties into their tertiary structures. To produce such proteins by recombinant DNA methods, techniques will have to be devised to synthesize these moieties and to correctly combine them with polypeptide sequences. For example, the oxygen-carrying ability of hemoglobin depends upon the occupation of the proteins "heme pocket" by a tetrapyrrole ring.

New techniques must be developed to separate the desired peptide sequences from the bacteria that produced it. Presently, insulin and somatostatin are cleaved from bacterial protein products that are hybrids comprised of the hormone sequence and bacterial enzyme. The hormone sequence and the bacterial enzyme sequence are spliced together by a polymethionine strand. Since neither insulin nor somatostatin contain methionine residues, the hormones can be cleaved from the hybrids by treating the hybrids with a system that cleaves methionine residues from peptides. Should more complex proteins with more complex amino acid compositions be cloned, new methods will be required for either the bacterial production of the peptide product or the retrieval of the product.

The activities of hormones are mediated by receptor molecules. A specific sequence within a peptide hormone's total amino acid sequence binds to a specific region of its target receptor; and it is by this binding that the hormone transfers information. As more becomes known of the chemistry and geometry of the active sites of receptor molecules, it may become possible to design peptide sequences that can activate specific receptors. Thus knowledge of hormone-receptor binding activity may eventually allow the designing by draftsman of compounds possessing specific biological activities. That is, an increased understanding of the mechanisms by which biologically active molecules convey information may eventually allow the development by applied genetics of novel molecules possessing specific therapeutic activity. Such an approach to develop a substance possessing a specific therapeutic activity contrasts with the approach in which biologically active molecules are sought by screening organisms such as fungi, bacteria, and plants for heretofore undiscovered compounds.

Many hormones cannot be extensively used for research because they are either unavailable or too expensive. But until the physiological properties of a hormone are understood, the therapeutic values of that hormone will remain unknown. Recombinant DNA techniques are being used to overcome this circular problem. In one laboratory somatostatin is being used as a research tool to study the regulation of the hormonal milieu of burn patients. One experiment may use as much as 25 mg of the hormone, which as a product of solid state synthesis costs as much as $12,000. A reduction in the cost of the hormone would allow for more extensive research on the physiological and therapeutic qualities of somatostatin. Genentech's production of this hormone by recombinant DNA techniques may result in a commercially available product, the cost of which may be less than that of the solid state synthesis product. Recombinant DNA techniques can be used to produce peptides of unknown therapeutic value at costs that would not prohibit investigations of the therapeutic values of these peptides.

Genes coding for peptide products have been shown to contain silent sequences of DNA that are not expressed as amino acids of the peptide product. Therefore industries wishing to use rDNA techniques to produce marketable therapeutic products will pursue the development of technologies whereby gene sequences conceived by draftsman can be expeditiously synthesized. Gene synthesis may become automated as, by analogy, the amino acid sequencing of peptides is now automated.

This report first considers the use of recombinant DNA technology in producing therapeutic substances used to treat or control disease states. This discussion is followed by a consideration of the potential application of such technology in the field of basic medical research. Finally, the possible application of recombinant DNA techniques to diagnose and treat errors of metabolism is discussed.

II. Therapeutic Substances

Recombinant DNA technology is being used to produce products useful in the treatment and control of disease. A disease state characterized by a missing or ineffective gene product can be treated by the administration of the required gene product, which has been harvested from bacterial cells into which the particular gene has been cloned. A gene coding for insulin has been cloned into bacteria; insulin has been harvested from the medium of this clone. The treatment or control of disease by gene products harvested from bacterial fermentations is then possible. The use of recombinant DNA technology to treat and control disease by enhancing the body's own defense mechanism is also a possibility. The production of interferon and vaccines by the use of genetic manipulations will be discussed in this section. The applicability of current techniques of applied genetics to produce products of blood plasma is reviewed. Finally, the production of substances valuable in the therapeutic management of pain, the endogenous opiates,will be considered as a potentially important process using recombinant DNA technology.

II.A. The Use of Recombinant DNA Technology to Supply a Missing or Otherwise Ineffective Gene Product.

II.A.1. Insulin

Insulin is a polypeptide hormone synthesized by pancreatic beta cells. It is essential for the proper regulation of carbohydrate and fat metabolism, especially glucose and fat oxidation. In addition, it stimulates amino acid and glucose transport into cells, protein synthesis, and glucogen and mucopolysaccharide formation.

Diabetes mellitus is a metabolic disorder characterized by an inadequate supply of effective insulin and subsequent disturbances of carbohydrate, fat, and protein homeostasis. Diabetics, therefore, require daily intravenous injections of insulin. 2,000 kilograms of insulin are used in the world each year, half of this being used in the United States.[1] The total number of diabetics in the United States is rising by 6% annually; the number of diabetics worldwide is increasing.[2] This reported increase is due in part to improved detection and treatment of diabetes in developing countries. The cost of insulin is rising. World insulin production from animal sources is tending to plateau,while the demand for the hormone may exceed supplies from animal sources by the year 2000.[3] Finally, one complication of animal-derived insulin is that its injection into some diabetic patients triggers immune-response reactions.[4]

Insulin is first synthesized by the pancreatic beta cells as a long precursor molecule. This precursor must be processed by the cell before it is biologically active. This involves deleting an amino acid sequence referred to as the C-chain from the middle of the protein. The two remaining peptides, the A and B chains, are then joined by disulfide linkages to form the active peptide hormone.

The cloning of the gene sequence for the precursor molecule, proinsulin, of the rat has been accomplished.[5] The use of proinsulin for the production of marketable insulin would involve performing the necessary posttranslational modifications of the protein, a function not performed by the gene's host bacterium, *E. coli*. Workers at Genentech in South San Francisco, California, have succeeded in cloning synthetic genes for the alpha and beta chains of insulin into bacteria.[6] The two gene producers are harvested separately and then combined *in vitro* to yield the biologically active form of the hormone. The harvesting of the insulin chains involves clipping the insulin sequences from the bacterial gene product, the enzyme B-galactosidase. For insulin derived from such process to be marketable with FDA approval would require that this clipping process be proven accurate.

II.A.1. Insulin

1. Steiner, D.F. et al. *Diabetes* 27(8), 877-878 (1978).
2. Gunby, P.*J. Am. Med. Assoc.* 240(16), 1697-1698 (1978).
3. Steiner, D.F. et al. *Diabetes* 27(8), 877-878 (1978).
4. Gunby, P. *J. Am. Med. Assoc.* 240(16), 1697-1698 (1978).
5. Ullrich, A. et al. *Science* 196, 1313-1319 (1977).
6. Gunby, P. *J. Am. Med. Assoc.* 240(16), 1697-1698 (1978).

II.A.2. Growth Hormone (GH) (Somatotropin)

Human growth hormone (hGH) is an unbranched polypeptide containing 188 amino acid residues and two internal disulfide bridges. Its physiological functions are the promotion of skeletal growth, protein anabolism, fat and carbohydrate metabolism, water and salt metabolism. Excess concentrations of the hormone in childhood result in gigantism, while excesses in adulthood result in acromegaly. Pituitary dwarfism and hypopituitarism are diseases of hGH deficiency which are treated with injections of the hormone. Usually, intramuscular injections are administered three times weekly over months or years to achieve desired growth rates.[1] One complication is that chronic administration may result in the development of circulating antisomatotropin antibodies.[2] Also, response to somatotropin is generally inversely proportional to the duration of treatment[3], implying the possible induction of a "down-regulation" mechanism.

The sole source of hGH for children deficient in the hormone is hGH extracted from pituitary glands obtained from autopsies. For the past 15 years the National Pituitary Agency in Baltimore, Maryland, has been supplying this hGH to patients free of charge.[4] On the average, each child treated for hGH deficiency requires at least 120 pituitary glands each year. Recently demand for the hormone has begun to exceed supply.[5]

Two groups have cloned hGH into bacteria.[6] One group at the University of California San Francisco Medical School was headed by Drs. Baxter and Goodman. Drs. Goeddel and Seeburg of Genentech of South San Francisco headed the second group. The feasibility of commercially producing hGH fermentation would in part depend on the demonstrable existence of a significant market for the hormone. A knowledge of the National Pituitary Agency's plans for the future, as well as the rate of the increase in need for the hormone, would be valuable in assessing the advisability of commercially producing hGH. The presently available hGH is distributed under the name "Asellacrin."[7]

II.A.2. GH

1. Govoni, L.E. and Hayes, J.E. Drugs and Nursing Implications (3rd Ed.) 663 (Appleton-Century-Crofts, New York, 1978).
2. Ibid. pg. 663.
3. Ibid. pg. 664.
4. Spilker, B. N.E.J.M. 298(11) 634 (1978).
5. Ibid. pg. 634.
6. Gonzalez, E.R. J. Am. Med. Assoc. 242(8), 701-702.
7. Govoni, L.E. and Hayes, J.E. Drugs and Nursing Implications (3rd Ed.) 663 (Appleton-Century-Crofts, New York, 1978).

II.A.3. Somatostatin (Somatotropin Release Inhibiting Factor, SRIF)

Somatostatin (SRIF) is a tetradecapeptide hormone with one internal disulfide bridge. It is widely distributed in the gastrointestinal tract and has been shown to be localized in the delta cells of the pancreas. SRIF has also been shown to be present in the brain, dorsal root ganglia, and the autonomic plexi of the intestine.[1] Cultured pancreatic delta cells have been demonstrated to release somatostatin into the surrounding media.[2] A multitude of effects of SRIF have been reported. It inhibits the secretion of insulin, glucogeon, thyrotropin, GIP, VIP, pancreozymin, secretin, motilin, gastrin, and growth hormone.[3] Studies using rats have shown that SRIF administration can reduce serum amylase and lipase, as well as reduce both hormonally and non-hormonally stimulated pancreatic volume and enzyme secretion in man.[4] Patients with SRIF secreting tumors develop diabetes and alterations in carbohydrate metabolism.[5]

The demonstration of SRIF's effects on the secretion of particular hormones has resulted in proposals that SRIF may be of therapeutic value in the control of acromegaly, acute pancreatitis, and insulin-dependent diabetes.[6] However, the serum half-life of SRIF is less than one minute.[7] This fact coupled with the fact that SRIF has such diverse actions restricts the potential of this hormone as a marketable therapeutic product. The cloning of a synthesized gene coding for human somatostatin has been accomplished. The gene sequence was synthesized and inserted into E. coli by Itakura and co-workers.[8]

II.A.3. Somatostatin

1. Couper, J.R. et al The Biochemical Basis of Neuropharmacology (3rd Ed.) 276-277 (Oxford University Press, New York, 1978).
2. Rizza, R.A. et al. Med. Clin. No. Amer. 62(4) 747-752 (1978).
3. Ibid.
4. Lankisch, P.G. et al. Gut 18(9) 713-716 (1977)
5. Rizza, R.A. et al. Med. Clin. No. Amer. 62(4) 747-752(1978
6. Itakura, K. et al. Science (198) 1056-1062 (1977).
7. Rizza, R.A. et al. Med. Clin. No. Amer. 62(4) 747-752 (1978).
8. Itakura, K. et al. Science (198) 1056-1062 (1977).

II.A.4. Human Chorionic Gonadotropin (hCG)

Human chorionic gonadotropin (hCG) is a multimeric hormone synthesized in the syncytial trophoblast of the placenta. It is detectable in the urine within a few days after the embryo's implantation and is known to support corpus luteal function in early pregnancy and may affect placental steroid hydroxylation and aromatization.[1] It has been used therapeutically in conjunction with other gonadotropins to bring about ovulation in women with hypopituitarism or amenorrhea/anovulation syndromes.[2] Its administration to women results in elevated serum levels of estrogen and progesterone,indicating its effects on ovarian steroid metabolism.[3] hCG has been administered to men with resultant increased testosterone production by the testicular interstitial cells.[4] It has been used therapeutically to maintain fertility in cryptorchid men.[5] The therapeutic effects of hCG may be due to its ability to exert luteinizing hormone (LH)-like effects as hCG has chemical, biological, and immunological properties of LH.[6]

The alpha subunit of hCG has been cloned into a bacterial plasmid by Drs. Fiddes and Goodman at UCSF.[7] This subunit is thought to be identical with the subunits for LH, FSH, and TSH. A major problem with the idea of cloning hCG for use in a medical therapy is that hCG is a glycoprotein.[8] Should glycosylation of the amino acid sequences of hCG prove necessary for biological activity, a problem arises in that *E. coli* are not capable of properly glycosylating the hCG that might be synthesized.

Presently commercially available hCG (Android-HCG, Antuitrin, A.P.L., Chorex, Follutein, Gonadex, Libigen, Pregnyl, Stemutrolin) is extracted from gravid human urine.[9]

II.A.4. hCG

1. Kase, N.G. in *Duncan's Diseases of Metabolism (Endocrinology)* (ed. Bondy and Rosenberg) 1621 (W.B. Saunders Co., Phila., 1974).
2. Ney, R.L. in *Duncan's Diseases of Metabolism (Endocrinology)* (ed. Bondy and Rosenberg 995 (W.B. Saunders Co., Phila., 1974).
3. *Ibid.*, pg. 975.
4. *Ibid.*, pg. 975.
5. Lipsett, M.B. in *Duncan's Diseases of Metabolism (Endocrinology)* (ed. Bondy and Rosenberg) 1574 (W.B. Saunders Co., Phila., 1974).
6. Ney, R.L., pg. 975.
7. Fiddes, J.C. and H.M. Goodman, *Nature* 281 351-356 (1979).
8. Govoni, L.E. and Hayes, J.E. *Drugs and Nursing Implications* (3rd Ed.) 139 (Appleton-Century-Crofts, New York, 1978).
9. *Ibid.*, pg. 139.

II.A.5. Human Placental Lactogen (hPL, Human Chorionic Somatomammatropin)

One group has recently reported the production of DNA from messenger for human placental lactogen (hPL).[1,2] This hormone is synthesized by the syncytial trophoblast of the placenta. Its activity is characterized as lactogenic, luteotropic, and growth promoting. It affects pregnancy maintenance, lactation, and fetal/maternal nutrition.[3] It is related chemically and biologically to growth hormone, and prolactin. Specific therapeutic uses for this hormone have not been assigned.

II.A.5. hPL

1. Shine, J. et al. Nature 270(8) 494-499 (1977).
2. Seeburg, P.H. et al. Cell 12 157-165 (1977).
3. Blank, M.S. J. Steroid Biochem. 8 403-414 (1977).

II.A.6. Future Prospects

A large market exists for enzymes capable of activating plasminogen in order to effect fibrinolysis, the dissolving of intravascular blood clots. One such enzyme currently being investigated is Urokinase, an enzyme presently obtained from human urine. Varidase is a presently marketed combination of proteolytic enzymes derived from certain strains of hemolytic streptococci. One of the enzymes, Streptokinase, activates plasminogin,while another enzyme Streptodornase dissolves the viscous nucleoprotein of dead cells.[1] Recombinant DNA technology could be applied to the production of these enzymes for which a market already exists. The market in the U.S. is estimated to be $\$10 \times 10^6$/year presently and is reported to presently approach $\$100 \times 10^6$/year in Japan.

Heparin is an acidic carbohydrate that is used chemically to prevent thrombosis. It is commercially prepared from bovine lung and porcine intestinal mucosa and marketed under the names Depo-Heparin, Hepathrom, Hepriar, Lipo-Hepin, Liquaemin Sodium, and Panheprin.[2] It is used in the prophylaxis and treatment of blood clots arising from surgery, blood transfusions, dialysis, and laboratory tests.[3] Heparin not only binds coagulation factors but also gamma globulin, B-lipoprotein, histones and globins. The heparins derived from beef lung and pork lung have been shown to possess many components.[4] One commercial method of preparation uses enzymes of *Flavorbacterium heparinum* to produce a more homogenous product.[5] Amplification of these enzymes by recombinant DNA techniques could lower the cost of production of heparin by this method. More than six metric tons of heparin was marketed in the U.S. in 1976.[6]

Several possibly marketable products of recombinant DNA technology, such as hCG and the interferons, are glycoproteins. However, present recombinant DNA methods do not include techniques for performing this posttranslational modification of the peptide chains. Should methods for performing such glycosylations be developed the possibility of using recombinant DNA technology to produce marketable amounts of therapeutic glycoproteins would be a more attractive one. Recently, Dr. Creeger has reported the cloning of genes coding for glycosyltransferases from *E. coli* into *S. typhimurium*.[7] Further developments in the genetic manipulations of enzymes that transfer carbohydrate moieties may result in a technology capable of performing controlled glycosylations of specific peptides.

II.A.6. Future Prospects

1. Govoni, L.E. and Hayes, J.E. *Drugs and Nursing Implications* (3rd Ed.) 667-668 (Appleton-Century-Crofts, New York, 1978).
2. *Ibid.* Pgs. 326-329.

3. Ibid. Pgs. 326-329.
4. Jacques, L.B. Science 206 528 (1979).
5. Ibid., pg. 528.
6. Ibid., pg. 528.
7. Creeger, E.S. and Rothfield, L.I. J. Biol. Chem. 254(3) 1979).

II.B. The Use of Recombinant DNA Technology to Produce Interferons

II.B.1. Interferons

Interferons are low molecular weight glycoproteins produced by fibroblasts, leukocytes, and lymphoblasts. They are known to stimulate antiviral activity. Interferons have also been shown to cause modulations of the immune system, inhibition of cell growth, and regulation of interferon synthesis. These glycoproteins, therefore, possess potential as therapeutic agents against many viral diseases such as rabies and herpes.[1] Their effects on the immune system and cell growth render them attractive as potential antitumor agents.[2] Presently much available interferon is derived from processing human leukocytes; but this method is limited by available starting material and the susceptibility of the preparations to pathogens.[3] Methods are being developed for culturing diploid fibroblast cells in order that their interferon be harvested.[4] Drawbacks to this approach exist, but the methodology at present already produces a relatively cost-efficient interferon. Recombinant DNA technology could be applied to the production of interferon peptide sequences. However, the most active forms of the interferon are glycosylated. The present recombinant DNA methods are unable to properly glycosylate cloned gene products. mRNA coding for human interferon has been isolated by Cavalieri and co-workers.[5] This mRNA was injected into *X. laevis* oocytes and subsequently translated into a protein product.

II.B.1. Interferons

1. Galasso, G.J. and J.K. Dunnick: Interferon an antiviral Drug for use in Man, Texas Reports on Biology and Medicine 35 p. 478 (1977).
2. H. Strander, Anti Tumor Effects of Interferon and its possible use as an anti-neoplastic agent in man, Texas Reports on Biology and Medicine 35 429-435 (1977).
3. Cantell, K., S. Hirvonen, K.E. Mogensen, and L. Pyhalu: Human Leukocyte interferon: production, purification, stability and animal experiments In Vitro 35-38 (1974).
4. Vilcek, J., Havell, E.A., Kohase, A., Superinduction of interferon with metabolic inhibitors: possible mechanisms and practical applications J. Infect. Diseases 133 A22-A29 (1976).
5. Cavalieri, R.L., E.A. Havel, J. Vilcek, S. Pestka, Synthesis of human interferon by *Xenopus laevis* oocytes: Two structural genes for interferon in human cells, PNAS US 74 3287-91 (1977).

II.B.2. Vaccines Against Parasitic Diseases

Presently over a billion people, primarily in developing countries, suffer from tropical diseases and parasitic infections.[1] Hookworm, trachoma, schistosomiasis, malaria, and filariasis affect more than 1.6 billion people.[2] Millions of cattle die each year from parasitic anaplasmosis, and East Coast fever.[3] A major health problem is to engender resistance or immunity to these parasites. One possible approach to this problem is to clone gene sequences coding for the surface antigen of parasites into bacteria or yeast. The antigen could thus be amplified independently of the parasite. Purified antigen could then be used to develop vaccines against specific parasites.

The most common infectious disease in man is malaria.[4] There are four malarial parasites: *Plasmodium vivax*, *Plasmodium ovale*, *Plasmodium malariae*, and *Plasmodium falciparum*. These parasites are transmitted by mosquitos; and the subsequent infection is treated with alkaloid quinine or its derivatives. A vaccine to malaria is reportedly being developed by Cetus Corporation of California.

II.B.2. Parasitic Diseases

1. Bloom, B.R. *Nature* (279) 21-26 (1979).
2. *Ibid*.
3. *Ibid*.
4. *Burrows Textbook of Microbiology*, 21st Edition. 833 (W.B. Saunders Co., Phila., 1979).

II.B.3. Vaccine Against Infectious Diseases

Of blood donors in Western Europe and the U.S., between 3% and 5% show evidence of having been infected with hepatitis B virus (HBV).[1] In African and Asian countries, 5% and 10% of the people may be chronically infected with HBV while the majority of the adult population may have been infected.[2] The development of a vaccine against HBV would be a marketable therapeutic product. Burrell *et al*. have reported the cloning of HBV sequences, some sequences coding for HBV antigens, into plasmids taken up by *E. coli*.[3] This is a major step in the development of quantities of HBV antigen sufficient to consider the production of a vaccine against HBV.

Newcastle disease virus is a pathogen that infects poultry and can infect the human handlers of the infected poultry.[4] Diseased birds are highly infectious. The development of a vaccine against surface antigens of the virus by means of recombinant DNA methods or hybridoma technology is a possibility.

II.B.3. Vaccine Against Infectious Diseases

1. Burrell, C.J. et al *Nature* *279* (5708) 43.
2. *Ibid*., pg. 43.
3. *Ibid*., pg. 43-47.
4. *Burrows Textbook of Microbiology* (21st Edition) 833 (W.B. Saunders Co., Phila., 1979).

II.C. Products of Blood Plasma

Presently there are fourteen products of human plasma

that are licensed by the FDA for parenteral use.[1] These include albumin, immunoglobulins, coagulation factors, and anticoagulants. The currently licensed method of extracting these products from human plasma is an ethanol-water fractionation process. Fifteen establishments hold FDA licenses to fractionate plasma.[2] Eleven of these are pharmaceutical companies in the U.S. The products of one foreign company are licensed by the FDA, as are the products of four nonprofit organizations in the U.S. A conservative FDA policy of licensing procedures of fractionation, that is, inertia against licensing new procedures of production, is reportedly linked to the fact that products of human blood other than albumin and immunoglobulins can transmit diseases.[3] Hepatitis is a disease that can be transmitted by blood components and plasma derivatives.

Albumin is a single polypeptide chain of 584 amino acids and a molecular weight of 66,290.[4] It has been completely sequenced.[5] It constitutes about 60% of total serum proteins[6] and is used therapeutically as a plasma volume expander and in the correction of hypoproteinemia.[7] Situations in which it is commonly used in the clinic are shock, excessive bleeding, and severe burns. Much of the principal work on albumin extraction was done during World War II in order that battlefield requirements for a plasma expander be met. The present U.S. market for albumin is about $25 million, and the protein is marketed under the trade names Albuminar, Albumisol, Albuspan, Buminate, Pro-Bumin, and Proserum.[8] The present production of albumin requires that between two and three million liters of human plasma be fractionated annually.[9]

The only reported recombinant DNA work done on albumin to date is an analysis of mRNA populations using cDNA probes.[10] That albumin is an unglycosylated single chain peptide of 584 amino acids allows for the possibility of fermentation production of albumin by bacteria that host an artificial gene sequence coding for albumin. The ease with which albumin is presently fractionated from plasma and the conservative licensing practices of the FDA are two considerations that diminish the prospect of albumin production by recombinant DNA techniques. As the needs for albumin increase[11] and as recombinant DNA techniques become more economically feasible, albumin produced by fermentations may be introduced into the marketplace.

Blood clotting is a process that is characterized as a proteolytic cascade involving approximately twenty primary factors. Six of these factors are currently licensed as marketable products of human plasma.[12] They are Factor VIII (Antihemophilic Factor, AHF), Fibrinogen, and Factor IX complex,which is a mixture of Factors II, VII, IX, and X. Fibrinogen is a large protein consisting of three interlinked polypeptide chains that are associated with carbohydrate.[13] It is used clinically as a coagulant and its present U.S. market is less than $10 million. Factor IX complex is used as a hemostatic. Christmas disease or hemophilia B is a congenital bleeding disease characterized by the absence of an active form of Factor IX.[14] AHF is a very large glycoprotein that is absent in patients with the sex-linked recessive disease hemophilia A.[15] It is used clinically as an antihemophilic factor. The markets for AHF and Factor IX complex are each less than $10 million. These marketed coagulation factors are, then, large complex proteins associated with carbohydrate. The markets for these factors are relatively small and limited. It would be not only very difficult to produce these factors with the present recombinant DNA technology, but also economically unfeasible considering the nature of the market for these proteins.

Plasmin, or fibrinolysin, is a molecule of two polypeptide chains linked by a single disulfide bond.[16] Its main function *in vivo* is to digest the fibrin in blood clots, and it is used clinically as a thrombolytic agent. It can be activated by streptokinase or urokinase, as is discussed under "Future Prospects" in this report. Plasmin is an interesting candidate to be cloned for production of a thrombolytic agent, as are its activators streptokinase and urokinase. Plasmin molecules are approximately 10 times larger than insulin. Its production by fermentation processes would probably require that two separate chains be individually produced, isolated, and then joined.

Antibodies are proteins produced by B-lymphocytes, and plasma cells are either bound to the cell membrane or secreted as extracellular products. Antibodies bind their targets, antigens, with highly specificity; and it is this property of antibodies that renders them valuable. As foreign elements, such as bacteria and viruses, invade an organism, that organism generates antibodies specific for the surface antigens of these foreign particles. These antibodies bind the antigens, and this binding is part of the organism's immune response system.

Until recently, specific antibodies were only obtained for research by injecting purified antigens into laboratory animals. The animal was then allowed time to elaborate antibodies specific for the antigen. Blood was then collected from these animals and antibody was extracted from the blood. The final product may be quite specific, but often the "antibody" product is actually an assortment of antibodies with various specificities. This can pose problems for a researcher attempting to avoid cross-reactivities between his antibody probe and antigens for which he is not testing.

Recently a new technique has been developed allowing the isolation of highly specific and pure antibodies. Mice are injected with purified antigen; but rather than collect the animal's blood, the animal's spleen cells are collected. These cells are fused with myeloma tumor cells and then cultured *in vitro*. The fused cells, called hybridomas, are screened to determine which cells are producing the desired antibody; and selected cells may then be grown in mass culture or injected into mice to boost antibody production.

Highly specific antibodies from hybridomas can be used for such precise work as the "mapping" of cell surface proteins to determine the molecular anatomy of developing cells and tissues. Studies with lab animals suggest that monoclonal antibodies may have important medical applications. Mice treated with monoclonal antibodies have been rendered resistant to challenges by rabies virus and melanoma tumor cells.[17] Sera-Lab in Crawly Down, England, is now selling monoclonal antibodies.[18] Now several U.S. firms are marketing them and it has been reported that several U.S. firms are developing hybridoma technology or seriously considering doing so.[19] Hybritech Inc. of La Jolla, California, has announced its expectations to be marketing a diagnostic product for hepatitis infection this year.[20] At present the Salk Institute in La Jolla, California, and the Institute for Medical Research in Camden, New Jersey, have facilities for the maintenance and distribution of donated hybridomas.

Cellular growth in cell culture medium may require serum supplementation. Factors present in serum may fulfill nutritive as well as regulatory needs of the cells. One relatively well characterized constituent of serum supplement is platelet-derived growth factor (PDGF).[21] This polypeptide hormone is derived from end-cells called platelets which circulate in the plasma and perform a major function in the prevention of bleeding. The platelets are derived of megakaryocytes in the bone marrow. These megakaryotes, themselves derived of multi-potential stem cells, have been grown in tissue culture.[22] Platelets are capable of recognizing traumatized tissue, such as ruptured epithelium of vascular tissue; and upon recognition of damage, the platelets undergo morphological and biochemical changes that ultimately result in the discharge of secretory granules. The PDGF so released stimulates local connective tissue to initiate wound repair. Growth promoting activity of platelets has been demonstrated for Balb/C 3T3 cells,[23] glial cells,[24] rat epithelial mammary cells, and Swiss 3T3 cells.[26] Other tissues, as the submaxillary gland and pituitary, have been shown to possess growth factor activity for cultured fibroblasts.

PDGF is proving to be a useful tool with which to investigate the ordered biochemical events of the fibroblasts, and the pathogenesis of atherosclerosis. It may prove to be an important commercial product, the availability of which would circumvent the expensive requirement of sera for cell culture work. PDGF appears to be a polypeptide containing between 100 and 350 amino acids. The most promising means of production of PDGF would be the isolation of the factor from cell culture or the isolation of the hormone from outdated blood.

II.C. Products of Blood Plasma

1. Proceedings of the International Workshop on Technology for Protein Separation and Improvement of Blood Plasma Fractionation (Ed. by H. E. Sandberg) 5 (DHEW Publication No. (NIH) 78-1422).
2. Ibid., pg. 4.
3. Ibid., pg. 8.
4. Putnam, F.W. (ed.) The Plasma Proteins (2nd Ed.) 64 (Academic Press, New York, 1975).
5. Putnam, F.W. op. cit. pg. 65.
6. Putnam, F.W. op. cit. pg. 162.
7. Windholz, M. (ed.) The Merck Index (9th Ed.) 8212 (Merck and Co., Inc., Rahway, N.J., 1976).
8. Windholz, M. op. cit. pg. 8212.
9. Sandberg, H.E. op. cit. pg. 7.
10. Hastie, N.D. et al. PHAS 75(3) 1217-1221.
11. Sandberg, H.E. op. cit. pg. 6.
12. Ibid., pg. xi.
13. Windholz, M. op. cit. pg. 3994.
14. Putnam, F.W. op. cit. pg. 115.
15. Putnam, F.W. op. cit. pg. 114.
16. Putnam, F.W. op. cit. pg. 110.
17. Fox, J.L. Chemical and Engineering News 57(1) 17.
18. Fox, J.L. op. cit. pg. 17.
19. Fox, J.L. op. cit. pg. 17.
20. Schmeck, H.M., New York Times, January 27, 1980.
21. Scher, C.D., et al. (1979). Biochem. Biophys. Acta 560, 217-241.
22. Metcalf, D., et al., (1975). PNAS 72, 1744-1748.
23. Antoniades, H.N., et al., (1979). PNAS U.S. 76, 1809-1813.
24. Heldin, C.H. et al., (1977). Exp. Cell Res. 109, 429-437.
25. Eastman, C.E. and Sirbasku, D.A. (1978). J. Cell Physiol. 97, 17-28.
26. Kepner, N. et al., (1978). Platelets: A Multidisciplinary Approach (Gaetano, G., Garattini, S., eds.), pp. 205-212, Raven Press, New York.

II.D. Therapeutic Management of Pain

Recently Nakanishi and his co-workers have reported the construction of bacterial plasmids that contain the nucleotide sequence for bovine corticotropin-beta-lipotropin precursor.[1] Within this precursor molecule synthesized by the pituitary are amino acid sequences which constitute the poly-peptides referred to as Met-enkaphalin, α-endorphin, β-endorphin, and γ-endorphin. These small polypeptides have been collectively referred to as endogenous opiates, endogenous morphine-like substances, and endorphins.[2] It is thought that these endogenous opiates are derived from the larger precursor molecule by selective proteolytic cleavage. The construction of a plasmid containing gene sequences coding for the endogenous opiates raises the possibility of using recombinant DNA technology to produce narcotic analgesics.

Morphine sulfate, an opium derivative, is presently used for relief of severe pain.[3] It is also used as a preanesthetic medication, for relief of dyspnea of heart valve failure and pulmonary edema, and for relief of pain from myocardial infarction. It has been used for obstetric analgesia and for management of cough and gastrointestinal disorders.[3] Opiates are used to alleviate the pain suffered by terminally ill cancer patients.

Morphine is thought to act by binding "opiate" receptors distributed nonrandomly throughout the brain. The endogenous opiates have been shown to bind these receptors, with β-endorphine having a higher affinity for the receptors than morphine.[4] The endogenous opiates produce analgesia when injected intraventricularly into experimental animals.[5] However, these small peptides effect little if any analgesia when administered peripherally, presumably because the endogenous opiates have difficulty in passing from the systemic circulation into the brain.[6] This property of the endogenous opiates minimizes their attractiveness as useful therapeutic narcotics as does their addictive quality. The amino acid sequences of these peptides are short enough that gene sequences coding for them could be synthesized.

II.D. <u>Pain</u>

1. Nakanishi, S. *et al.* *PNAS* (USA) *75* 6021-6025 (1978).
2. Marx, J.L. *Science* *193* 1228-1229 (1976).
3. Govoni, L.E. and Hayes, J.E. *Drugs and Nursing Implications* (3rd Ed.) 478-481 (Appleton-Century-Crofts, New York, 1978).
4. Marx, J.L. *Science* *193* 1227-1229 (1976).
5. *Ibid.*
6. *Ibid.*

III. The Use of Recombinant DNA Technology for Basic Medical Research

III.A. Comment

The technology of genetic manipulation will benefit basic medical research as tools are developed for more sophisticated study of the regulation of genomic expression. Cloning gene sequences into bacteria will allow amplification of gene product formation allowing the study of gene products at different stages of their maturation. cDNA probes will make it possible to quantitate transcription and to quantitate the movement of transcripts throughout intracellular compartments.

III.B. The Use of Recombinant DNA Technology in the Study of DNA Replication

An understanding of the mechanisms of DNA replication is necessary for an understanding of processes involving cell division. Gametogenesis, embroyogenesis, postnatal development, tissue regeneration, immune response, and oncogenesis are processes that will be more thoroughly understood as more becomes known of DNA replication.

Eukaryotic genomes are comprised of 3 species of DNA.[1] This was determined by recombinant DNA techniques that measured the rate at which DNA sequences formed hybrid molecules. Unique sequences are present only once per haploid genome. These constitute more than half of the total DNA. 10^2 to 10^5 copies of each intermediately repetitive sequence is present. Cellular RNA is apparently transcribed from only unique and intermediately repetitive DNA called replicons.[2] There are thought to be 35,000 replicons in each diploid human nucleus.[3] Replication is initiated in the middle of each replicon and proceeds as bidirectional forks to the two termini of the replicon. Evidence has been presented that threads of DNA may become attached to the nuclear membrane; and the hypothesis has been presented that this anchoring may be a part of the replication process.[4] Many questions of eukaryotic gene replication remain unanswered: How does the DNA uncoil for replication? Why are sequences replicated early in the cycle rich in guanine and cytosine residues, while those sequences replicated later are rich in thymidine and adenine residues? Is the anchoring of DNA to the nuclear envelope a mechanism by which the DNA is ordered for replication? Is the initiation of replication ordered or random? How are the replicons related to one another?

As it has been estimated that only a fraction of the DNA in the mammalian genome codes for gene products, functions for "silent" sequences have yet to be determined. Perhaps these silent regions are, at least in part, regulatory sequences that participate in the control of gene replication, transcription, and structure.

The DNA coding sequences for gene products are sometimes interrupted internally by silent DNA. This has been found to be true of the ovalbumin gene,[5] the immunoglobulin genes,[6] and the globin genes.[7] The recent reports by Tucker[8] demonstrate how recombinant DNA techniques can be applied to study the structure and function not only of genes coding for gene products but also of silent introns as well. His group has cloned the complete nucleotide sequence of the gamma-2b immunoglobulin heavy chain of the mouse. The nucleotide sequence of the DNA was determined and consequently sequences of introns and exons of silent DNA were determined. As more intronic and exonic sequences are catalogued, similarities among silent DNA will contribute to the understanding of the functions of these sequences. The functions may include the regulation of replication, transcription, and chromosomal structure.

Medical researchers are interested to understand the mechanisms of cell division and the problems that arise from improper cell division. Trisomy 21 (Down's Syndrome, Mongoloidism) results from improper chromosome separation during gametogenesis. Embryologists may further understand the processes of differentiation, development, and teratogenesis when the mechanisms of cell division are more thoroughly understood. The processes of tissue regeneration, such as the constant renewal of skin and mucosa and the process of wound healing, are processes of cell division. Finally, understanding the regulation of cell division may provide clues to the understanding of uncontrolled and undesirable cell division as in oncogeny.

III.B. rDNA/Replication

1. Lewin, B. Gene Expression-2 193 (John Wiley and Sons, London, 1975).
2. Ibid., pg. 54.
3. Ibid., pg. 54.
4. Ibid., pg. 60.
5. Garrapin, A.C. et al. Nature 273 349 (1978).
6. Tonegawa, S. et al. PNAS (USA) 75(3) 1485-1489 (1978).
7. Gilbert, W. Nature 271 501 (1978).
8. Tucker, P.W. et al Science 206 1299-1306.

III.C. The Use of Recombinant DNA Technology in the Study of DNA Transcription

All cells of an organism, except the germ line, contain the same genetic material. Differentiation and differential gene expression must then be functions of the regulation of gene expression. An understanding of the controls of transcription would further understanding of embryogenesis, postnatal development, aging, and the body's genetic response to epigenetic factors.

The human chromosomes are made of chromatin which contains the duplex DNA structure, basic proteins called histones, more acidic proteins known as nonhistone proteins,[1] and RNA molecules of cRNA.[2] Chromatin also possesses some proteins with enzyme activity such as RNA polymerase.[3] Histones are thought to participate in the structuring of DNA. Nonhistone proteins and cRNA may participate in regulating the accessibility of DNA to nonchromatin factors such as transcriptional enzymes. One model of eukaryotic gene expression postulates four types of DNA sequences that constitute a positive control system.[4] Sensor sequences are activated by specific agents, hormone-protein complexes for example. This activation initiates transcription of an integrator sequence. The resulting transcript then binds to a receptor sequence. This binding initiates transcription of the structural gene and the gene product is subsequently produced. The transcript may be a sequence containing a tRNA sequence, an rRNA sequence, or a heterogeneous nuclear RNA (hnRNA) to be processed into messenger RNA (mRNA).

Recombinant DNA technology allows the medical researcher the opportunity to study the structure of gene sequences. An understanding of the structure of sensor, integrator, receptor, and structural sequences in turn may provide information helpful towards an understanding of the functioning of these sequences. The genes coding for the hemoglobins offer an opportunity to study concerted gene regulation as well as the process of differentiation. Cloning of human alpha, beta, gamma, and epsilon globin genes has been accomplished.[5,6,7] In the course of normal development, the human ceases to produce gamma globin in the perinatal period. Gamma and alpha globin are the constituents of fetal hemoglobin of HbF.[8] Adult hemoglobin, HbA, is comprised of alpha globin and beta globin. The hereditary persistence of fetal hemoglobin,

HPFH, is a benign abnomality that affects about 1 in a thousand American Blacks.[9] The normal developmental changes in hemoglobin gene expression do not occur in individuals with persistent fetal blood types. An understanding of the structure and function of the globin gene family is essential for further understanding of the developmental changes in gene expression of the globins. The globin genes are genes characterized by intronic sequences. Recombinant DNA technology can facilitate the isolation and sequencing of globin gene sequences allowing investigation of the intronic and exonic sequences that may participate in differential gene expression.

The immunoglobulins also constitute a multigene family characterized by intronic sequences. Of the immunoglobulins, much remains to be understood of gene maturation, gene product formation, and lymphocyte differentiation. Recently several groups have reported cloning gene sequences coding for various heavy and light, constant and variable regions of the immunoglobulins.[10,11,12,13,14,15] Evidence suggests that variable and constant gene sequences, separated by long stretches of DNA, may be brought together at the DNA level so that transcription of the joined sequences results in a transcript carrying information from two previously separated gene sequences. Recombinant DNA technology is being used to study the structure of the immunoglobulin genes in order to better understand the mechanisms by which lengths of DNA are looped out to produce a gene product coding for a specific antibody. Such work may further understanding of agammaglobulinomia, the transient hypoglobulinemias, the development of autoimmunities, and leukemia.

Steroid hormones are regulators of gene transcription. They are thought to bind protein receptors in the cytosol, and the steroid-peptide complex is then translocated into the nucleus. The hormone-receptor complex then interacts with chromatin in such a way as to promote the transcription of specific gene sequences. Do gene sequences under the control of any one specific steroid have structural similarities that might help account for their response to the steroid-receptor complex? Ovalbumin and lysozyme are proteins whose synthesis in the hen oviduct is controlled by estrogen. The gene sequences coding for these two proteins have been cloned.[16,17,18,19] The ovalbumin gene has been found to be internally interrupted by lengths of silent DNA. Studies of other gene sequences under estrogenic control may reveal similar intronic or exonic sequences that may further the understanding of the mechanism by which a steroid alters gene transcription. More broadly, the mechanisms by which the adrenals, gonads, and placenta exert their effects on development, homeostasis, behavior, and the body's genetic response to epigenetic factors will be more thoroughly understood.

III.C. rDNA/Transcription

1. Lewin, B. Gene Expression-2 101 (John Wiley and Sons, London, 1975).
2. Ibid., pg. 357.
3. Ibid., pg. 101.
4. Ibid., pg. 368-369.
5. Wilson, J.T. et al. Nuc. Acids. Res. 5(2) 563-581 (1978).
6. Little, P. et al. Nature 273 640-643 (1978).
7. Proudfood, N.J. PNAS 76(11) 5435-5439 (1979).
8. Harris, H. The Principles of Human Biochemical Genetics 138 (North-Holland Pub. Co. Amsterdam, 1975).
9. Willimas, R.A. (ed.) Textbook of Black-Related Diseases 223 (McGraw Hill, N.Y., 1975).
10. Seidman, J.G. et al. Nature 271 582 (1978).
11. Tonegawa, S. et al. PNAS (USA) 74(8) 3518-3522 (1977).
12. Tonegawa, S. et al PNAS (USA) 75(3) 1485-1489 (1978).
13. Tucker, P.W. et al. Science 206 1303-1306 (1979).
14. Tucker, P.W. et al. Science 206 1299-1303 (1979).
15. Hozumi, N. et al. Nuc. Acids Res. 5(6) 1779-1799 (1978).

16. Humphries, P. et al. Nuc. Acids Res. 4(7) 2389-2405 (1977).
17. Woo, S.L.C. et al. Biochem 16(26) 5670-5675 (1977).
18. Garapin, A.C. et al. Nature 273 349 (1978).
19. Baldacci, P. et al. Nuc.Acids Res. 6(8) 2667-2680 (1979).

III.D. The Use of Recombinant DNA Technology in the Study of Posttranscriptional Processing

The RNA transcribed from gene sequences is in some cases altered by deletions or additions of sequences before it is transported to the cytoplasm. It may, then, be that the gene expression of developing and differentiated cells is under a form of negative control or censorship. That is, sequences not destined to be expressed in their mature form by a cell may be transcribed but not fully processed. An understanding of posttranscriptional processing is necessary to thorough understanding of development, homeostasis, aging, and pathogenesis.

Messenger RNA is a product of posttranscriptionally processed hnRNA, a larger precursor molecule that is the direct transcript of a gene sequence.[1] The processing that hnRNA undergoes may include the deletion of internal sequences, the degradation of sequences at the 5' end, the addition of polyadenylate (polyA) sequences to the 3' end of the molecule, and capping. The polyA tail is thought to function in mRNA conservation, nucleocytoplasmic transport, and the regulation of the molecule's translation. Two classes of mRNA are found outside the nucleus. These are membrane bound mRNA's thought to code for proteins that will be eventually exported out of the cell and cytosolic mRNA's thought to code for proteins to be retained within the cell. mRNA may be sequestered from translation as in embryonic cells in which mRNA is produced and sequestered. The cells are, then, preprogrammed in anticipation of a future stage of development.

45S RNA is transcribed from genes associated with the nucleolus.[2] Portions of the 45S molecule are then methylated. Intranucleolar processing of the 45S RNA involves degrading the unmethylated sequences of the molecule. The 28S and 18S sequences that remain after this degration are then transported into the cytoplasm to participate in ribosome assembly.

4S tRNA and 5S rRNA are thought to be transcribed in their mature form rather than as partial sequences of larger sequences.[3]

The presence of masked messenger RNA's (mmRNA) is established in echinoderms and amphibia.[4] The existence of untranslated mRNA species in mammalian myoblasts has been reported,[5] this untranslated mRNA being a ribonucleoprotein particle. cDNA probes specific for particular messages will be used to quantitate amounts of preprogrammed message cells. Such probes will also be used to determine the relative amounts of a particular message present in the various intracellular compartments - nuclear, polysomal, and sequestered. The process of myogenesis will be more fully understood as more details of mRNA processing become known. Myosin is a protein present in muscle fibrils. Studies of dystrophic mouse muscle suggest that the myosin of the diseased muscle is embryonic in nature.[6] The embryonic form of myosin is more vulnerable to proteolysis than the adult form. Therefore, a perturbation of the genetic program for myosin expression may participate in the pathogenesis of muscular dystrophies. Such a perturbation could exist at any of several levels of gene regulation. A knowledge of the structures of the myosin genes will contribute to the understanding of the myogenesis as well as further understanding of transcriptional control and posttranscriptional processing. Recently the cloning of sequences coding for myosin light chain into *E. coli* by Deshpande and co-workers has been reported. This will allow for an examination of the structure of a myosin gene. Such study may reveal mechanisms by which the developmental expression of this gene is regulated.

The looping out of internal sequences of DNA prior to the transcription of immunoglobulin genes has been discussed previously in this paper. Tucker and coworkers report that the primary mRNA transcript of a cloned gene region of mouse gamma immunoglobulin heavy chain has internal sequences of RNA the functions of which are not known.[7,8] It is speculated that these regions are excised by ribonuclease activity, and that the remaining sequences are then spliced together, the result being the mature mRNA. This then is another example of recombinant DNA technology being used to discover new details of the structure and function of eukaryotic genes and their products.

III.D. rDNA/Posttranscription

1. Lewin, B. Gene Expression-2 271 (John Wiley and Sons, London, 1975).
2. Ibid., pg. 298.
3. Ibid., pg. 329.
4. Epstein, C.J. Biol. Reprod. 12 82-105 (1975).
5. Buckingham, M.E. et al. Biochem Soc. Trans. 5 474-477 (1977).
6. Stracher, A. et al. Ann. N.Y. Acad. Sci. 317 349-355 (1979).
7. Tucker, P.W. et al. Science 206 1303-1306 (1979).
8. Tucker, P.W. et al. Science 206 1299-1303 (1979).

III.E. The Use of Recombinant DNA Technology for Gene Mapping

The positions of more than 200 human genes have been assigned to particular chromosomes.[1] These include the sequences coding for immunoglobulins, interferons, alpha-1-antitrypsin, compement components, coagulation factors, hemoglobins, procollagen, and hCG. Knowledge of the particular gene may prove to be valuable to those attempting to clone, excise, alter, transfer, or otherwise manipulate a gene sequence. Some hemoglobin thalasemmias are results of point mutations. If normal genes could be cloned into the bone marrow cells of thalasemmic patients and these cells then reimplanted into the *in vivo* bone marrow, the expression of proper globin in otherwise thalasemmic patients might be accomplished. Such genetic surgery could benefit homozygotes for HbS (sickle cell anemia), the birth frequency of homozygotes among U.S. Blacks being 1:625.[2] Such genetic surgery may be facilitated by a sophisticated knowledge of the human chromosomal gene map. Radiolabelled cDNA probes or radiolabelled transcripts of cloned gene sequences could be used to hybridize with human chromosomal DNA. Autoradiography could then be used to determine the exact location or locations of the hybridized sequence(s) coding for the particular gene product.

III.E. rDNA/Gene Mapping

1. Ruddle, F.H. in *Human Genetics* (ed. by Armendares, S. and Lisker, R.) 200 (Excerpta Medica, Amsterdam, 1977).
2. Willimas R.A. (ed.) *Textbook of Black-Related Diseases* 223 (McGraw Hill, N.Y. 1975).

III.F. Diagnosis and Treatments of Inborn Errors of Metabolism

Hope has been expressed that genetic engineering may aid in the prevention, diagnosis, prognosis, and treatment of inherited disease. In 1900, congenital malformations constituted 3.3% of all causes of infant mortality in the U.S.[1] In 1979, the U.S. infant mortality rate was 13.6 per 1,000 live births, a new low.[2] The infant mortality rate due to congenital anamolies was 2.5 per 1,000 live births or about 18% of all infant mortalities.[3] As infant mortality from infectious disease declines, congenital malformations are accounting for a greater percentage of total infant mortality. There are over 1,000 recognized single gene disorders in man.[4] In one hospital 15% of pediatric beds were reportedly occupied by genetically determined diseases or patients whose problems had significant genetic influence.[5] 7% of the beds in another hospital were so occupied.[6]

Since not all genes are expressed by the fetus in utero, not all gene disorders can be determined by *in vitro* culture of cells derived from aminocentesis. The defective genes of 45 inborn errors are expressed in utero and, of these, assays are available to detect 35.[7] Techniques may be developed in which cDNA probes are used to determine the presence or absence of normal alleles of particular genes in fetal cells. Early detection of missing or ineffective genes would allow for termination of the conceptus or preparation of early treatment of the metabolic deficiency.

III.E. Diagnosis

1. Williams, R.A. *Textbook of Black-Related Diseases* pg. 32 (McGraw Hill, New York, 1975).
2. Wegman, M.E. *Pediatrics* 64(6) 835-842 (1979).
3. *Ibid.*, pg. 835-842.
4. Kirkman, H.N. *Ped. Ann.* 7 54-61 (1978).
5. Williams, R.A. *Textbook of Black-Related Diseases* pg. 32 (McGraw Hill, New York, 1975).
6. *Ibid.*, pg. 32.
7. Emery, A.E.H. in *Human Genetics* (ed. by Armedares, S. and Lisker, R.) 170 (Excerpta Medica, Amsterdam, 1977).

IV. Case Study on the Role of Applied Genetics with Regards to Gentamicin

IV.A. Introduction

According to Perlman,[8] in a period of a little over 35 years, over 40,000 antibiotics have been discovered. Of these, only a small percentage (about 2%) are produced on a commercial scale. In addition, some, including 23

---

[8] D. Perlman, 1979. In "Microbial Technology" (H.J. Peppler & D. Perlman, ed.) Chap. 8. P. 241-280. Academic Press, NY.

penicillins, 13 cephalosporins, dihydrostreptomycin, clindamycin, 6 tetracyclines and 2 rifamycins are prepared by combination of microbial synthesis and chemical modification. Only chloramphenicol and pyrrolnitrin are manufactured on a commercial scale by synthesis. The total chemical synthesis for penicillins, cephalosporins, polymycins, gramicidin, tetracyclines, kanamycins, griseofulvin and cycloheximide has been completed on a laboratory scale.

Listed in Table IVa. are the 90 commercially important antibiotics and presented in Table IVb. the antibiotics produced by major manufacturers are listed.

The 1975 value of pharmaceutical preparations, based on data from a variety of sources, was summarized by SRI, Menlo Park, California.[1] In the class of compounds called antiinfectives and antiparasitics, the total value of shipments for 1975 is reported as 1,165 millions of dollars. Systemic antibiotics accounted for 791 millions of dollars or 67% of this total. The penicillins accounted for 244.5 millions of dollars or 30% of the systemics (21% of all the antifectives and antiparasitics). Graphical representation of production statistics and values in dollars per kilogram, as summarized by SRI from 1965 to 1979, is presented in Fig. IVa.

The role that fermentation plays in production of the penicillins and semisynthetic penicillins is summarized in Fig. IVb. In addition, a precursor to the very important class of antibiotics, the cephalosporin series, can be produced via the penicillin fermentation, i.e., 7-amino-desacetoxy cephalosporinic acid. Mutation and selection processes have had a major impact in improving penicillin production yields. For example, the yield of benzylpenicillin

---

[1]SRI. 1977, In "Chemical Economics Handbook" Medicinal Products, 566.100A.

Table IVa: Some Antibiotics Produced on a Commercial Scale

Antibiotic	Microbial source	Antibiotic spectrum[b] G+	G-	My	AF	AT	Other	Chemical type	Therapeutic or other use	Some commercial source[c]
Amphomycin	*Streptomyces canus*	+						Peptide	Topical	42
Amphotericin B	*Streptomyces nodosus*				+			Polyene	Oral or parenteral	(58)
Avoparcin[a]	*Streptomyces candidus*	+	+					Glycopeptide	Animal growth promotant	(4)
Azalomycin F[a]	*Streptomyces hygroscopicus*	+	+		+				Topical (AF)	54
Bacitracin	*Bacillus subtilis*	+						Peptide	Topical; also animal growth promotant	7,(33),42,45 (47),(48)
Bambermycins	*Streptomyces bambergenesis*	+						Phosphoglycolipid	Animal growth promotant	(31)
Bicyclomycin[a]	*Streptomyces sapporonensis*	+	+						Topical	27
Blasticidin S[a]	*Streptomyces griseochromogenes*				+			Nucleoside	Agricultural (AF)	35
Bleomycins	*Streptomyces berticillus*	+	+			+		Peptide	Parenteral (AT)	44a
Cactinomycin[a]	*Streptomyces chrysomallus*	+				+		Peptide	Parenteral (AT)	22
Candicidin B	*Streptomyces griseus*				+			Polyene	Topical	42,(47)
Candidin[a]	*Streptomyces viridoflavus*				+			Polyene	Topical	(47)
Capreomycin	*Streptomyces capreolus*			+				Peptide	Parenteral	21

○ = U.S. Companies

(continued)

Table IVa: (continued)

Antibiotic	Microbial source	Antibiotic spectrum[b] G+	G-	My	AF	AT	Other	Chemical type	Therapeutic or other use	Some commercial source[c]
Cephalosporins	Cephalosporin C is produced by Cephalosporium acremonium and converted to 7-ACA, which is used for prep. of semisynthetic cephalosporins	+	+					Peptide	Oral and parenteral	5,12,14,18 23,26,27, 29,41,45, 49,52,58, 59,60
Chloramphenicol	Streptomyces venezuelae; commercial manufacture is by chemical synthesis	+	+				Rickettsia		Oral or parenteral	46
Chromomycin A_3[a]	Streptomyces griseus	+				+			Parenteral (AT)	59
Colistin	Bacillus colistinus		+					Peptide	Parenteral	8,37,39,52
Cycloheximide	Streptomyces griseus				+				Agricultural (AF)	35,61
Cycloserine	Streptomyces orchidaceus			+				Amino acid	Parenteral (TB)	5,33
Dactinomycin	Streptomyces antibioticus					+		Peptide	Parenteral (AT)	44
Daunorubicin	Streptomyces peucetius					+			Parenteral (AT)	23,52
Doxorubicin	Streptomyces peucetius					+			Parenteral (AT)	23
Enduracidin[a]	Streptomyces fungicidus	+		+				Peptide	Animal growth promotant	59
Erythromycin	Streptomyces erythreus	+						Macrolide	Oral and parenteral; animal growth promotant	1,5,13,14, 18,19,21, 41,48,49, 52,61

(continued)

Table IVa: (continued)

Antibiotic	Microbial source	Antibiotic spectrum[b] G+	G-	My	AF	AT	Other	Chemical type	Therapeutic or other use	Some commercial source[c]
Fortimicins[a]	*Micromonospora olivoasterospora*	+	+					Aminoglycoside	Parenteral	39
Fungimycin[a]	*Streptomyces coelicolor* var. *aminophilus*				+			Polyene	Topical	42
Fusidic Acid[a]	*Fusidium coelcineum*	+						Steroid	Parenteral	40
Gentamicins	*Micromonospora purpurea*	+	+					Aminoglycoside	Parenteral	14,15,56
Gramicidin A	*Bacillus brevis*	+						Peptide	Topical	42,47,62
Gramicidin J[a] (S)	*Bacillus brevis*	+						Peptide	Topical	43
Griseofulvin	*Penicillium griseofulvum*				+			Spriolactone	Oral	14,29,34, 40,44$_n$,59
Hygromycin B	*Streptomyces hygroscopicus*	+	+				Helminths	Aminoglycoside	Animal feed suppl.	11,41,59
Josamycin[a]	*Streptomyces narbonesis*	+						Macrolide	Oral and parenteral	55
Kanamycins	*Streptomyces kanamyceticus*	+	+	+				Aminoglycosides	Parenteral	12,14,43, 44a,52
Kasugamycin[a]	*Streptomyces kasugaensis*	+	+					Aminoglycoside	Agricultural antibacterial	8,43,55
Kitasatamycin[a]	*Streptomyces kitasatoensis*	+						Macrolide	Oral and parenteral	60
Lasalocid	*Streptomyces hazelensis*	+					Coccidia	Polyether	Agricultural use as coccidiostat and growth promotant	22
Lincomycin	*Streptomyces lincolnensis*	+							Oral and parenteral	61
Lividomycin[a]		+	+							36,52

(continued)

Table IVa: (continued)

Antibiotic	Microbial source	Antibiotic spectrum[b] G+	G-	My	AF	AT	Other	Chemical type	Therapeutic or other use	Some commercial source[c]
Macarbomycins[a]	*Streptomyces phaechromogenes*	-						Phosphoglycolipid	Animal growth promotant	43
Mepartricin[a]					+			Polyene	Topical	57
Midecamycin[a]		-						Macrolide	Oral and topical	43
Mikamycins[a]	*Streptomyces mitakaensis*	+						Peptide	Animal growth promotant	8
Mithramycin	*Streptomyces* species	+		+		+			Parenteral (AT)	48
Mitomycin C	*Streptomyces caespitosus*	+	+		+				Parenteral	12,39
Mocimycin[a]		+							Animal growth promotant	28
Monensin	*Streptomyces cinnamonensis*	+					Coccidia	Polyether	Animal growth promotant	41
Myxin	*Chromobacterium iodinum* plus chemical modification				+			Phenazine	Topical in veterinary use	32
Neocarzinostatin	*Streptomyces carzinostaticus*	+				+		Peptide	Parenteral (AT)	
Neomycins	*Streptomyces fradiae*	+	+					Aminoglycoside	Oral and topical	6,7,11,14, 23,42,44a, 47,48,52, 53,58,59,61
Nosiheptide[a]	*Streptomyces actinosus*	+						Peptide	Animal growth promotant	52
Nisin	*Streptococcus cremoris*	+						Peptide	Food preservative	
Novobiocin	*Streptomyces niveus*	+							Oral and topical	43,52,61
Nystatin	*Streptomyces noursei*				+			Polyene	Oral and topical	4,14,15,52, 58

(continued)

Table IVa: (continued)

Antibiotic	Microbial source	G+	G-	My	AF	AT	Other	Chemical type	Therapeutic or other use	Some commercial source[c]
		Antibiotic spectrum[b]								
Oleandomycin	*Streptomyces antibioticus*	+						Macrolide	Oral and parenteral	48
Paromomycin	*Streptomyces rimosus*	+	+				Protozoa	Aminoglycoside	Oral	46
Penicillin G[d]	*Penicillium chrysogenum*	+						Peptide	Oral and parenteral; also as animal growth promotant	2,3,6,8,9, 10,12,14, 22,25,28, 29,31,41, 43,44,45, 48,52,58, 60,63
Penicillin V[d]	*Penicillium chrysogenum*	+						Peptide	Oral	1,2,3,8,10, 11,12,21, 25,28,29, 31,41,43, 48,52,58, 60,63
Streptomycin	*Streptomyces griseus*	+	+	+				Aminoglycoside	Parenteral; use in agriculture to control bacteria	6,14,28,29, 43,44,48, 52,58,63
Dihydrostreptomycin	*Streptomyces humidus* (Also chemical reduction of *streptomycin*)	+	+	+				Aminoglycoside	Parenteral	6,14,28,43, 45,48,52,63
Tetracyclines										
Chlortetracycline	*Streptomyces auerofaciens*	+	+				Rickettsia	Tetracycline	Parenteral and oral; animal growth promotant	4,5,14,20, 23,49,52, 50,57,

(continued)

Table IVA: (continued)

Antibiotic	Microbial source	Antibiotic spectrum[b] G+	G-	My	AF	AT	Other	Chemical type	Therapeutic or other use	Some commercial source[c]
Tetracyclines										
6-Demethyl-7-chlortetracycline	Streptomyces aureofaciens	+	+				Rickettsia	Tetracycline	Parenteral and oral	4,18,49,52
5-Hydroxy-tetracycline	Streptomyces rimosus	+	+				Rickettsia	Tetracycline	Parenteral and oral animal growth promotant	5,11,14,15, 18,28,34, 38,48,49,50
Tetranactin[a]	Streptomyces flaveolus	+					Insects	Macrotetralide	Insecticide	17
Thiopeptin[a]	Streptomyces tateyamensis	+						Peptide	Animal growth promotant	27,35
Thiostrepton	Streptomyces azureus	+						Peptide	Animal growth promotant	58
Tobramycin[a]	Streptomyces tenebrarius	+	+					Aminoglycoside	Parenteral	11,41
Trichomycin[a]	Streptomyces hachijoensis				+		Trichomonas	Polyene	Topical	27
Tylosin	Streptomyces fradiae	+					PPLO	Macrolide	Veterinary; animal growth promotant	21,41
Tyrothricin	Bacillus brevis	+	+					Peptide	Topical	10,42,47,62
Tyrocidine	Bacillus brevis	+	+					Peptide	Topical	42
Validamycin[a]	Streptomyces hygroscopicus var. limonesis	+	+					Aminoglycoside	Parenteral	59
Vancomycin	Streptomyces orientalis	+						Glycopeptide	Parenteral	41
Variotin[a]	Paecilomyces varioti				+				Topical	40,44a
Viomycin	Streptomyces floridae	+	+	+				Peptide	Parenteral	48,60
Virginiamycin	Streptomyces virginiae	+						Peptide	Animal growth promotant	51

(continued)

Table IVa: (continued)

[a] Not distributed in the United States (1978).

[b] G+, gram-positive bacteria; G-, gram-negative bacteria; My, mycobacteria; AF, antifungal; AT, antitumor.

[c] Key to sources (from Perlman, 1979): O is U.S. firm

1. Abbott Laboratories (U.S.A.)
2. Aktiebolaget Astra (Sweden)
3. Aktiebolaget Fermenta (Sweden)
4. American Cyanamid (U.S.A.)
5. Ankerfarm S.p.A. (Italy)
6. Antibioticos S.A. (Spain)
7. Apothekernes Laboratorium fur Special praeparater A/S (Norway)
8. Banyu Pharmaceutical Company (Japan)
9. Beechan Pharmaceutical Company (England)
10. Biochemie GmbH (Austria)
11. Biogal (Hungary)
12. Bristol-Myers Company (U.S.A.)
13. Chemibiotic Ltd. (Ireland)
14. China National Chemicals Import and Export Corporation (China)
15. Chinoin (Hungary)
16. Chong-Kun-Dong Corporation (S. Korea)
17. Chugai Pharmaceutical Company (Japan)
18. Companhia Industrial Produtora de Antibioticos S.A.R.L. (CIPAN) (Portugal)
19. Compania Espanola de la Penicillina y Antibioticos S.A. (Spain)
20. Diaspa S.p.A. (Italy)
21. Dista Products Ltd. (England)
22. Farbenfabriken Bayer AG (W. Germany)
23. Farmitalia S.p.A. (Italy)
24. Fermentfarma S.p.A. (Italy)
25. Fermion Oy (Finland)
26. Fervet S.p.A. (Italy)
27. Fujusawa Pharmaceutical Company (Japan)
28. Gist-Brocades n.v. (Holland)
29. Glaxo Laboratories Ltd. (England)
30. Gruppo Lepetit S.p.A. (Italy)
31. Hoechst AG (W. Germany)
32. Hoffmann-LaRoche Inc. (U.S.A.)
33. IMC Chemicals Group, Inc. (U.S.A.)
34. Imperial Chemical Industries Ltd. (England)
35. Kaken Chemical Company (Japan)
36. Kowa Company (Japan)
37. Kayaku Antibiotics Research Company Ltd. (Japan)
38. Krakow Pharmaceutical Works ("Polfa")(Poland)
39. Kyowa Hakko Kogyo Company (Japan)
40. Leo Pharmaceutical Products (Denmark)
41. Eli Lilly Company (U.S.A.)
42. H. Lundbeck and Company (Denmark)
43. Meiji Seika Kaisha Ltd. (Japan)
44. Merck and Company, Inc. (U.S.A.)

44a. Nihon Kayaku Company (Japan)

45. Novo Industri A/S (Denmark)
46. Parke, Davis and Company (U.S.A.)
47. S.B. Penick and Company (U.S.A.)
48. Pfizer, Inc. (U.S.A.)
49. Proter S.p.A. (Italy)
50. Rachelle Laboratories Inc. (U.S.A.)
51. Recherche et Industrie Therapeutique (Belgium)
52. Rhone-Poulenc S.A. (France)
53. Roussel-UCLAF (France)

(continued)

Table IVa: (continued)

54. Sankyo Company Ltd. (Japan)
55. Sanraku Ocean Company, Ltd. (Japan)
56. Schering Corporation (U.S.A.)
56a. Shionogi & Co., Ltd. (Japan)
57. Societa Prodotti Antibiotici S.p.A. (Italy)
58. E.R. Squibb and Sons (U.S.A.)
59. Takeda Chemical Industries Ltd. (Japan)
60. Toyo Jozo Company Ltd. (Japan)
61. The Upjohn Company (U.S.A)
62. Wallerstein Laboratories, Inc. (a division of G.B. Fermentation Industries, Inc., USA)
63. Wyeth Laboratories (U.S.A.)

d. Both penicillins G and V are used as sources of 6-aminopenicillanic acid, an intermediate in the synthesis of semisynthetic penicillins, also.

Table IVb: Antibiotics Produced by Major Manufacturers

Company and Location	Chemical
Abbott Laboratories Chemical and Agricultural Products Division North Chicago, IL	erythromycin; erythromycin lactobionate; erythromycin phosphate; erythromycin stearate; fumagillin
Universal Enterprises, Ltd., subsidiary Barceloneta, PR	erythromycin
American Cyanamid Company Agricultural Division Hannibal, MO Willow Island, WV	chlortetracycline
Lederle Laboratories Division Fine Chemicals Department Pearl River, NY	chlortetracycline hydrochloride; demeclocycline hydrochloride; minocycline hydrochloride; nystatin; tetracycline hydrochloride
American Home Products Corporation Wyeth Laboratories, Inc., division Paoli, PA	ampicillin; ampicillin, sodium; dicloxacillin, sodium; nafcillin, sodium; penicillin G, benzathine; penicillin G, potassium; penicillin G, procaine; phenoxymethylpenicillin, potassium
Anderson Development Company Adrian, MI	m-cresyl acetate
Beecham, Inc. Beecham Pharmaceuticals — U.S.A., division Beecham Laboratories, division Piscataway, NJ	amoxicillin; ampicillin; carbenicillin; cloxacillin, sodium; dicloxacillin, sodium; methicillin, sodium; ticarcillin
Bristol-Myers Company Bristol Laboratories, division Syracuse, NY	amoxicillin trihydrate; amikacin; ampicillin, sodium; ampicillin trihydrate; cephapirin; cephapirin, sodium; cloxacillin, sodium; dicloxacillin, sodium; hetacillin; hetacillin, potassium; kanamycin sulfate; methicillin, sodium; oxacillin, sodium; phenethicillin, potassium; phenoxymethylpenicillin, potassium; rolitetracycline nitrate; tetracycline; tetracycline hydrochloride; tetracycline phosphate complex
CPC International Inc. Penick Corporation Newark, NJ	candicidin; gramicidin; neomycin; neomycin palmitate; tyrocidine hydrochloride; tyrothricin
Diamond Shamrock Corporation Biosciences Unit Nutrition and Animal Health Division Louisville, KY	n.a.

(continued)

Table IVb: (continued)

Company and Location	Chemical
International Minerals & Chemical Corporation Chemical Group Terre Haute, IN	bacitracin (medicinal and nonmedicinal); bacitracin, zinc (medicinal and nonmedicinal)
International Rectifier Corporation Rachelle Laboratories, Inc. subsidiary Long Beach, CA	chloramphenicol; chlortetracycline; doxycycline; oxytetracycline hydrochloride; tetracycline hydrochloride
Eli Lilly and Company Tippecanoe Laboratories Lafayette, IN	cephalexin monohydrate; cephaloridine; cephalothin; cephalothin, sodium; erythromycin; hygromycin B; penicillin G, potassium; penicillin G, procaine (medicinal and nonmedicinal); phenoxymethylpenicillin; phenoxymethylpenicillin, potassium; tylosin; vancomycin
Merck & Co., Inc. Merck Chemical Manufacturing Division Danville, PA	penicillin G, procaine (nonmedicinal)
Pfizer Inc. Chemicals Division Brooklyn, NY	bacitracin; dihydrostreptomycin; oleandomycin; viomycin
Groton, CT	doxycycline; methacycline; neomycin; neomycin sulfate (nonmedicinal); oleandomycin; penicillin G; penicillin G, potassium; penicillin G, procaine; penicillin G, sodium; penicillin O, sodium; phenethicillin, potassium; phenoxymethylpenicillin; polymyxin B sulfate; streptomycin (medicinal and nonmedicinal); tetracycline; troleandomycin; viomycin
Terre Haute, IN	oxytetracycline (nonmedicinal); streptomycin (nonmedicinal)
SmithKline Corporation Smith Kline Chemicals Division Swedeland, PA	cefazolin; cephradine
Squibb Corporation E.R. Squibb & Sons, Inc., subsidiary U.S. Pharmaceutical Company Operations	amphotericin B; neomycin; nystatin (medicinal and nonmedicinal); penicillin G, potassium; penicillin G. procaine (nonmedicinal); thiostrepton
Ersana, Inc. subsidiary Humacao, PR	cephradine
Squibb Manufacturing, Inc., subsidiary Humacao, PR	amphotericin B; nystatin
Trade Enterprises, Inc., subsidiary Humacao, PR	amoxicillin; ampicillin; ampicillin trihydrate; dihydroampicillin
The Upjohn Company Fine Chemical Division Kalamazoo, MI	erythromycin; erythromycin stearate; lincomycin hydrochloride; neomycin sulfate; novobiocin; novobiocin, calcium; novobiocin, sodium; spectinomycin dihydrochloride; spectinomycin sulfate; tetracycline hydrochloride
Upjohn Manufacturing Company, subsidiary Arecibo, PR	clindamycin; clindamycin palmitate; clindamycin phosphate; lincomycin

Fig. IVa : From SRI. 1979, In "Chemical Economics Handbook" Antibiotic Chemicals, S65.100A

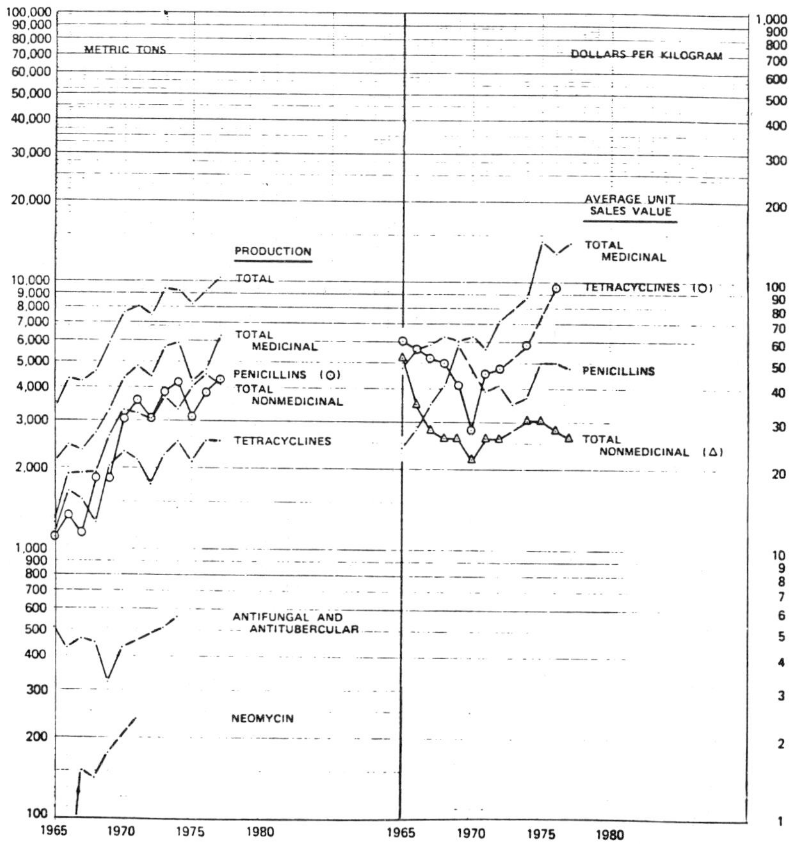

Legend for Fig. IVa

TETRACYCLINES: Effective for the treatment of diseases caused by gram-positive and gram-negative bacteria, and a variety of microbial infections. Chlortetracycline and oxytetracyclines are used extensively as animal feed supplements to protect animals against disease and hasten weight gain.

NEOMYCIN: Neomycin is active against gram-positive, gram-negative, and mycobacteria, and effective in the treatment of bacterial infections of the skin and digestive tract. Data reported for 1955-1964 are for human or veterinary use only; 1965-1977 data are for all uses, both medicinal and nonmedicinal.

TOTAL MEDICINAL: Data include total antibiotic production for human and veterinary medicine in the treatment of diseases caused by microorganisms (bacteria, yeasts, fungi). In addition to their action against microorganisms, certain antibiotics have demonstrated some effectiveness in the treatment of cancers, and have been approved for such use in the United States: Blenoxane (bleomycin sulfate), Adriamycin (doxorubicin hydrochloride), Mutamycin (mitomycin), Cosmegan (dactinomycin), and Mithracin (mithramycin).

TOTAL NONMEDICINAL: Data include total antibiotic production for nonmedicinal uses. The largest market is in animal feeds, to stimulate growth and enhance feed utilization (as well as to treat and prevent diseases) in cattle, swine, and fowl. The size of this market is certain to decrease in the next five years due to FDA plans to severely restrict or ban use of penicillin and tetracyclines in animal feeds.

Fig. IVb: Flowchart for production of Penicillins from SRI 1979. In "Chemical Economics Handbook."

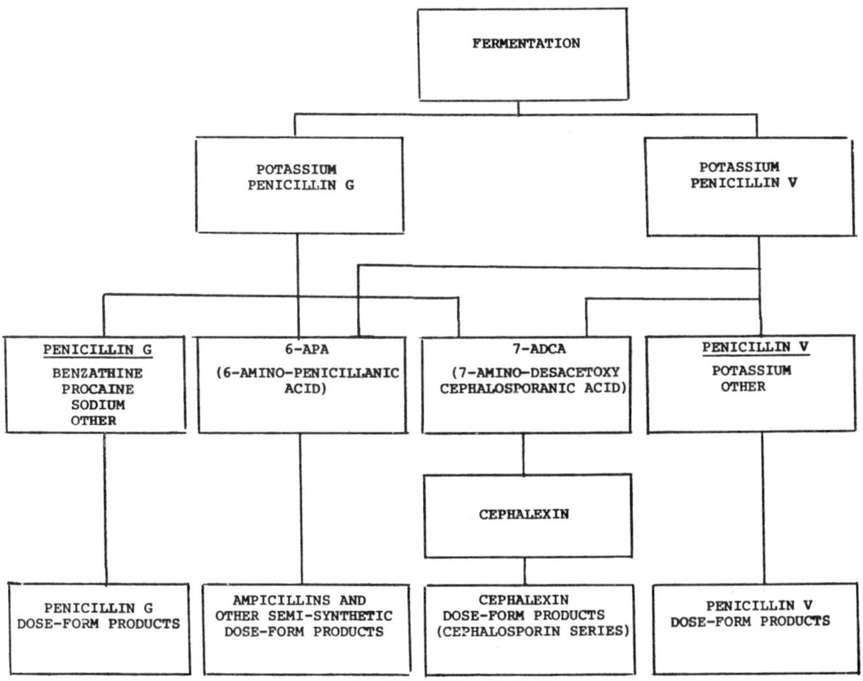

was improved to well over 20 mg/ml by sue of strains selected from Penicillium chrysogenum NRRL 1951, a culture producing only 120 μg/ml.[1]

Elander and Chang[2] have summarized large scale strain improvement programs concerned with the induction, selection and utilization of superior penicillin-producing variants of P. chrysogenum, as carried out in government, university and industrial laboratories over the past 30 years. (Most recent strain improvement work has been done in industrial laboratories).

Obviously during this period of intensive selection work, attention was focused on searching for strain characteristics that may be correlated with high yield. A correlation between increased productivity and reduced sporulation and growth was noted by the Wisconsin group.[2] The genetic basis for this correlation has not yet been established, for lack of fundamental information. Very little is known about the genetics of penicillin production.

Heterokaryon experiments have determined that penicillin yield is determined by nuclear genes, and there is evidence from both heterokaryons and parasexual recombinants that several different genes are involved. Future improvements in penicillin titers may be brought about by more extensive application of the parasexual cycle in breeding.[3,4]

Many factors influence the economics of fermentation processes including raw material costs, cell yields, batch vs. continuous operation and other types of biochemical processing pro-

---

[1] D. Perlman, 1979. In "Microbial Technology" (H.J. Peppler & D. Perlman, ed.) Chap. 8. P. 241-280. Academic Press, NY.

[2] Elander, R.P. & Chang, L.T., 1979. In "Microbial Technology (H.J. Peppler & D. Perlman, ed.), Vol. II., Chap. 12 p. 243-302, Academic Press, NY.

[3] Merrick, M.J. 1976. Proc-Int. Symp. Genet. 2nd Microorg. 2nd 1974. pp. 229-242.

[4] Hopwood, D.A., 1977. Dev. Ind. Microbiol. 18, 9-21.

cedures. Swartz[1] has recently reported on a detailed economic analysis of penicillin G manufacturing costs in establishing priorities for fermentation process improvement.

The study is based upon information from two sources. The first is Dr. Joseph Lein of Panlabs, who has provided results of pilot and production scale data for series of cultures developed at Panlabs. In addition, A.G. McKee & Co. provided a breakdown of manufacturing and production costs for a new bulk, nonsterile penicillin G production facility. A four year strain improvement program tripled average potency (gm free acid/liter) from 8.0 in 1973 to 26.3 in 1976. However, as is clearly pointed out by Swartz,[1] the important concept from a processing point of view is productivity (kg product produced per total liter total tank capacity per day). Thus cycle time becomes accounted for. Using the above definition, strain improvement increased productivity to 3.2 (kg produced/1000 liters total tank capacity/operations day) from an initial value of 0.72, a more than four-fold increase.

Another key set of data required, and one that can be a measure of strain improvement, is overall yield from carbon and energy sources and from other major raw materials such as precursor molecules (in penicillin G - sodium phenylacetate) or nitrogen sources. Over the course of the Panlabs program, the yield value was doubled from 0.05 gmK-Pen G/gm glucose equivalent to 0.12.

As pointed out by Swartz,[1] since yield from oxygen and heat can be expected to follow yield from carbon and energy sources, considerable economies can result from improvements in yield. Since in the example given by Swartz approximately 30% of production costs are raw material costs, improvement in yield should have a significant economic impact.

---

[1] Swartz, R.W. 1979. In "Annual Reports on Fermentation Process" (D. Perlman) Vol. 3 p. 75-110. Academic Press. N.Y.

Furthermore, Swartz[1] then summarized approaches to penicillin G process improvement, for the details of which the reader is referred to the original article. The most desirable is the genetic approach, since in his case study an improvement of 18% in penicillin productivity would result in the greatest saving. Another important way to decrease cost is to obtain higher cell mass to utilize cheaper raw materials.

The difficulty with the genetic approach is one of predictability. There is as yet no basis for expecting an increase of 18% in penicillin production via genetic means. Efforts to improve penicillin yields by recombination will increase in the antibiotic industry for three main reasons: the availability of divergent genetic lines from independent firms such as Panlabs, Inc.; the applicability of protoplast fusion techniques to *P. chrysogenum*; the economic incentive.

Although from a historical point of view significant accomplishments have been made regarding penicillin, it was thought that some perspective may be obtained by conducting a more detailed analysis on gentamicin. The reasons include the following:

a) it is a different class of antibiotic;

b) the drug is a mixture of three gentamicins, and thus one would expect strain improvement approaches to be very challenging;

c) the patents on gentamicin are due to expire soon.

IV.B. Aminocyclitol Antibiotics

The discovery of streptomycin by Waksman in 1944 began a new era, extending chemotherapy to the Gram-positive *Mycobacterium tuberculosis* and to the Gram-

1 Swartz, R.W. 1979. In "Annual Reports on Fermentation Process" (D. Perlman) Vol. 3 p. 75-110. Academic Press. N.Y.

negative bacteria. Following on the heels of the discovery of this broad-spectrum antibiotic were the announcements of the discoveries of neomycin (1949), paromomycin (1956), kanamycin (1957), and gentamicin (1963), among others. Tobramycin was later discovered (1967), followed by the semi-synthetic amikacin in 1972.

These aminocyclitol antibiotics (also called "aminoglycosides") are typical secondary metabolites. They are low molecular weight natural products which are bactericidal and inhibit protein synthesis via an interaction with the 30S ribosomal subunit. They can be divided into two categories depending on whether the aminocyclitol moiety is 2-deoxystreptamine (DOS) or some other cyclic alcohol. The latter group includes streptomycin and spectinomycin which posses aminocyclitol moieties of streptidine and actinamine,respectively. By far, the largest category is the DOS-containing group which can be further divided into a subgroup in which DOS is substituted on 2 adjacent hydroxyls (4,5-disubstituted group), a subgroup in which the substitution is on 2 non-adjacent hydroxyls (4,6-disubstituted group), and a third subgroup in which DOS is monosubstituted.

4,5-disubstituted	4,6-disubstituted	Monosubstituted
Neomycins	Kanamycins	Destomycins
Paromomycins	Tobramycin	Hygromycin B
Lividomycins	Seldomycins	Apramycin
Ribostamycin	Gentamicins	
Xylostasin	Sisomicin	
Butirosins	Verdamicin	

Streptomycin, discovered in 1944, was the first clinically useful broad-spectrum antibiotic. Its use immediately produced a remarkable decrease in mortality from tuberculosis. It is bactericidal in action. Today it is used mainly in combination with other antibiotics, although it is still the drug of choice for tularemia. Two of its drawbacks are ototoxicity and potential nephrotoxicity. In addition, resistance has developed in the clinic. Neomycin, which was discovered in 1949, extended the antibacterial spectrum of the aminoglycosidic aminocyclitols to *Proteus* and was very effective against staphylococci, the tubercle bacillus, *Klebsiella* and *Escherichia coli*. However, its use has been limited to topical applications because of its greater toxicity as compared to streptomycin.

In 1957, the introduction of kanamycin brought renewed interest in the aminocyclitol antibiotics. It became the drug of choice against serious Gram-negative infections, being less toxic than neomycin and possessing a wider antibacterial spectrum than streptomycin. Kanamycin has found use against refractory urinary infections,especially those due to staphylococci, *Escherichia coli* and gonococci, and in newborn infants with sepsis. For years it was the most effective drug known for *Klebsiella*, *Enterobacter*, *Proteus* and *Shigella*. It is also effective against the tubercle bacillus and *Neisseria*. On the debit side, kanamycin lacks good activity against *Pseudomonas*, *Bacteroides*, streptococci, enterococci, pneumococci and clostridia. Other disadvantages are ototoxicity, potential nephrotoxicity and resistance development. These deficiencies provided an opening for gentamicin development in the 1960's.

Aminocyclitol antibiotics are highly polar and relatively insoluble in lipids. As a result, they are minimally absorbed from the gut and are usually administered intramuscularly. Despite this lack of oral activity and potential toxicity (see below), they are extremely important chemotherapeutic agents because of their broad-spectrum activity. Gram-negative bacteria are responsible for two-thirds of all infections which result in hospitalization. Aminocyclitol antibiotics are also valuable because they exert a synergistic effect with other antibiotics against Gram-positive bacteria and are useful in tuberculosis. As a result, 13 aminocyclitol antibiotics are on the market with total world sales in 1978 of a half billion dollars.

IV.C. Importance of Gentamicin

The most important liability of aminocyclitol antibiotics is their vestibular and auditory toxicity and the necrosis of kidney tubules. One of the important advances made by the discovery of gentamicin was its lower degree of ototoxicity and nephrotoxicity, as compared to the earlier streptomycin and kanamycin. Indeed, nephrotoxicity is not a real problem today with gentamicin; however, extreme care must be exercised to avoid ototoxicity. A second advantage was its greater activity, compared to other aminocyclitol antibiotics, against the Gram-positive *Staphylococcus aureus*. A very major advance was its activity against the Gram-positive genera *Proteus* and *Pseudomonas*. Indeed it was the first commercial antibiotic with activity against *Pseudomonas aeruginosa*. It is also the most active aminocyclitol antibiotic against enterobacteria. Gentamicin is the drug of choice for septicemia and serious infections of the central nervous system, urinary tract, gastrointestinal tract, skin, bone, and soft tissue (including burn-related infections) caused by Gram-negative bacteria. Another advantage was its activity against bacterial strains which developed resistance to streptomycin and kanamycin. The resistance was found to occur via reduced uptake of the antibiotic (permeability mutations), ribosomal changes (target mutations) and by R-factor (plasmid) coded enzymatic inactivation of the antibiotics by phosphorylation, nucleotidylylation and acylation. Although hospital strains have developed some resistance to gentamicin, this has occurred slowly and is not yet a major problem.

Another advance made by Schering Corporation's discovery of gentamicin in 1963 was the utility of actinomycetes (i.e. filamentous bacteria) other than Streptomyces for antibiotic production. Up until 1963, all important antibiotics from antinomycetes came from the genus Streptomyces. With Schering's discovery of the usefulness of Micromonospora as an antibiotic producer, the whole field of antibiotic discovery has shifted to genera other than Streptomyces. Indeed, this probably was the reason that companies such as Bristol-Myers (i.e., the Bristol-Banyu Research Laboratories in Tokyo) and Parke-Davis looked even at non-filamentous bacteria for new antibiotics, resulting in both companies discovering the aminocyclitol antibiotic, butirosin. Although butirosin did not become commercially important, its α-hydroxy-γ-aminobutyric acid side chain became very important. When the side-chain was chemically attached to kanamycin A by Bristol-Banyu chemists, the important semi-synthetic antibiotic, amikacin, resulted. This indeed is the best aminocyclitol antibiotic known today for use against aminocyclitol-resistant pathogens.

IV.D. Aminocyclitol Antibiotic Fermentations

Fermentations of aminocyclitol antibiotics are usually carried out in fermentors of 20,000 to 100,000 gallons capacity with agitation and aeration. Fermentation media usually contain a combination of soybean meal as protein source and dextrin, starch and/or glucose as carbon and energy sources. Temperature is usually controlled at 25° to 37°C and fermentationslast from 3 to 6 days. In some cases, titers reach 15g/l. The antibiotic is found both in the extracellular broth and associated with mycelia.

The bound aminocyclitol antibiotic can be removed by acidification. Isolation is carried out by filtration, followed by extraction with an acidic ion-exchange resin. Elution with acid or base gives the crude antibiotic. Purification is achieved by chromatography, crystallization or precipitation as a salt.

IV.E. Aminocyclitol Antibiotics at Schering

The discovery of gentamicin was announced by Schering workers in 1963 and it has been used clinically ("Garamycin, Schering-Plough") in the U.S. since 1969. The drug is produced by *Micromonospora purpurea* as a complex ("gentamicin C complex") containing gentamicin C_1, C_{1a}, and C_2 plus a large number of trace components. The three members of the C complex contain the unusual sugar, garosamine, in glycosidic linkage with the C-6 hydroxyl of DOS. Commercial preparations of the drug usually contain 28%C_1, 35%C_{1a}, and 37%C_2. The gentamicin C complex differs from the earlier-discovered neomycins, kanamycins and paromomycins in its possession of both C-methyl and N-methyl substituents. In the laboratory, the complex many be produced at 28°C in shaken flasks (280 rpm) with a medium of 5% potato dextrin, 0.5% glucose, 3.5% soybean meal, 0.7% $CaCO_3$ and 0.00013% cobalt chloride; the duration is 6 to 7 days.

Biosynthetic experiments at Schering have shown D-glucose to be incorporated into the antibiotic but D-glucosamine to be a better precursor. The methyl groups arise from the methyl of L-methionine.[1] The biosynthetic path involves two cobalt-dependent steps, both of which are C-methylations. The N-methylations do not appear to be cobalt-dependent.

The importance of gentamicin to Schering (now Schering-

1. B.K. Lee, R.T. Testa, G.H. Wagman, C.M. Liu, L. McDaniel and C. Schaffner. 1973. J. Antibiot. 26:728

Plough) cannot be overemphasized. In 1974 they were the 5th largest U.S. antibiotic producer even though they marketed only a single antibiotic. Their worldwide sales of gentamicin were about 155 million dollars,which represented approximately 19% of total sales. In 1978, sales of gentamicin reached 226 million dollars, 21% of total sales. Even more dramatic is the fact that gentamicin alone provides about 45% of the company's earnings.

Schering-Plough has continued their thorough examination of the genus *Micromonospora* for new antibiotics. In 1970 they reported the discovery of a dehydro derivative of gentamicin C_{1a} produced by *Micromonospora inyoensis*[1]. This compound (antibiotic 6640) was named sisomicin. It is more active than the gentamicin C complex against certain Gram-negative organisms, but twice as toxic in acute toxicity tests in animals. Chronic toxicity in cats is about 30% greater than gentamicin. Although Bayer in West Germany is commercializing this compound, Schering-Plough has placed their emphasis on a semi-synthetic derivative, Sch 20569 (1-N-ethylsisomicin), which is called netilmicin[2]. Netilmicin is somewhat more active *in vitro* than sisomicin and is active against most sisomicin-resistant strains. Although more acutely toxic than gentamicin in mice, netilmicin was surprisingly less nephrotoxic in rats[3], and showed less chronic toxicity in cats. Because of its high potency, broad spectrum and lowered chronic toxicity in laboratory animals, it has reached clinical tests in man. Recent clinical reports from various investigators indicate, however, that

---

1. M.J. Weinstein, J.A. Marquez, R.T. Testa, G.H. Wagman, E.M. Oden, and J.A. Waitz. 1970. J. Antibiot. 23:551.

2. G.H. Miller, G. Arcieri, M.J. Weinstein, and J.A. Waitz. 1976. Antimicrob. Agents Chemother. 10:827.

3. F.C. Luft, M.N. Yum, and S.A. Kleit. 1976. Antimicrob. Agents Chemother. 10:845.

its low toxicity advantage in animals is not carrying over into man[1] (also compare Snydman *et al.*, 1979[2], with Miller *et al.*, 1976[3]). It is doubtful that netilmicin will be pushed as a commercial antibiotic if there is no safety advantage in man as compared to gentamicin.

Another Schering-Plough discovery is verdamicin, a natural analogue of sisomicin produced by *Micromonospora grisea* and first described in 1975[4]. Its antibacterial spectrum *in vitro* and *in vivo* is similar to gentamicin and sisomicin, although it is 1.5 to 3 times as active as gentamicin *in vivo* against Gram-positive and Gram-negative bacteria with the exception of *Pseudomonas*. Against *Pseudomonas*, verdamicin is equivalent to gentamicin. Along with the higher activity, however, is a higher (almost twice) acute toxicity of verdamicin, about the same as sisomycin. However, verdamicin appears less ataxic in cats than gentamicin.

Mention should also be made of the discovery of non-aminocyclitols such as rosaramicin (previously called rosamicin) by the Schering-Plough group[5]. This is a macrolide antibiotic, produced by *Micromonospora rosaria*, which is thought to have some advantages (e.g. better Gram-negative activity) over erythromycin. Another discovery has been that of the everninomycins by *Micromonospora carbonaceae*[6,7].

1. C.E. Cox. 1979. Curr. Ther. Res. Clin. Exp. 25:603.
2. D.R. Snydman, F.P. Tally, S.H. Landesman, M. Barza and S.L. Gorbach. 1979. Antimicrob. Agents Chemother. 15:50.
3. G.H. Miller, G. Arcieri, M.J. Weinstein, and J.A. Waitz. 1976. Antimicrob. Agents Chemother. 10:827.
4. M.J. Weinstein, G.H. Wagman, J.A. Marquez, R.T. Testa and J.A. Wartz. 1975. Antimicrob. Agents Chemother. 7:246.
5. G.A. Wagman, J.A. Waitz, J. Marquez, A. Murawski, E.M. Oden, R.T. Testa and M.J. Weinstein. 1972. J. Antibiot. 25:641.
6. G.H. Wagman, G.M. Luedemann and M.J. Weinstein. 1965. Antimicrob. Agents Chemother. 1964. p. 33.
7. A. K. Ganguly and S. Szmulewicz. 1975. J. Antibiot. 28:710

IV.F. Use of Genetics for Development of Gentamicin

Genetic procedures, especially mutagenesis, have had a major effect on process development costs of fermentation products. Indeed mutation has been the main factor responsible for the survival of the fermentation industry in its competition with the chemical industry for manufacture of natural compounds. In various fermentation processes, hundred to thousand-fold increases in broth titer have been achieved over the years by mutation. Indeed, mutation to higher productivity is the cheapest way of improving the economics of a biological process. The most common method used to obtain high-yielding mutants is that of treating a population of uninucleate vegetative cells or spores with a mutagenic agent until a certain "desired" degree of killing is obtained, plating out the survivors and testing each resulting colony or a randomly selected group of colonies for product formation in flasks. Another common procedure is to select obvious mutants with altered colony morphology or color for testing in flasks. Other sub-populations that are sometimes chosen for liquid fermentations are auxotrophic mutants, their revertants, mutants resistant to toxic end-products, to toxic precusors, or to toxic analogues of precursors, revertants of non-producing mutants and mutants producing higher antibiotic concentrations on agar plates.[1] All of these screening methods are rather indirect since no successful selection method has ever been devised for producers of secondary metabolites (i.e. metabolites not required for growth). Furthermore, they all depend on effective mutation and the ability to separate the mutated nuclei from all the other nuclei in the population.

During the early years of gentamicin at Schering,

---

1. A.L. Demain. 1973. Adv. Appl. Microbial. 16:177.

strain improvement did not play a major role. As a result, gentamicin has been one of the most expensive antibiotic products on the market (ca.$3.00/gram of bulk material and $50.00/gram in parenteral form). The major reason appears to be the lack of techniques available in the 1960's for mutation of filamentous and poorly sporulating bacteria such as *Micromonospora purpurea*. This lack of a uninucleate cell's cellular stage (i.e. individual vegetative cells or spores), coupled with extremely low growth rates, made strain improvement by mutagenesis very difficult. Even if a mutation to higher antibiotic production had occurred in such a multinucleate filament, it probably was masked by the other parental titer nuclei in the strand. To make the mutated nucleus the only viable nucleus would probably have required excess mutagenesis which could cripple the vigor of the culture. Apparently strain improvement efforts at Schering had yielded only marginal results into the early 1970's. Facing the end of gentamicin patent protection in the U.S.A. in May 1980 and the major importance of this single antibiotic to the profits of the corporation, Schering management opted for outside genetic help. Fortunately, a new company named Cetus Corporation, had just been formed in 1971 in Berkeley, California. Cetus was dedicated to improving existing biological processes and devising new biological processes with an emphasis on genetics. The raison d'etre of this new company was to fill a vacuum created by the lack of government funding of academic work in applied biology. For years, there had been major public funding of applied physics, applied mathematics, applied electronics and applied chemistry, and as a result, there was no real gap between academic discoveries in these fields and their translation into industrial processes. The same situation did not exist in biology and the founders of Cetus saw this need and moved into the gap. According to Dr. Peter Farley, President of Cetus, a contract was signed in 1973 between Schering-Plough and Cetus to genetically develop a number of Micromonospora organisms. At this point, the young Cetus Corporation (1) devised methods of properly handling Micromonospora and (2) used their automated large-scale screening capabilities. The aspect of proper handling is exemplified by U.S. Patent 3,950,224 ("Method of Cloning Filamentous Organisms"; inventors: Calvin B. Ward and Robert D. Bruner) awarded to Cetus on April 13, 1976. As stated in this patent "For sporulating microorganisms...., it is usually possible to use a single spore as the colony-forming unit in genetic experiments. Since most spores are presumably uninucleate, it is relatively easy to obtain genetically pure colony-forming units". In non-sporulating filamentous organisms "each colony-forming unit may, therefore, contain many tens or hundreds of nuclei which may or may not be genetically identical. If a filamentous culture is treated with a mutagen to introduce genetic inhomogeneity into the culture, any genetic variations having significantly better productivity in the culture are typically masked." In the patented procedure, the filaments are fragmented by homogenization with or without glass beads and those of the fragments which are less than a predetermined size are obtained by filtration through a filter of known pore size (1 to 6 microns). These small fragments are of such a size that colonies grown from such filaments will be predominantly genetically pure. These fragments are then subjected to normal mutagenesis procedures. The patent points out that such homogenization can produce a 100-fold increase or more in the viable count of Micromonospora purpurea (the species which makes gentamicin). Since all of the data in the patent deal with Micromonospora purpurea, it is obvious that this work was done during the Cetus/Schering-Plough relationship and was important in the improvement of gentamicin production. The second aspect of the Cetus effort dealt with automated large-scale screening. This is best understood by quoting from Lewin[1] (1978). "For the first five years life at Cetus was something of a struggle. Because the pharmaceutical industry was receptive to the idea of exploiting modern biology, Cetus chose to concentrate its initial efforts in this sector. Research labs in the industry spend much of their time searching for new strains of microorganisms that, for instance, will give better yields of antibiotics. By conventional techniques, the best equipped labs can screen up to 500 colonies a week, and since mutation rates giving rise to the hoped-for new strain may be as low as one in 10,000, or even one in ten million, such a search can take a very long time. With Don Glaser's inspiration, Cetus developed an automated system that can zip through 50,000 tests at a time... And so, slow though it was to get under way, the screening technique established itself as the essential financial support of the company."

It is of interest that although there had not been significant funding of the screening aspect of microbiology, government funding of a very basic project led in an indirect way to Cetus' ability in this respect. For a number of years, the government funded the development of Dr. Donald Glaser's "Dumbwaiter", a huge instrument (40 feet long and two stories high) which is capable of detecting "needles in haystacks", i.e. detecting a few mutant colonies of a particular type out of millions of colonies. The Dumbwaiter, described by Bylinsky[2], "is

1. R. Lewin. 1978. New Scientist, October 5, p. 18.
2. G. Bylinsky. 1974. Fortune, February, p. 96.

designed to process automatic surveillance and screening in accordance with instructions programmed in a computer." This instrument was used by Glaser and his group to isolate many new thermosensitive DNA replication mutants in *Escherichia coli* despite their extremely infrequent occurrence[1]. The machine is in Glaser's laboratory at the University of California at Berkeley and is available to academic investigators from other institutions but has not been used by industry. Although the machine was never used by Cetus, Glaser is a founder of Cetus and Chairman of its Scientific Advisory Board. Many of Glaser's students and post-doctoral associates have obtained permanent employment at Cetus. Although the Dumbwaiter is many times more costly than anything that Cetus could afford in the early 1970's, the know-how and experience of its builders became available at Cetus and was put to work on the Schering-Plough project. This episode shows how government funding can stimulate innovations in industry even though it was rather indirect in this case. At any rate, Cetus' yield improvements had a significant impact on corporate planning at Schering-Plough as evidenced by the following report of the Schering-Plough-New York Analyst's Meeting which occurred on December 7, 1977, at the end of the 4-year Cetus-Schering relationship. In this report, Ronald M. Nordmann of Blyth Eastman Dillon & Co., Inc. summarizes the remarks of Richard J. Bennett, President and Chief Operating Officer of Schering-Plough, about the consequences of the Cetus yield improvements. "Fermentation yield improvements were characterized as 'dramatic' in recent years reducing the company's need for additional manufacturing capacity. For example, the elimination of the proposed plant in Ireland

1. C.G. Sevastopoulos, C.T. Wehr and D.A. Glaser. 1977. Proc. Natl. Acad. Sci. USA, 74:3485.

resulted from a decision that yield improvements in existing facilities would cover near term future needs in Europe. Schering-Plough has been working with Cetus Corporation, a California-based operation, in an attempt to achieve yield improvements. We believe that Schering invested $1-2 million annually with Cetus Corporation on this project. We understand that the relationship with Cetus Corporation has now been severed, and the company expects to do this work in-house." Since a new fermentation plant costs $60-80 million, the knowledge and performance of Cetus presumably was of great importance to Schering-Plough.

IV.G. Genetic Aspects of Gentamicin Research Unrelated to Strain Improvement

The Sterling-Winthrop Research Institute of the Sterling Drug Company has used a genetic approach to produce a promising new derivative of gentamicin. This approach is based on the work of Shier, Rinehart and Gottlieb[1], and has been termed mutational biosynthesis or mutasynthesis. Shier, Rinehart and Gottlieb produced a new group of aminoglycosidic aminocyclitol antibiotics which they called hybrimycins. They started with *Streptomyces fradiae* which produces neomycins B and C, a pair of antibiotics possessing DOS. They mutated this strain to non-production and showed that it did not produce neomycin unless supplemented with DOS. Such mutants are called "idiotrophs"[2]. Thus their mutant was blocked in the biosynthesis of the DOS. By substituting streptamine for DOS in the medium, they obtained hybrimycins A1 and A2; by substituting with *epi*streptamine in the medium,

1. W.T. Shier, K.L. Rinehart, Jr. and D. Gottlieb. 1969. Proc. Natl. Acad. Sci. USA 63:198.
2. K. Nagaoka and A.L. Demain. 1975. J. Antibiot. 28:627.

they obtained hybrimycins B1 and B2. The hybrimycins were less active than the corresponding neomycins.

The Sterling-Winthrop workers prepared a DOS-idiotroph of *Micromonospora purpurea* and when it was fed streptamine, it made the novel compound Win 42122-2 (deoxygentamicin)[1]. At present, deoxygentamicin has been tested in animals and although somewhat less active than gentamicin, it is considerably less toxic and thus appears to have a higher therapeutic index.[2] According to Dr. Paul E. Came, Director of Microbiology at Sterling-Winthrop, the compound will soon enter clinical tests in man to determine whether it is a safer gentamicin.

The Schering-Plough workers have also carried out mutational biosynthesis but chose to work on the sisomicin-producer, *Micromonospora inyoensis*.[3] At least 9 new sisomicins have been prepared by feeding idiotrophs streptamine, 2,5-dideoxystreptamine, 5-*epi*streptamine, 2-deoxy-5-*epi*streptamine, (+)-2-deoxy-N-methylstreptamine, and (+)-2,5-dideoxy-N-methylstreptamine.[4] From statements made by Schering-Plough scientists at scientific meetings, it is doubtful that any of these will be commercialized.

These *Micromonospora inyoensis* idiotrophs have also been used at Schering-Plough to provide knowledge of the biosynthetic pathway to sisomicin. The feeding and incorporation of pseudotrisaccharides and pseudodisaccharides into sisomicin has allowed the development of a proposed

1. D. Rosi, W.A. Goss and S.J. Daum. 1977. J. Antibiot. 30:88.
2. P.E. Came, J.R. O'Connor, R.A. Dobson, R.B. Wagner and R.J. Fabian. 1979. Antimicrob. Agents Chemother. 16:813.
3. R.T. Testa, G.H. Wagman, P.J.L. Daniels and M.J. Weinstein. 1974. J. Antibiot. 27:917.
4. P.J.L. Daniels and D.F. Rane. 1979. In "Microbiology - 1979". D. Schlessinger, ed., pp. 314-317. American Society for Microbiology., Washington D.C.

pathway of sisomicin biosynthesis.[1] In addition, the Schering-Plough scientists have used blocked mutants of *Micromonospora purpurea* to develop a proposed pathway of gentamicin biosynthesis.[2] In this case, they fed proposed intermediates such as gentamicin A and observed production of the gentamicin C complex.

In 1974, workers at the Kyowa Hakko Kogyo Company in Japan disclosed their discovery of antibiotic XK-62-2 (called "sagamicin") produced by *Micromonospora sagamiensis* along with gentamicins C_1, C_2 and C_{1a}.[3] They also reported the production of sagamicin by *Micromonospora purpurea* and *Micromonospora echinospora*. In 1975, Daniels *et al.* of Schering-Plough reported that mutant J1-33 of *Micromonospora purpurea* no longer produced gentamicin C_1 and C_2, but did produce C_{1a} and a new component C_{2b} which was identical to sagamicin.[4] Daniels *et al.* (1975) show that the *in vitro* spectrum of sagamicin (C_{2b}) is qualitatively very similar to the gentamicin C complex and close examination of their Table 3 shows sagamicin to be considerably more active than gentamicin against *Staphylococcus aureus*, some *Escherichia coli* strains, *Pseudomonas aeruginosa*, *Klebsiella aerogenes* and *Proteus mirabilis*.[5] We have been told that Kyowa Hakko is proceeding along with the commercialization of sagamicin.

---

1. R.T. Testa and B.C. Tilley. 1975. J. Antibiot. 28:573.
2. B.K. Lee, T.L. Nagabhushan, R.G. Condon, G. Shimonaski, M.G. Kalyanpur, M. Patel and J.A. Waitz. 1979. Antimicrob. Agents Chemother. 16:589.
3. R. Okachi, I. Kawamoto, S. Takasawa, M. Yamamoto, S. Sato, T. Sato and T. Nara. 1974. J. Antibiot. 27:793.
4. P.J.L. Daniels, C. Luce, T.L. Nagabhushan, R.S. Jaret, D. Schumacher, H. Reimann and J. Ilavsky. 1975. J. Antibiot. 28:35.
5. M.J. Weinstein, G.M. Luedemann, E.M. Oden, G.H. Wagman, J.P. Rosselet, J.A. Marquez, C.T. Coniglio, W. Charney, H.L. Herzog and J. Black. 1963. J. Med. Chem. 6:463.

IV.H. Public Support of Gentamicin Development

We have failed to find any evidence that the development of gentamicin or the application of genetics to this effort has ever been supported by U.S. government funding.

V. Case Study on Interferon

V.A. INTRODUCTION

Interferons are a class of proteins produced by the body in response to a viral infection. An important component of the body's own natural defense system, interferons inhibit the ability of viruses to replicate in protected cells. In clinical and animal studies interferons have shown promise in the treatment of viruses, such as hepatitis, herpes and rabies, and more recently have also been shown to be beneficial in the treatment of cancer. To date only several hundred patients have been treated with interferon because of the very limited supplies available. The major source of the interferon for these trials thus far has been the "buffy cell coats" from blood donations. Before any conclusions as to the efficacy of interferon in treating cancer or viral disease can be reached, many more thousands of patients will have to undergo clinical trials. This is essential for obtaining knowledge of proper dosages and schedules, dangers and side effects.

V.B. HISTORICAL BACKGROUND

Since the 1930's it has been known that the infection of one virus in animals and in cell cultures interferes with the subsequent infection of a second virus. Investigations into this phenomenon led Isaacs and Lindenmann (1) to the

(1) A. Isaacs and J. Lindenmann: Virus Interference I. The Interferon, *Proc. R. Soc. Ser B.* *147*:258 (1957).

discovery of interferon. They found that the first virus did not simply block other viruses from access to the cells, but rather induced the cells to secrete a substance into the surrounding fluids, making uninfected cells resistant to subsequent viral challenge (figure 1).

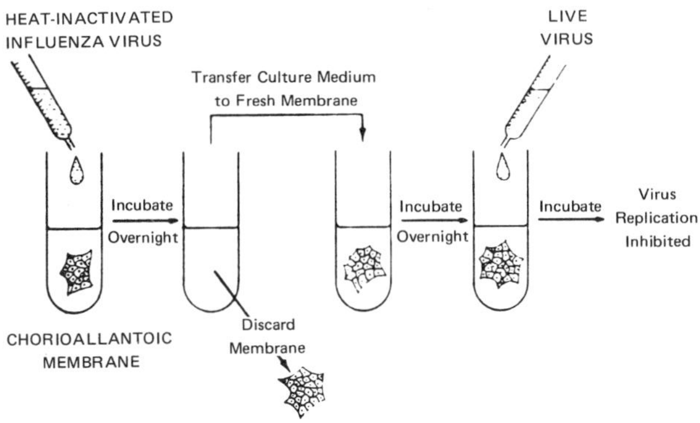

Fig. 1. The discovery of interferon. A schematic representation of the original experiment of Isaacs and Lindenmann (1957).

It has been established that this substance, interferon, is a relatively low molecular weight protein, but remarkably non-specific in its action, i.e. that interferon induced by one virus is effective against a broad range of other viruses. However, interferon is species specific in that interferon produced by one species generally stimulates antiviral activity only in cells of that or closely related animal species. For example, human interferon protects human and monkey cells, but not mouse or chicken cells.

There is substantial evidence that interferon plays an important role in the natural recovery from viral infections. It is perhaps the earliest of the known host responses, operative within hours after infection. Probably the most compelling evidence comes from work in animal studies. Gresser et al (1) have shown that mice treated with antisera against mouse interferon are killed by injections containing amounts of herpes simplex virus several hundred fold less than the LD50 for mice not treated with antisera. Additionally, in man there is evidence strongly supporting a causal relationship between the interferon system and natural recovery from many established viral infections (see figure 2).

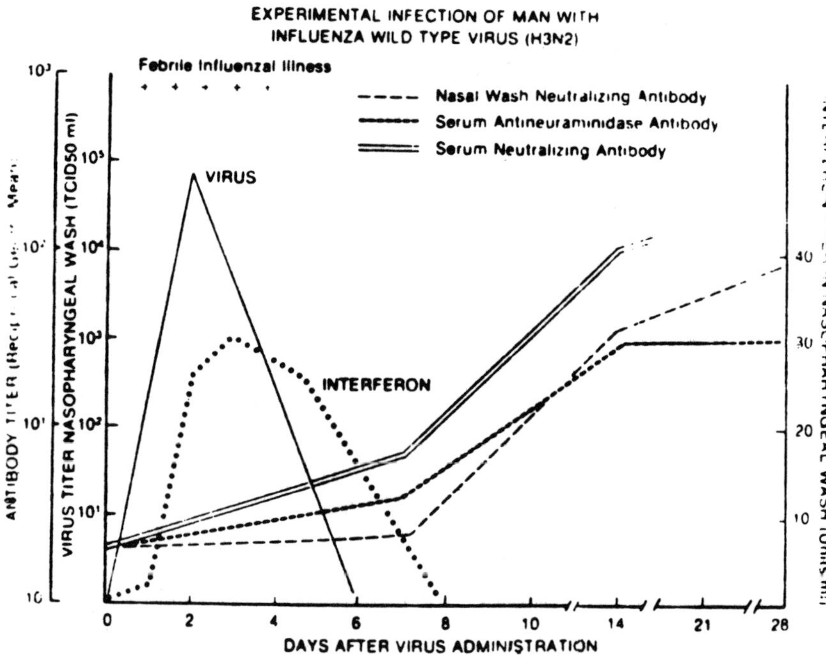

Figure 2. Production of virus, interferon, and antibody during experimental infection of man with influenza virus (from a study by Drs. B. Murphy, E. Chaihub, S. Nusinoff, J. Kasel, C. Uhlendorf, S. Baron, and R. Chanock, National Institutes of Health).

---

(1) J. Gresser, M.G. Tovery, C. Maury, and M.T. Bandu: ..le of Interferon in the pathogenesis of virus diseases in mice as demonstrated by the use of antiinterferon serum II. Studies with Herpes simplex, Maloney sarcoma, vesicular stomatitis, Newcastle Disease and Influenza viruses. J. Infect. Dis. 135: 846-851 (1977)

In recent years it has become increasingly clear that interferon produces a number of cellular alterations in addition to the antiviral state. Most notable are modulations of the immune system, inhibition of cell growth, and regulation of interferon synthesis. The medical potential of interferon has also received considerable attention. In addition to the possibility of treating a number of viral diseases, interferon may prove to be a useful antitumor agent. Figure 3 is an attempt to summarize the interferon system, from the induction and synthesis of interferon to the production of the antiviral state.

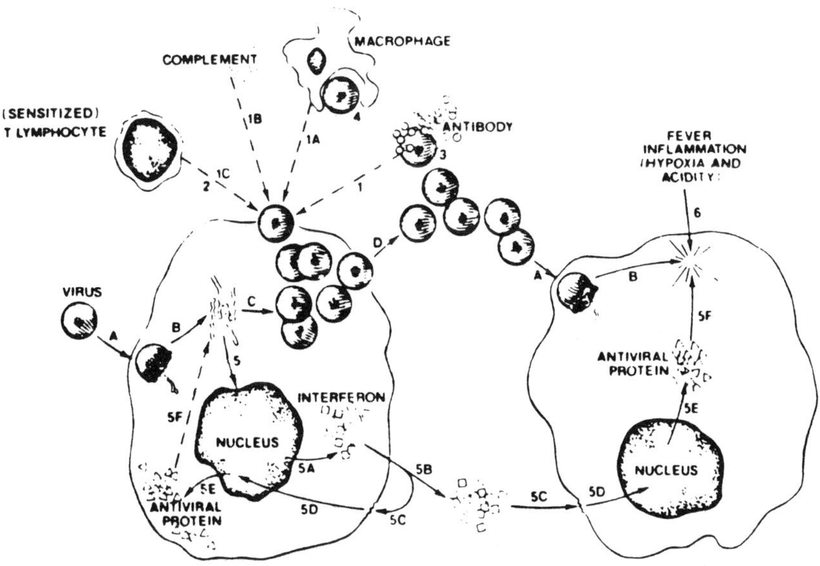

Figure 3. Host defenses against virus infection. The extracellular defenses include antibody (1 & 3) and the phagocytic macrophage (4) which may combine with and inactivate virus. Defenses against intracellularly replicating virus include: cytolysis of virus-infected cells bearing virus surface antigens by synthesized T lymphocytes (2) acting directly or through their mediators; antibody plus macrophages (1A); antibody plus complement (1B); antibody plus T lymphocytes (1C); and antibody plus K cells (not shown). Additional inhibition of intracellular viral replication (D and C) can be caused by the nonspecific defense mechanisms of interferon (5-5F). local acidity, local hypoxia (6), and interfering virus (not shown).

V.C. THERAPEUTIC POTENTIAL

A) Antiviral agent

Although interferon was identified in 1957, not until the mid-1970's was there enough material to begin clinical trials. Supplies consisted mainly of human leukocyte interferon. The results of these trials, although limited, are encouraging. The drug is relatively non-toxic and has been effective in animal and clinical trials, in the treatment of influenza, rhinoviruses,rabella, hepatitis B, cytomegalovirus, herpes, varicella-zorter, and vaccina.

Interferon is most effective when administered as a prophylactic agent, i.e., given 24 hours prior to infection, but by necessity most clinical trials have dealt with it as a therapeutic agent.

The dosages of interferon administered in the treatment of viruses are dependent upon a number of factors; These include: the disease treated, the route of administration, sensitivity of a particular virus to interferon, the amount of interferon likely to be produced by the patient in response to the infection, and accessibility of the infected organ to the interferon administered. In addition, because interferon is degraded in the body itself, daily doses are needed (1).

Generally the acute diseases require large doses over a relatively short period of time, whereas the chronic infections are best treated by lower doses for sustained periods of time. In terms of actual numbers, these doses range from 3.5 X 10E5 units/kg body weight per

---

(1) G.J. Galasso and J.K. Dunnick: Interferon, an Antiviral Drug for Use in Man, _Texas Reports on Biology and Medicine, 35_: 478 (1977).

day, as in the treatment of cytomegalovirus in infants (1), to doses of about 100 fold less in the treatment of chronic hepatitis B (2).

Interferon may be used also as a vaccine supplement. In principle, interferon can be used to control a viral infection until the body developes antibodies against the virus. Support for this idea comes from animal studies where vaccines against rabies were more effective when supplemented with interferon(3,4). It should be noted that interferon has some side-effects. These will be discussed later.

B) Antitumor agent

Because some tumors are believed to be induced by viruses, interferon was examined as an antitumor agent. In animal studies when the tumor was induced by a viral agent, interferon effected a cure if the viral or tumor load could be reduced (5). In human trials, interferon was first used to treat Hodgkin's disease and osteogenic sarcomas because of their believed viral association. In these cases, although interferon has been shown to be somewhat effective, it is not clear that the action is a result of

---

(1) A.M. Arvin, A.S. Yeager, T.C. Merigan: Effect of Leukocyte Interferon on Urinary Excretion of Cytomegalovirus by Infants, J. Infect. Dis. 133(suppl) : A205-210 (1976).

(2) H.B. Greenberg, R.B. Pollard, L.I. Lutwick, P.B. Gregory, W.S. Robinson, and T.C. Merigan. Infection in Patients with Chronic Active Hepatitis. N. Eng. J. Med. 295: 517-522 (1976).

(3) J. Hilfenhaus, H.E. Karges, E. Weinmann, and R. Barth: Effect of Administered Human Leukocyte Interferon on Experimental Rabies In Monkeys, Infect. Immun. 4: 1156-1162 (1975).

(4) G.M. Baer, J.H. Shaddock, V.A. Moore, P.A. Yager, S.S. Baron, and H.B. Levy: Successful Prophylaxis against Rabies in Mice and Rheus Monkeys. The Interferon System and Vaccine, J. Infect. Dis. 136: 286-294 (1977).

(5) G.J. Galasso and J.K. Dunnick: Interferon, an Antiviral Drug for Use in Man, Texas Reports On Biology and Medicine. 35: 478 (1977).

its antiviral activity, and may in part be due to some of its other activities such as growth inhibition and enhancement of macrophage and T-cell activity.

The most extensive clinical trials to date have been with patients with osteogenic sarcoma(1,2), a virulent form of bone cancer which, in the normal course of the disease, spreads to the lungs. On average, only 20% of the patients with no evidence of metastases at the time of the diagnosis survive an additional five years. In 1971 researchers in Stockholm began a program to determine the effectiveness of interferon against this disease. They found that of those patients treated with interferon 65% remained free of metastases 2.5 years after the diagnosis. In constrast, among patients treated in other Swedish hospitals not receiving interferon, only 30% remained disease-free over the same period of time.

Since then, clinical trials on other forms of cancer have been initiated. Those that have shown responsiveness to interferon include Hodgkins's disease, multiple myeloma, melanoma, breast cancer, certain forms of leukemia and lymphoma, and bladder cancer. It should be stressed that these results are preliminary in nature and that much more work is needed. It is also likely that in the future interferon will be used in combination with conventional chemotherapy and radiation treatment.

C) Interferon induction *in vivo*

Because of the scarcity of interferon, a second

---

(1) H. Strander: Antitumor Effects of Interferon and Its Possible Use as an Anti-neoplastic Agent in Man. *Texas Reports on Biology and Medicine.* *35* : 478 (1977).
(2) H. Strander, K. Cantell, P.A. Jakobson, U. Nilsonne, and G. Soderberg: Interferon Treatment of Osteogenic Sarcoma: A Clinical Trial. *Fogarty Intern. Center Proc.* *28*: 377-380 (1977).

approach to its production has been pursued. This approach involves stimulating the patient's own capacity to produce interferon by treatment with a suitable inducer. A variety of inducers have been identified, the most prominent being inactivated viruses, bacterial endotoxins, and synthetic ribonucleotides.

The use of these interferon inducers in the prevention and treatment of viral diseases has met with two major problems. The first is the hyporesponsive period, during which the organism (or even cells in culture) is unable to produce adequate levels of interferon in response to repeated inductions. This is the primary reason why researchers have focused on the use of exogenous interferon to maintain the antiviral state. The second obstacle is the toxicity of the inducers, which are generally pyrogenic. Among the toxic effects noted are inhibition of liver regeneration, damage to nervous tissue, and embryo toxicity (1).

D) Other Uses

Should interferon continue to show efficacy in the treatment of viral infection, the drug is likely to play a large role in the prevention of viral infections among highly immunosuppressed patients. Two groups in need of such treatment are cancer patients and recipients of organ transplants.

Radiation therapy as well as many forms of chemotherapy severely compromise immune function in cancer patients. This is dramatized by the fact that infection is the single largest cause of death in cancer patients (about 35% of all

---

(1) W.E. Stewart II. The Interferon System. Springer-Verlag, Wein, Austria. pp. 27-55 (1979).

cancer related deaths (1). Although the majority of such infections are nonviral, viral infections and the susceptibility of these patients to viral infections is a serious matter.

In organ transplant patients, the immunosuppressed state is intentionally produced to reduce the risk of organ rejection. Thus, viral infections such as cytomegalovirus plague these individuals. A study of kidney transplants has shown that, if interferon is administered before and for six weeks after the transplant, the incidence of infection is reduced. Cytomegalovirus infections are most troublesome during the first three months after the operation (2).

E) Pharmokinetics and Side Effects

The lifetime of interferon in the organism depends in part on the form of administration, although once in the blood it has a half life of about 4 hours. This is believed to be more due to adsorption of the protein to cell surfaces than to its degradation. With intramuscular injections, peak interferon levels in the blood are reached in 5-8 hours, whereas with subcutaneous injections peak levels are reached in 8-10 hours. In addition, because of the size of the interferon molecule, a number of organ barriers such as the gut are not easily penetrated (3). Thus it is not surprising that interferon is poorly adsorbed by weanlings from mother's milk (less that 1%), and is not likely to be

---

(1) J.L. Ambrus, C.M. Ambrus, J.B. Mink, and J.W. Pickorer: Causes of Death in Cancer Patients. J. Medicine 6: 61-71 (1975).

(2) G.J. Galasso and J.K. Dunnick: Interferon, an Antiviral Drug for Use in Man, Texas Reports On Biology and Medicine. 35: 478 (1977).

(3) G.W. Jordan, R.P. Fried, T.C. Merigan: Administration of Human Leukocyte Interferon in Herpes Zoster. I. Safety, Circulating Antiviral activity, and Host Response to Infection. J. Infect. Dis. 130: 56-62 (1974).

readily adsorbed by the skin (1). Experiments with humans and chimpanzees show that most of the interferon applied intranasally is rapidly cleared, although some of the dose remains (2,3). Perhaps most important is the difficulty in penetrating the blood/brain barrier (4,5), which complicates treatment of infections of the central nervous system. Consequently, some attempts using direct intrathecal administration of interferon have been made. Although the interferon appears to be tolerated, distribution of the drug is difficult and considerable work is needed before any conclusions can be formed.

The dose-dependent side effects associated with interferon include: fever for several hours after receiving injection; nausea and vomiting; malaise and myalgias, which occur during onset of fever; marrow suppressive effects; and decreased feedings and weight gain in infants. All of these effects are relieved when the treatment is stopped (6). Additionally, it should be emphasized that many of the clinical trials have used a very crude form of interferon. Upon further purification a number of the side effects, in particular the nausea and lassitude, have

---

(1) T.W. Schafter, M. Lieberman, M. Cohen, et al.: Interferon Administered Orally, Protection of Neonatal Mice from Lethal Virus Challenge. Science 176: 1326-1327 (1972).
(2) P.E. Johnson, S.B. Greenberg, M.W. Harmon, et al.: Recovery of Applied Human Leukocyte Interferon from the Nasal Mucosa of Chimpanzees and Humans. J. Clin. Microbiol. 4: 106-107 (1976).
(3) T.C. Mergian, S. Reed, T.S. Hall et al.: Inhibition of Respiratory Virus Infection by Locally Applied Interferon. Lancet 1: 563-567 (1973).
(4) D.V. Habif, R. Lipton and K. Cantell: Interferon Crosses Blood cerebiospinal Fluid Barrier in Monkeys. Proc. Soc. Exp. Biol. Med. 149: 287-289 (1975).
(5) M. Ho, C. Nash, C.W. Morgan. Interferon Administered in the Cerebiospinal Space and its Effect on Rabies in Rabbits. Inf. and Immun. 9: 286-293 (1974)
(6) T.C. Mergian: Pharmacokinetics and Side Effects of Interferon in Man. Texas Reports on Biology and Medicine 35: 541-547 (1977).

diminished. The marrow suppresive effects (demonstrated by depression of white cell, platelet and reticulocyte counts), although reversible, are not diminished with further purification. In general, these complications have not prevented continued therapy, if the dosages are properly managed. However, the lassitude and marrow suppressive effects limit the amount of interferon which can be administered at any one time.

V.D. MEASUREMENT OF INTERFERON

Interferon is an extremely potent inhibitor of viral growth and the great majority of interferon assays are based on quantitating this activity. Generally speaking, the interferon assays can be divided into three groups;

a) measurement of the cytopathic effect (CPE)

b) plaque reduction assays

c) virus yield reduction methods.

The CPE inhibition assay involves determining the concentration of interferon at which 50% of the cells are protected against virus destruction (1). It entails the selection of a virus that causes a characteristic cell damage easily identified by low power microscopy. The method is considered to be simple, reliable, and reproducible. However, a considerable amount of subjectivity is involved in evaluating the 50% endpoint. The dye uptake method is a potential modification that can considerably improve the objectivity of this asssay. This method involves the evaluation of cell damage by measuring the uptake by the cell of a vital dye (neutral red). Cells

---

(1) N.H. McMannus: Microtiter assay for interferon, Microspectrophotometric Quantitation of Cytopathic effect. App. and Envir. Micro. 31: 35-38 (1976).

taking the dye can be enumerated, or the dye can be extracted from the cells and evaluated colorimetrically.

The plaque reduction method involves determination of the sample dilution at which the number of plaques in a monolayer of cells is reduced by 50%. The method is similar to the CPE method but is a bit more accurate. However, it is much more expensive with regards to time and labor (1).

In the yield reduction methods, the reduction in viral growth is measured over one cycle of virus replication. The growth of the viruses is measured by a number of differing methods: titration of infectious virus, hemeagglutinin measurements, measurement of viral neuraminidase, determination of hemeadsorption, and measurement of viral nucleic acid.

V.E. INTERFERON INDUCERS

The first step in the production of interferon involves the recognition of some chemical species leading to the derepression of genetic material coding for interferon. These chemical species are referred to as inducers. The first inducers identified were viruses. Subsequently, "foreign" nucleic acids (RNA) were also found to be capable of inducing interferon. To date a tremendous number of different substances have been reported to be inducers. For *in vitro* studies, the two most potent and commonly used inducers are probably the viral and synthetic polynucleotides (particularly poly rI - poly rC).

Viruses from every major grouping have been shown to induce interferon, albeit with varying degrees of

---

(1) V.B. Campbell, T. Grunberger, M.S. Kockman, and S.L. White: A Microplaque Reduction Assay for Human and Mouse Interferon. *Can. J. Microbiol.* *21*: 1247-1253 (1975).

effectiveness. Newcastle Disease Virus (NDV) is considered to be an example of a "good" inducer. It is consistently capable of inducing relatively high titers of interferon, both *in vivo* and *in vitro*. Many factors affect the ability of the virus to induce interferon, a number of which are related to the metabolic changes that the virus causes in the infected cells. Many "poor" interferon inducers can be significantly improved by treatments which slightly damage the virus (such as heat or UV light).

In 1967 double stranded synthetic polyribonuleotides were discovered to be efficient inducers of interferon (1). Poly rI - poly rC was used for this study and, despite considerable efforts to improve it, is still the most potent of the polynucleotide inducers. The following characteristics are essential for efficient inducer activity:

a) high molecular weight (for poly rI-poly rC, a molecular weight of greater than 300,000 is necessary) (2)
b) the 2'-hydroxyl group (Apparently any modification of this group results in a decrease of activity. Consistent with this decrease, the deoxyribose polymers are completely inactive) (3)
c) double strandedness (4)
d) nucleolytic resistance (In many systems the effectiveness of the polynucleotides as inducers has been related to their resistance to nucleolytic degradation. Formation of a complex of the polynucleotide with cationic polymers,

---

(1) A.K. Field, A.S. Tytell, G.P. Lampson, and M.R. Hilleman: Inducers of Interferon and Host Resistance II. Multistranded Synthetic Polynucleotide Complexes. *Proc. Natl. Acad. Sci.* *58*: 1004-1009 (1967).
(2) P.M. Pither, and D.W. Hutchinson: The Mechanism of Interferon Induction By Synthetic Polyribonucleotides. In: *Interferons and Their Actions* (W.E. Stewart II ed.) *1*, pp. 13-36 Cleveland, Ohio CRC Press Inc, 1977.
(3) E. DeClerq, B. Zmudzka, and D. Shugar: Antiviral Activity of Polynucleotides, Role of the 2-Hydroxyl and a Pyrimidine 5-Methyl. *FEBS letters* *24*: 137-140 (1972).
(4) C. Colby: The Induction of Interferon by Natural and Synthetic Polynucleotides. *Prog. Nucleic Acid Res. Mol. Biol.* *11*: 1-32 (1971).

such as polylysine or DEAE dextran, has improved their stability and in some systems their effectiveness as inducers.) (1,2) In this respect, improvements have made poly rI-poly rC more effective as an inducer.

It is also worthwhile to mention that in human fibroblast cells, the kind of interferon produced is dependent upon the nature of the inducer. Studies by Havell et al. (3) have shown that fibroblasts produce only one form of interferon , the "fibroblast" interferon, when exposed to poly rI-poly rC. Upon exposure to Newcastle Disease Virus, both "fibroblast" and "leukocyte" interferons are produced (in about an 80/20 ratio). It is reasonable to speculate that if these studies are pursued, human fibroblasts might also become a major source of leukocyte interferon.

V.F. INTERFERON PRODUCTION

Typically cell cultures begin to synthesize interferon approximately 2 hours after exposure to poly rI-poly rC. The rate of synthesis reaches a maximum in about 3 hours and then decreases at a rate as rapidly as it rises. This rapid increase and decrease is usually accompanied by a corresponding increase and decrease in biologically active mRNA in the induced cells (see figure 4). The generally accepted view is that the termination of synthesis is due to the irreversible inactivation of mRNA A repressor protein also induced by poly rI-poly rC is believed to be responsible for the post-transcription control of interferon synthesis. The evidence for this model is three-fold:

(1) J. Rice, W. Turner, M.A. Chirigos, and N.R. Rice: Enhancement by Poly I:C - Induced Interferon Production in Mice. Applied Microbiol. 19: 867-869 (1970).
(2) F. Dianzani, P. 'Cantagalli, S. Gagnoni, and G. Rita: Effect of DEAE Dextran on Production of Interferon Induced by Synthetic Double Stranded RNA in L-Cell Cultures. Proc. Soc. Exp. Biol. Med. 128: 708-710 (1968).
(3) E.A. Havell, T.G.Hayes and J. Vilcek: Synthesis of Two Distinct Interferons by Human Fibroblasts. Virology 89: 330-334 (1978).

a) cells induced by poly rI-poly rC become temporarily refractive to repeated induction, perhaps due to the residual levels of the repressor protein (1).
b) selective treatment of induced cells with various metabolic inhibitors is capable of greatly enhancing and sustaining the synthesis of interferon. This treatment is called "superinduction," and the success of this process may be due to the preferential sensitivity of the repressor protein mRNA to these antimetabolites (2).
c) the amount of mRNA in cells induced with poly rI-poly rC and those superinduced at their peak synthesis rates differs by no more than a factor of two, even though the amount of interferon produced by these cells differ by a factor of 20-fold or more, as has been demonstrated using various mRNA translation systems, such as injection into <u>xenopus oocytes</u>, to quantitate the amount of interferon mRNA (3).

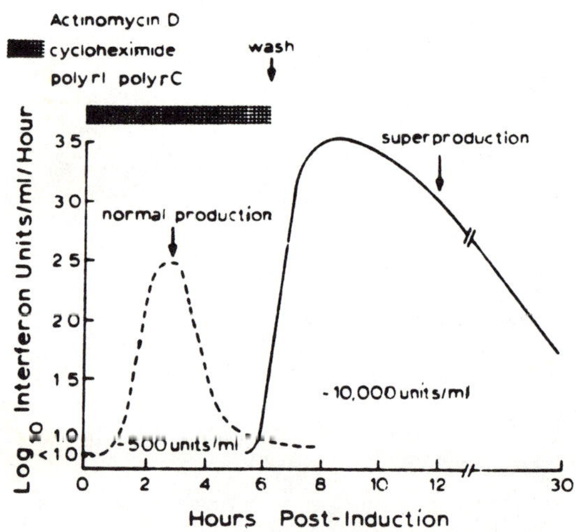

Fig. 4. Superproduction of interferon. The system illustrated is the human fibroblast method described by Havell and Vilček (1972).

---

(1) M. Kohase and J. Vilcek: Regulation of Human Interferon Production Stimulated with Poly I - poly C Correlation between Shutoff and Hyporesponsiveness to Reinduction. <u>Biology</u> <u>76</u>: 47-54 (1977).
(2) J. Vilcek, E.Z. Havell, and M. Kohase: Superinduction of Interferon with Metabolic Inhibitors: Possible Mechanisms and Practical Applications. <u>J. Infect. Dis.</u> <u>133</u>: A22-A29 (1976).
(3) N.B.K. Raj and P.M. Pitha: Relationship Between Interferon Production and Interferon Messenger RNA Synthesis in Human Fibroblasts. <u>PNAS</u> <u>79</u>: 1483-1487 (1977).

V.G. CHEMISTRY

There are three known types of human interferon: leukocyte, fibroblast, and immune interferons (1). Fibroblasts produce predominantly "fibroblast" interferon and leukocytes produce predominantly "leukocyte" interferon. Although the two interferons produce an identical antiviral state, they are antigenically very dissimilar. Additionally there is a so-called lymphoblastoid interferon produced in the "Namalva" cell line, derived from a Burkitt's lymphoma. This cell line seems to produce a mixture relatively rich in both leukocyte and fibroblast interferon.

Immune interferon is a relatively recent discovery. It is produced by lymphoblasts in response to various mitogens and antigens (2). As yet not much is known about this type of interferon. It is not stable at pH 2, as are the others. It may be more involved in modulation of the immune system than in the production of an antiviral state, which suggests that it is more closely related to the lymphokines than the "classical" interferons.

Both fibroblast and leukocyte interferons are believed to be glycoproteins; they are about 20,000 in molecular weight, and stable at pH 2. Leukocyte interferon is stable under non-reducing SDS conditions and appears to be less resistant to mechanical stress than fibroblast interferon. Both are extremely potent chemicals with specific activities estimated at 10E9 to 10E10 units/mg. (A unit is defined somewhat arbitrarily as the amount necessary to protect 50% of the cells against viral damage.) Only recently have these

---

(1) K.H. Fantes: Interferons: Chemical Properties. Texas Reports in Biology and Medicine 35: 173-180 (1977).
(2) J.S. Younger: Properties of Interferon Induced by Specific Antigens. Texas Reports on Biology and Medicine 35: 17-23 (1977).

proteins been purified to near homogeneity (1,2). The problems are that, first and foremost even crude interferon is scarce and, secondly, that once purified these proteins are not particularly stable in dilute solutions.

An area in need of better definition is the role of the carbohydrate moiety in the activity of interferon. For example, the carbohydrate moiety appears to influence the species specificity of interferons. Another question is the effect the carbohydrate moeity has on the various other activities of interferon.

In studies with leukocyte interferon, treatment with either deglycosylating enzymes or periodate markedly reduced the antiviral activity (3). The same appears to be true with fibroblast interferon (4). In addition, "buffy coat" production of interferon in the presence of inhibitors of glycosylation produced either less interferon or an interferon with reduced activity. If Biogen's announced results with recombinant DNA techniques are correct, interferon without the carbohydrate moiety retains some activity _in vitro_. The questions still to be answered are how much the antiviral activity is reduced, and how the _in vivo_ stability is altered.

---

(1) K. Cantell, S. Hirvonen, K.E. Morgensen, and L. Pitha: Human Leukocyte Interferon: Production, Purification, Stability and Animal Experiments. _In Vitro_ _10_:35-38 (1974).

(2) E. Knight: Interferon: Purification and Initial Characterization from Human Diploid Cells. _PNAS_ _73_: 520-523 (1976).

(3) W.E. Stewart II., L.S. Lin, M. Wiranowska-Stewart, K. Cantell: Elimination of Size and Charge Heterogenetic of Human Leukocyte Interferon by Chemical Cleavage. _PNAS_ _74_: 4200-4204 (1977).

(4) S. Bose, D. Gurari-Rotman, V.T. Ruegg, L. Corley, and C.B. Anfinsen: Apparent Dispensibility of the Carbohydrate Moiety of Human Interferon for Antiviral Activity. _J. Biol. Chem._ _251_: 1659-1662 (1976).

V.H. GENETIC CONTROL OF THE INTERFERON SYSTEM

Using hamster-human somatic cell hybrids, the genes responsible for the production of human fibroblast interferon have been assigned. Only those hybrids which retained chromosome 5 as the only human chromosome were inducible for interferon, whereas hybrids containing other chromosomes but lacking 5 were not inducible (1). However, genes located on chromosomes 2 and 18 facilitate and/or are responsible for the activation of the gene on chromosome 5. Subsequent work has shown that cells containing multiple copies of chromosome 5 longarm produce considerably more interferon than the normal diploid cultures. Additionally, those with multiple copies of chromosome 5 shortarm are very poor producers. Such evidence would indicate that the gene for the interferon mRNA is on the 5 longarm and the repressor gene is on the 5 shortarm.

Similar somatic cell hybrid studies have indicated that the gene responsible for the antiviral state is located on chromosome 21. Furthermore, the genetic control of the interferon induced cell growth inhibition is located on chromosome 21 (2).

V.I. POTENTIAL OF LARGE-SCALE PRODUCTION

At present the biggest impediment to further research and clinical testing with interferon is its lack of availability. Historically, the first major source of interferon was established by Dr. Kari Cantell in

---

(1) Y. H. Tan, R.P. Creagan, and R. H. Ruddle: The Somatic Cell Genetics of Human Interferon: Assignment of Human Interferon Loci to Chromosome 2 and 5. _PNAS_ _71_: 2251-55 (1974).

(2) Y.H. Tan, J.A. Tischfield and R.H. Ruddle: The Linkage of Genes for the Human Interferon Induced "Antiviral Protein" and Indophendoxidase B Traits to Chromosome 21. _J. Exp. Med._ _137_: 317-330 (1973).

cooperation with the Finnish Red Cross (1). In 1978 the American Cancer Society purchased 40 billion units of interferon from the Finnish Red Cross for $2 million. This constituted the first significant supply of interferon in the United States and stimulated further interest in interferon in this country. The following year, Congress made a special appropriation so that the National Cancer Institute (NCI) might also purchase interferon. Following this action, the NCI a announced request of proposals to supply 50 billion units each of leukocyte, fibroblast, and lymphoblastoid interferon.

Human Leukocyte Interferon

Much of the interferon used to date has been produced from leukocytes under the direction of Dr. Kari Cantell in Finland. Dr. Cantell utilized the most available source of normal human diploid cells, "buffy coat" leukocytes. Through cooperation with the Finnish Red Cross, which operates a single blood bank for the entire country in Helsinki, this process has been scaled up to an annual production of about 200 billion units. In Cantell's scheme the "buffy coats" are first primed and then induced with Sendai Virus at a high multiplicity. Twenty hours later the interferon is removed and partially purified. This method is presently able to produce about 2 million units per donor, which is equivalent to about a single daily dose in the treatment of some forms of cancer. Thus, the supply has obvious limitations.

A facility of similar size has been established at the Memorial Sloan-Kettering Cancer Center, but many more

---

(1) K. Cantell, and S. Hirvonen: Preparation of Human Leukocyte Interferon for Clinical Use. Texas Reports on Biology and Medicine 35: 138-144 (1977).

centers will be neeeded to provide for clinical trials.

There are two obvious problems involved in the scale-up of leukocyte interferon. First, there is only a modest potential for scale-up with "buffy coats", because of the low yield of interferon from a single donation. Secondly, keeping these preparations pathogen free is difficult as human blood donations are often contaminated with hepatitis and other viruses.

It should also be mentioned that interferon produced from buffy coats contains a small fraction of fibroblast type interferon (generally about 2-5%).

In this country Warner-Lambert, Revlon (via the blood products division), and National Patent Development Corp. have begun or have announced their intention to produce leukocyte interferon. Together these companies should be producing about 500 billion units annually within the next year or two.

Human Leukocyte Interferon from Lymphoblastoid Cell Lines

Many of the problems associated with increasing interferon production from "buffy coat" leukocytes can be avoided by the use of suspension cell lines (1). The Namalva lymphoblastoid cell line is presently perceived by some as the answer to the large scale production of interferon. The cells grow to relatively high cell densities in suspension and, upon induction with a number of paromyxoviruses, produce a mixture of leukocyte and fibroblast interferon (though predominantly leukocyte, about 80%).

---

(1) H. Strander, K.E. Morgensen and K. Cantell: Production of Human Lymphoblastoid Interferon. J. Clin. Microbiol. 2: 116-117 (1975).

Two difficulties arise from the production of interferon from lymphoblastoid cells. First is the derivation of the cell line from a Burkett's lymphoma (1), as both the Food and Drug Administration and the Bureau of Biologics presently object to the use of malignant cell lines for the production of drugs or other biological materials. Secondly, the yield of interferon using this method is relatively low, typically about 10-20 fold less (on a per cell basis) than that obtained with buffy coat leukocytes or fibroblast cells. This means that the scale of operation has to be radically increased to produce any real gains.

In England, Burroughs-Welcome is presently growing this line on a 1000 liter scale and in this country work conducted by NIH involves growth on a similar scale.

Human Fibroblast Interferon

Fibroblasts offer the greatest potential for the large-scale production of interferon from normal diploid human cells. At present, a number of cell lines isolated from neonatal foreskin fibroblasts can be induced to produce acceptable titers of fibroblast interferon (2). In addition to the potential for scale-up, it is considerably simpler to monitor and maintain these cell lines free from pathogens.

Fibroblasts require a surface for growth and the traditional method of cell culture has involved the use of

---

(1) G. Klein, L. Dombos, and B. Gothoskar: Sensitivity of Epstein-Barr Virus (EBV) Producer and Non-producer Human Lymphoblastoid Cell Lines to Superinduction with EB-Virus. Int. J. Cancer 10: 44-57 (1972).

(2) J. Vilcek, E.A. Howell, M.L. Gradoville, M. Mika-Johnson, and W.H.J. Douglas: Selection of New Human Foreskin Fibroblast Cell Strains for Interferon Production. Adv. In Exp. Med. And Biol. 110: 101-118 (1978).

rotating roller bottles containing a minimal amount of the growth medium. A variety of modifications have been devised to increase the efficiency of cell culture. The method offering the greatest scale-up potential for the growth of anchorage dependent cells is the use of charged microcarriers (as developed at the Massachusetts Institute of Technology) (1). These microcarriers are suspended in the growth medium by gentle stirring, while the cells attach and grow on their surface. Because the microcarriers offer an increased surface to volume ratio, the number of cells per unit volume of growth medium is significiantly increased.

Once the cells have covered the surface, they are induced to produce interferon by exposure to a synthetic polyribonucleotide and certain antimetabolites. The cycle time from inoculation to the collection of interferon is about 10 days (2).

Fibroblast interferon is somewhat different in character from the leukocyte interferon (3). The protein appears to be more heavily glycosylated and the purified material also appears to be more sensitive to mechanical stress (4). These differences have prompted some researchers to wonder if the two types would have equal therapeutic value. It would appear from the clinical trials of its antiviral efficacy that both forms of interferon are

---

(1) W.G. Thilly and D.W. Levine: Microcarrier Culture: A Homogeneous Environment For Studies of Cellular Biochemistry. Methods In Enzymology 58: 184-194 (1979).

(2) D.Giard and R. Fleischaker: Examination of Parameters Affecting Human Interferon Production with Microcarrier-grown Fibroblast Cells. (in press) (1980).

(3) K.H. Fantes: Interferons: Chemical Properties. Texas Reports in Biology and Medicine 35: 173-180 (1977).

(4) J.J. Sedmak, D.P. Jameson, and S.E. Grossberg: Thermal and Vortical Stability of Purified Human Fibroblast Interferon. Adv. in Exp. Med. and Biol. 110: 133-152 (1978).

equally useful (1). Fibroblast interferon has also shown very promising results in the treatment of malignant melanomas in the work being conducted at Roswell Park.

In this country both Cal-Biochem and HEM Research produce and sell limited quantities of fibroblast interferon. Both firms presumably are producing interferon in roller bottles. Flow General plans to begin production of interferon on microcarriers in June of 1980. Additionally a number of large pharmaceutical firms with experience in cell culture have now begun to examine the potential for producing fibroblast interferon.

Interferon from Recombinant DNA

In principle, insertion of the gene coding for human interferon into a bacteria cell could prove to be the most economical solution to the large scale production of human interferon. A number of groups, not only in this country but also in Japan, Europe, and in Israel, are very actively involved in this endeavor. A number of obstacles confront this approach, the most serious being the question of the carbohydrate moiety. As interferon is a glycoprotein and interferon produced by recombinant DNA is not, a number of tests are needed to determine how much of the antiviral and antitumor activities are retained. In addition, non-glycosylated proteins tend to have much shorter life times *in vivo* than do the glycosylated ones, which may mean larger and more frequent doses than in treatments using the native protein.

Biogen recently announced the production of leukocyte

---

(1) G.J. Galasso and J.K. Dunnick: Interferon, an Antiviral Drug for Use in Man. *Texas Reports on Biology and Medicine* *35*: 478 (1977).

interferon in *E. coli* (1). The results are encouraging but the yields are very low (more than 1000 times less than those produced by human fibroblasts). It will probably be another 3-5 years before any of this material becomes commercially available. Biogen hopes to be able to market the material at between one and ten dollars a dose (presumably a million units).

Comparison of Fibroblast and Recombinant DNA Technologies

Should interferon ever be produced in large scale, human fibroblasts or bacteria via recombinant technology will probably be utilized. Surprisingly, both methods are potentially able to produce the material at a cost of between $1 and $10 a dose. Interferon from fibroblasts has the advantage of being derived from normal human cells and being the authentic protein. Interferon from bacteria may be easier to produce because the requirements to cultivate are technically much easier. However, it is an incomplete protein and must be purified much more extensively before use in humans, because of exotoxins and endotoxins that are associated with *E. coli* and other bacteria.

Market Potential for Interferon

There is considerable speculation concerning the market potential for interferon. Left unrestricted, interferon could probably be as available as chlorophyll once was. Realistically speaking, however, it is likely to remain a restricted drug because of its potentially serious side effects. Should preliminary results from the clinical trials be substantiated, interferon is likely to be used in combination with conventional chemotherapy for preventing

(1) S. Nagata, H. Taira, A. Hall et al.: Synthesis in *E. coli* of a Polypeptide with Human Leukocyte Interferon Activity. *Nature* *284*: 316-320 (1980).

infection in patients whose immune systems have been seriously repressed (cancer and organ transplant patients), as an antiviral agent (for the treatment of chronic Hepatitis B and a few other diseases), and as an antitumor agent itself. The actual size of these groups is at best uncertain, but about 500,000 people per year is a cautious estimate. About 10E14 units of interferon per year would be needed to treat this population, and at $1-10 per million units this would represent between 100 million and one billion dollars in sales. For a summary see Table II.

Table II: Summary of Potential Methods for Interferon Production

Means of Production	Types of Interferon Produced	Potential for Scale-up	Present Cost ($/10^6 units)	Projected Cost ($/10^6 units)	Problems	Potential for Improvement
"Buffy Coat" Leukocytes	leukocyte, $\geq$95% fibroblast, $\leq$5%	No	50	-	-lack of scale-up -pathogen contamination	- minimal
Lymphoblastoid Cells	leukocyte, 80% fibroblast,20%	Yes	-	≃ 25	-poor yields -drug approval (cells derived from tumor)	-improved yields -expression of fibroblast interferon
Fibroblasts	fibroblast	Yes	43-200	≃1-10	-cell culture -economic competition with recombinant DNA	-improved yields -improved cell culture technology -expression of leukocyte type interferon
Recombinant DNA	leukocyte or fibroblast	Yes	-	≃1-10	-does not produce authentic interferon -in vitro drug stability -poor yields -drug approval -possible economic competition with fibroblast cell production	-improved yields -modified interferons